Principles of Electrodynamics

MELVIN SCHWARTZ

Consulting Professor of Physics
Stanford University

DOVER PUBLICATIONS, INC.
New York

TO MY WIFE, MARILYN

Published in Canada by General Publishing Company, Ltd., 30 Lesmill Road, Don Mills, Toronto, Ontario.
Published in the United Kingdom by Constable and Company, Ltd.

This Dover edition, first published in 1987, is an unabridged, corrected republication of the work first published by the McGraw-Hill Book Company, New York, 1972, in its International Series in Pure and Applied Physics.

Drawings by Judith L. McCarty

Manufactured in the United States of America
Dover Publications, Inc., 31 East 2nd Street, Mineola, N.Y. 11501

Library of Congress Cataloging-in-Publication Data

Schwartz, Melvin, 1932–
    Principles of electrodynamics.

    ". . . unabridged, corrected republication of the work first published by the McGraw-Hill Book Company, New York, 1972 . . ."T.p. verso.
    Includes index.
    1. Electrodynamics.   I. Title.
QC631.S38   1987   537.6   87-13607
ISBN 0-486-65493-1 (pbk.)

Contents

8 MULTIPOLE EXPANSION OF THE RADIATION FIELD; SOME FURTHER CONSIDERATIONS ON THE INTERACTION OF RADIATION WITH MATTER; INTERFERENCE AND DIFFRACTION 273

9 WAVEGUIDES AND CAVITIES 304

10 ELECTRIC AND MAGNETIC SUSCEPTIBILITY 321

Preface

Electromagnetic theory is beautiful! When looked at from the relativistic point of view where electric and magnetic fields are really different aspects of the same physical quantity, it exhibits an aesthetically pleasing structure which has served as a model for much of modern theoretical physics. Unfortunately this beauty has been all but buried as most textbooks have treated electricity, magnetism, Coulomb's law, and Faraday's law as almost completely independent subjects with the ground work always supplied by means of empirical or historical example. Occasionally a chapter is devoted to the relativistic coalescence of the various aspects of electromagnetism but use is rarely made of the requirement of Lorentz invariance in deriving the fundamental laws.

Our point of view here is quite different. Basically we have two purposes in mind—one is to exhibit the essential unity of electromagnetism in its natural, relativistic framework and the other is to show how powerful the constraint of relativistic invariance is. To these ends we shall show that all electromagnetism follows from electrostatics and the requirement that our laws be the simplest ones allowable under the relativistic constraint. The hope is that the student will make use of these new insights in thinking about theories that are as yet undeveloped and that the model we set here will be generally useful in other areas of physics.

A word about units. Unfortunately one of the results of the completely disconnected way in which electricity and magnetism have been taught in the past has been the growing acceptance of the mks over the cgs system of units. We have no special preference for centimeters over meters or of grams over kilograms. We do, however, require a system wherein the electric field **E** and the magnetic field **B** are in the *same* units. Using the mks system, as it is presently constituted, for electromagnetic theory is akin to using a meterstick to measure along an East-West line and a yardstick to measure along a North-South line. To measure **E** and **B** in different units is completely antithetical to the entire notion of relativistic invariance. Accordingly we will make use of the cgs (gaussian) system of units exclusively. Conversion to practical units where necessary can be carried out with no difficulty.

The author would like to express his most profound appreciation to Miss Margaret Hazzard for her patient and careful typing of the text.

MELVIN SCHWARTZ

1
Mathematical Review and Survey of Some New Mathematical Ideas

It would be delightful if we could start right out doing physics without the need for a mathematical introduction. Unfortunately though, this would make much of our work immeasurably more laborious. Mathematics is much more than a language for dealing with the physical world. It is a source of models and abstractions which will enable us to obtain amazing new insights into the way in which nature operates. Indeed, the beauty and elegance of the physical laws themselves are only apparent when expressed in the appropriate mathematical framework.

We shall try to cover a fair bit of the mathematics we will need in this introductory chapter. Several subjects are, however, best treated within the context of our physical development and will be covered later. It is assumed that the reader has a working familiarity with elementary calculus, three-dimensional vectors, and the complex number system. All other subjects will be developed as we go along.

1-1 VECTORS IN THREE DIMENSIONS; A REVIEW OF ELEMENTARY NOTIONS

We begin by reviewing what we have already learned about three-dimensional vectors. As we remember from our elementary physics, there are a large number of quantities that need three components for their specification. Position is, of course, the simplest of these quantities. Others include velocity and acceleration. Even though we rarely defined what was meant by a vector in mathematically rigorous terms, we were able to develop a certain fluency in dealing with them. For example, we learned to add two vectors by adding their components. That is, if $\mathbf{r}_1 = (x_1, y_1, z_1)$ and $\mathbf{r}_2 = (x_2, y_2, z_2)$ are two vectors, then

$$\mathbf{r}_1 + \mathbf{r}_2 = (x_1 + x_2, y_1 + y_2, z_1 + z_2)$$

If a is a number, then

$$a\mathbf{r}_1 = (ax_1, ay_1, az_1)$$

We also found it convenient to represent a vector by means of an arrow whose magnitude was equal to the vector magnitude and whose direction was the vector direction. Doing this permitted us to add two vectors by placing the "tail" of one at the "head" of the other as in Fig. 1-1. We also learned how to obtain a so-called scalar quantity by carrying out a type of multiplication with two vectors. If $\mathbf{r}_1 = (x_1, y_1, z_1)$ and $\mathbf{r}_2 = (x_2, y_2, z_2)$ are two vectors, then $\mathbf{r}_1 \cdot \mathbf{r}_2$ is defined by the equation

$$\mathbf{r}_1 \cdot \mathbf{r}_2 = x_1 x_2 + y_1 y_2 + z_1 z_2$$

It was also shown that $\mathbf{r}_1 \cdot \mathbf{r}_2$ could be obtained by evaluating $|\mathbf{r}_1| \, |\mathbf{r}_2| \cos \theta_{12}$, where $|\mathbf{r}_1|$ and $|\mathbf{r}_2|$ are, respectively, the magnitudes of $\mathbf{r}_1$ and $\mathbf{r}_2$ and θ_{12}

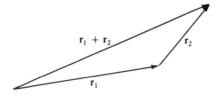

Fig. 1-1 The addition of two vectors can be accomplished by placing the "tail" of one at the "head" of the other.

is the angle between them. Another so-called vector was obtained by taking the cross product of $\mathbf{r}_1$ and $\mathbf{r}_2$. That is,

$$\mathbf{r}_1 \times \mathbf{r}_2 = (y_1 z_2 - y_2 z_1, z_1 x_2 - z_2 x_1, x_1 y_2 - y_1 x_2)$$

We shall have much more to say about the true nature of this beast very shortly. At the moment we just recall that it appears in some respects to be a vector whose magnitude is equal to $|\mathbf{r}_1|\,|\mathbf{r}_2|\sin\theta_{12}$ and whose direction, at right angles to both $\mathbf{r}_1$ and $\mathbf{r}_2$, is given by a so-called **right-hand rule** in going from $\mathbf{r}_1$ to $\mathbf{r}_2$. If we look from the head toward the tail of $\mathbf{r}_1 \times \mathbf{r}_2$, we would see the shortest rotation from $\mathbf{r}_1$ to $\mathbf{r}_2$ to be in the counterclockwise direction.

Unfortunately, we shall have to relearn much of the above within a more abstract framework if we are to make any progress beyond this point. We shall have to go back to our basic notions and see if we can define what we mean by vector in a more suitable, less intuitive manner. Only by doing so will we be prepared to say clearly which combinations of three numbers are vectors and which are not. We will also be able to define scalar in a reasonable way and will then see our way clear to an understanding of higher-rank tensors.

1-2 THE TRANSFORMATION PROPERTIES OF VECTORS UNDER SPATIAL ROTATION

To open the way for a more rigorous definition of vector, we proceed a bit further with our old intuitive notions. Let us consider a so-called **position vector**, that is, a vector from the origin of our coordinate system to the point (x,y,z). If we draw a unit vector along each of the three axes as shown in Fig. 1-2 and call them $\hat{\mathbf{i}}$, $\hat{\mathbf{j}}$, and $\hat{\mathbf{k}}$, respectively, we can write $\mathbf{r} = x\hat{\mathbf{i}} + y\hat{\mathbf{j}} + z\hat{\mathbf{k}}$. Now, we ask, what if we were to rotate our coordinate system to a new set of axes x', y', and z' with a new set of unit vectors $\hat{\mathbf{i}}'$, $\hat{\mathbf{j}}'$, and $\hat{\mathbf{k}}'$? How would $\mathbf{r}$ be expressed now? We answer this question very simply by expressing $\hat{\mathbf{i}}$, $\hat{\mathbf{j}}$, and $\hat{\mathbf{k}}$ in terms of the new unit vectors $\hat{\mathbf{i}}'$, $\hat{\mathbf{j}}'$, and $\hat{\mathbf{k}}'$. (This is possible

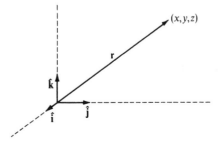

Fig. 1-2 The vector $\mathbf{r}$ can be expressed as $\mathbf{r} = x\hat{\mathbf{i}} + y\hat{\mathbf{j}} + z\hat{\mathbf{k}}$ where $\hat{\mathbf{i}}$, $\hat{\mathbf{j}}$, and $\hat{\mathbf{k}}$ are unit vectors along the x, y, z axes.

because any three-dimensional vector whatsoever can be expressed either in terms of $\hat{\mathbf{i}}$, $\hat{\mathbf{j}}$, and $\hat{\mathbf{k}}$ or in terms of $\hat{\mathbf{i}}'$, $\hat{\mathbf{j}}'$, and $\hat{\mathbf{k}}'$.) We write

$$\hat{\mathbf{i}} = a_{11}\hat{\mathbf{i}}' + a_{21}\hat{\mathbf{j}}' + a_{31}\hat{\mathbf{k}}'$$
$$\hat{\mathbf{j}} = a_{12}\hat{\mathbf{i}}' + a_{22}\hat{\mathbf{j}}' + a_{32}\hat{\mathbf{k}}' \qquad (1\text{-}2\text{-}1)$$
$$\hat{\mathbf{k}} = a_{13}\hat{\mathbf{i}}' + a_{23}\hat{\mathbf{j}}' + a_{33}\hat{\mathbf{k}}'$$

We note the obvious fact that

$$\hat{\mathbf{i}} \cdot \hat{\mathbf{i}}' = a_{11} \qquad \hat{\mathbf{i}} \cdot \hat{\mathbf{j}}' = a_{21} \qquad \hat{\mathbf{i}} \cdot \hat{\mathbf{k}}' = a_{31}$$
$$\hat{\mathbf{j}} \cdot \hat{\mathbf{i}}' = a_{12} \qquad \hat{\mathbf{j}} \cdot \hat{\mathbf{j}}' = a_{22} \qquad \hat{\mathbf{j}} \cdot \hat{\mathbf{k}}' = a_{32}$$
$$\hat{\mathbf{k}} \cdot \hat{\mathbf{i}}' = a_{13} \qquad \hat{\mathbf{k}} \cdot \hat{\mathbf{j}}' = a_{23} \qquad \hat{\mathbf{k}} \cdot \hat{\mathbf{k}}' = a_{33}$$

This, of course, permits us immediately to express the unit vectors $\hat{\mathbf{i}}'$, $\hat{\mathbf{j}}'$, and $\hat{\mathbf{k}}'$ in terms of $\hat{\mathbf{i}}$, $\hat{\mathbf{j}}$, and $\hat{\mathbf{k}}$, viz.,

$$\hat{\mathbf{i}}' = a_{11}\hat{\mathbf{i}} + a_{12}\hat{\mathbf{j}} + a_{13}\hat{\mathbf{k}}$$
$$\hat{\mathbf{j}}' = a_{21}\hat{\mathbf{i}} + a_{22}\hat{\mathbf{j}} + a_{23}\hat{\mathbf{k}} \qquad (1\text{-}2\text{-}2)$$
$$\hat{\mathbf{k}}' = a_{31}\hat{\mathbf{i}} + a_{32}\hat{\mathbf{j}} + a_{33}\hat{\mathbf{k}}$$

We realize that not all the nine quantities a_{ij} can be chosen independently. After all, only three angles are necessary to specify the rotation of one coordinate system into another. We expect then to have six equations linking the coefficients. We obtain these equations by requiring that $\hat{\mathbf{i}}'$, $\hat{\mathbf{j}}'$, and $\hat{\mathbf{k}}'$ form an orthogonal set of unit vectors.

$$\hat{\mathbf{i}}' \cdot \hat{\mathbf{i}}' = 1 = a_{11}{}^2 + a_{12}{}^2 + a_{13}{}^2$$
$$\hat{\mathbf{j}}' \cdot \hat{\mathbf{j}}' = 1 = a_{21}{}^2 + a_{22}{}^2 + a_{23}{}^2$$
$$\hat{\mathbf{k}}' \cdot \hat{\mathbf{k}}' = 1 = a_{31}{}^2 + a_{32}{}^2 + a_{33}{}^2$$
$$\hat{\mathbf{i}}' \cdot \hat{\mathbf{j}}' = 0 = a_{11}a_{21} + a_{12}a_{22} + a_{13}a_{23} \qquad (1\text{-}2\text{-}3)$$
$$\hat{\mathbf{j}}' \cdot \hat{\mathbf{k}}' = 0 = a_{21}a_{31} + a_{22}a_{32} + a_{23}a_{33}$$
$$\hat{\mathbf{k}}' \cdot \hat{\mathbf{i}}' = 0 = a_{11}a_{31} + a_{12}a_{32} + a_{13}a_{33}$$

Now to return to our original vector $\mathbf{r}$. We can write $\mathbf{r}$ in terms of its components in either of two ways:

$$\mathbf{r} = x\hat{\mathbf{i}} + y\hat{\mathbf{j}} + z\hat{\mathbf{k}} \qquad \text{or} \qquad \mathbf{r} = x'\hat{\mathbf{i}}' + y'\hat{\mathbf{j}}' + z'\hat{\mathbf{k}}'$$

Making use of Eqs. (1-2-1), we find immediately that

$$x' = a_{11}x + a_{12}y + a_{13}z$$
$$y' = a_{21}x + a_{22}y + a_{23}z \qquad (1\text{-}2\text{-}4)$$
$$z' = a_{31}x + a_{32}y + a_{33}z$$

We have traditionally used a right-handed coordinate system to specify the components of a vector. That is to say we have chosen $\hat{\mathbf{i}}, \hat{\mathbf{j}}$, and $\hat{\mathbf{k}}$ so that if we curl the fingers of our right hand from $\hat{\mathbf{i}}$ to $\hat{\mathbf{j}}$, our thumb will point along $\hat{\mathbf{k}}$. Expressing this in language somewhat more abstract and less anthropomorphic, we can say that $\hat{\mathbf{i}} \times \hat{\mathbf{j}} = \hat{\mathbf{k}}$ in such a system. Obviously there is nothing in nature that requires us to limit ourselves to right-handed coordinate systems, and we might ask if there is anything special about the set of numbers a_{ij} if the primed system should happen to be a left-handed system. For a left-handed system we can write

$$(\hat{\mathbf{i}}' \times \hat{\mathbf{j}}') \cdot \hat{\mathbf{k}}' = -1 \qquad (1\text{-}2\text{-}5)$$

Expressing $\hat{\mathbf{i}}', \hat{\mathbf{j}}'$, and $\hat{\mathbf{k}}'$ in terms of $\hat{\mathbf{i}}, \hat{\mathbf{j}}$, and $\hat{\mathbf{k}}$, we can rewrite this equation as follows:

$$[(a_{11}\hat{\mathbf{i}} + a_{12}\hat{\mathbf{j}} + a_{13}\hat{\mathbf{k}}) \times (a_{21}\hat{\mathbf{i}} + a_{22}\hat{\mathbf{j}} + a_{23}\hat{\mathbf{k}})]$$
$$\cdot (a_{31}\hat{\mathbf{i}} + a_{32}\hat{\mathbf{j}} + a_{33}\hat{\mathbf{k}}) = -1$$

Carrying out the indicated multiplications, we find

$$a_{11}(a_{22}a_{33} - a_{23}a_{32}) + a_{12}(a_{23}a_{31} - a_{21}a_{33})$$
$$+ a_{13}(a_{21}a_{32} - a_{22}a_{31}) = -1 \qquad (1\text{-}2\text{-}6)$$

The expression on the left of Eq. (1-2-6) is called the **determinant** of the matrix of numbers a_{ij} or det a_{ij} for short. It is often written in the notation

$$\det a_{ij} = \begin{Vmatrix} a_{11} & a_{12} & a_{13} \\ a_{21} & a_{22} & a_{23} \\ a_{31} & a_{32} & a_{33} \end{Vmatrix}$$

We see then that any transformation that takes us from a right-handed coordinate system to a left-handed coordinate system is characterized by having its determinant equal to -1. Indeed, as we can easily see, the determinant is equal to -1 whenever we change the handedness of our system and $+1$ if we keep it unchanged. By allowing transformations with either sign of determinant, we allow ourselves to deal with both rotations and reflections or with any combination of these transformations.

We have begun to think of our transformation as having an "identity" all its own. It is characterized by a set of nine numbers, which we have called a **matrix**. Furthermore we have seen in Eq. (1-2-4) that we can obtain the triplet (x', y', z') by "multiplying" the triplet (x, y, z) by this matrix, with the operation of multiplication being defined as

$$\begin{bmatrix} x' \\ y' \\ z' \end{bmatrix} = \begin{bmatrix} a_{11} & a_{12} & a_{13} \\ a_{21} & a_{22} & a_{23} \\ a_{31} & a_{32} & a_{33} \end{bmatrix} \begin{bmatrix} x \\ y \\ z \end{bmatrix} = \begin{bmatrix} a_{11}x + a_{12}y + a_{13}z \\ a_{21}x + a_{22}y + a_{23}z \\ a_{31}x + a_{32}y + a_{33}z \end{bmatrix} \qquad (1\text{-}2\text{-}7)$$

We can represent the above operation symbolically by writing

$$\mathbf{r'} = \mathbf{ar} \tag{1-2-8}$$

(In the future, a boldface sans serif symbol, such as $\mathbf{a}$, will mean that the symbol is a matrix and not a number.)

Suppose now that we wish to undertake two successive transformations, the first characterized by $\mathbf{a}$ and the second by another matrix $\mathbf{b}$. If we begin with the triplet $\mathbf{r}$, then the first transformation leads to the triplet $\mathbf{r'}$ and the second to the triplet $\mathbf{r''}$. That is,

$$\mathbf{r'} = \mathbf{ar}$$
$$\mathbf{r''} = \mathbf{br'}$$

Alternatively, we might have gone directly from the unprimed to the double-primed coordinate system by means of a transformation $\mathbf{c}$.

$$\mathbf{r''} = \mathbf{cr}$$

Writing out these transformations in detail will show that we could determine all the elements of $\mathbf{c}$ directly from $\mathbf{a}$ and $\mathbf{b}$ by means of the simple set of equations

$$c_{11} = b_{11}a_{11} + b_{12}a_{21} + b_{13}a_{31}$$
$$c_{12} = b_{11}a_{12} + b_{12}a_{22} + b_{13}a_{32}$$

or, in general,

$$c_{ij} = b_{i1}a_{1j} + b_{i2}a_{2j} + b_{i3}a_{3j}$$

We abbreviate this in the customary way by writing

$$c_{ij} = \sum_{k=1}^{3} b_{ik}a_{kj} \tag{1-2-9}$$

Thus the element c_{ij} can be obtained by taking the "scalar product," so to speak, of the ith row in $\mathbf{b}$ with the jth column in $\mathbf{a}$.

The operation which we have defined above in Eq. (1-2-9) is called the **product of two matrices a and b** and can be represented by the expression $\mathbf{c} = \mathbf{ba}$. Matrix multiplication, unlike the multiplication of two numbers, is *not* in general commutative, as the reader can very easily convince himself. That is to say the product $\mathbf{ab}$ is not in general equal to the product $\mathbf{ba}$. Multiplication is, however, associative. This means that we can in general write, for three transformations $\mathbf{a}$, $\mathbf{b}$, and $\mathbf{c}$,

$$\mathbf{a(bc)} = \mathbf{(ab)c} \tag{1-2-10}$$

To complete our picture we should point out that one of the possible transformations is the identity transformation which leaves the coordinate system unchanged. We write this matrix as $\mathbf{1}$ with the observation that

$$\mathbf{1} = \begin{bmatrix} 1 & 0 & 0 \\ 0 & 1 & 0 \\ 0 & 0 & 1 \end{bmatrix} \tag{1-2-11}$$

Returning back to Eqs. (1-2-1) and (1-2-2), we see that for every transformation $\mathbf{a}$ there is also an inverse transformation $\mathbf{a}^{-1}$ such that

$$\mathbf{a}\mathbf{a}^{-1} = \mathbf{a}^{-1}\mathbf{a} = \mathbf{1} \tag{1-2-12}$$

The inverse transformation is just given by the **transposed** matrix. That is to say

$$a_{ij}^{-1} = a_{ji} \tag{1-2-13}$$

(For those whose mathematical sophistication is just a bit above average, we might point out that the set of all transformations defined above constitute what is known in the trade as a **group**. The detailed properties of groups play an important role in the development of much of quantum mechanics and should be studied at the earliest possible moment by those who intend to extend their horizons in physics beyond the classical domain.)

We can now think in terms of the complete set of all transformations from one orthogonal coordinate system to another, including within our set both rotations (det $\mathbf{a} = +1$) and reflections (det $\mathbf{a} = -1$). The definition of scalar, vector, and various other entities is now best done in terms of this set of transformations.

Let us begin with what is intuitively the simplest of these entities, the scalar. Imagine that we are given a set of explicit instructions for determining some number. We follow these instructions scrupulously, coming up with a value for the number. We can now rotate our coordinate system or change its handedness (by means of the transformation $\mathbf{a}$). If the same set of rules for determining the number leads to the same result in the new system, regardless of the choice of rotation or reflection, then the number is a **scalar**.

Obviously there are innumerable trivial examples of scalars that we can readily cite. The number of cents in the dollar or the number of fingers on your hand have nothing to do with the coordinate system and hence are ipso facto scalars. Much less trivial, though, are numbers that are derived by means of rules which concern coordinates themselves. Let us take a simple example.

Suppose the rule tells us to take the x coordinate of a point, square it, add to that the square of the y coordinate of the same point, and add to the sum the square of the z coordinate of the point. We would have then a number equal to $x^2 + y^2 + z^2$. If we transform to a new system and follow the same prescription in the new system, we come up with $x'^2 + y'^2 + z'^2$. Unless we knew the Pythagorean theorem we would have no a priori expectation that the same rule applied in these different systems would give us the same result. Indeed it does because we have just determined the square of the distance from our point to the origin, and that quantity does not depend on the rotational orientation or the handedness of our system. Clearly then the number $x^2 + y^2 + z^2$ is a scalar.

Let us try a more difficult example now. Consider two points whose coordinates in one system are (x_1,y_1,z_1) and (x_2,y_2,z_2). We can form the expression $x_1x_2 + y_1y_2 + z_1z_2$ and evaluate it in this coordinate system. We can now transform coordinates and evaluate the same expression in the new system, obtaining $x_1'x_2' + y_1'y_2' + z_1'z_2'$. Again we have no a priori expectation that the two numbers will come out to be the same. Making use of Eqs. (1-2-4) and (1-2-3), the reader can easily convince himself that this is, however, the case—the numbers are the same and so the expression $x_1x_2 + y_1y_2 + z_1z_2$ is a scalar. (The result is not entirely unanticipated for we remember that this expression is the scalar product of $\mathbf{r}_1$ and $\mathbf{r}_2$ and can also be written as $|\mathbf{r}_1||\mathbf{r}_2|\cos\theta$. The latter formula does not depend on the coordinate system.)

There is a great temptation now to let every "constant" of nature, like charge and mass, be labeled a scalar. In fact we must be exceedingly careful since an attribute like charge is defined operationally in terms of forces by external fields, and we must investigate the behavior of the entire system under *both* rotation and reflection before we can conclude that the attribute is a scalar. We shall have more to say about this very shortly.

We go on now to the definition of another important entity, the **pseudoscalar**. The pseudoscalar differs from the scalar in only one important respect. The sign of the number we obtain by following our prescription in a left-handed coordinate system is *opposite* to that we obtain in a right-handed system. For pure rotations, scalars and pseudoscalars behave identically.

To find an example of a pseudoscalar is not difficult at all. Let us take three points in space which in one coordinate system have the components (x_1,y_1,z_1), (x_2,y_2,z_2), and (x_3,y_3,z_3). We can construct a determinant D out of these nine numbers:

$$D = \begin{Vmatrix} x_1 & y_1 & z_1 \\ x_2 & y_2 & z_2 \\ x_3 & y_3 & z_3 \end{Vmatrix}$$

$$= x_1(y_2z_3 - y_3z_2) + y_1(z_2x_3 - z_3x_2) + z_1(x_2y_3 - x_3y_2)$$

$$(1\text{-}2\text{-}14)$$

(It is quite clear that D is equal to $\mathbf{r}_1 \cdot (\mathbf{r}_2 \times \mathbf{r}_3)$ and has magnitude equal to the volume of the parallelopiped determined by $\mathbf{r}_1$, $\mathbf{r}_2$, and $\mathbf{r}_3$.) If we change coordinates by means of a pure rotation with det $\mathbf{a} = +1$, we find that D' as evaluated in the new system for the same three points is unchanged. On the other hand, if we change from a right-handed to a left-handed system, D' changes sign. (Take the simplest such transformation corresponding to $x' = -x$, $y' = -y$, and $z' = -z$ and substitute above. Then rotate to any other left-handed system.)

So far we have been dealing with prescriptions whereby we come up with single numbers. However, we have already discovered some entities which require three components for their specification, like the coordinates of a point. This brings us then to another class of mathematical objects which we shall call the **polar vector.**

Imagine that we have a prescription for calculating a triplet and obtain (v_1, v_2, v_3) as the result of following this prescription. Consider next that we transform to a new coordinate system by means of the transformation $\mathbf{a}$ and then apply the same prescription as before. We would obtain a new triplet (v'_1, v'_2, v'_3). These triplets are the components of a polar vector $\mathbf{v}$ if and only if, for any choice of $\mathbf{a}$, we have

$$
\begin{aligned}
v'_1 &= a_{11}v_1 + a_{12}v_2 + a_{13}v_3 \\
v'_2 &= a_{21}v_1 + a_{22}v_2 + a_{23}v_3 \\
v'_3 &= a_{31}v_1 + a_{32}v_2 + a_{33}v_3
\end{aligned}
\tag{1-2-15}
$$

Obviously the three coordinates of a point (x, y, z) constitute a polar vector. So do the three components of velocity and acceleration. Using the notation we have developed earlier, we can write

$$\mathbf{v}' = \mathbf{a}\mathbf{v}$$

An important characteristic of a polar vector is the fact that $\mathbf{v}$ changes sign under the pure inversion represented by

$$
\mathbf{a}_{\text{inv}} = \begin{bmatrix} -1 & 0 & 0 \\ 0 & -1 & 0 \\ 0 & 0 & -1 \end{bmatrix}
\tag{1-2-16}
$$

Under this transformation, of course,

$$
\begin{aligned}
v'_1 &= -v_1 \\
v'_2 &= -v_2 \\
v'_3 &= -v_3
\end{aligned}
\tag{1-2-17}
$$

On the other hand, we can imagine a triplet of numbers which behaves exactly like a polar vector under rotation but does *not* change sign under inversion. Such an entity comes under the classification of **axial vector.** To construct such a triplet we need only take the vector product of two polar vectors $\mathbf{v}$ and $\mathbf{w}$:

$$\mathbf{u} = \mathbf{v} \times \mathbf{w}$$
$$= [(v_2 w_3 - v_3 w_2), (v_3 w_1 - v_1 w_3), (v_1 w_2 - v_2 w_1)] \qquad (1\text{-}2\text{-}18)$$

Under an inversion,

$$\mathbf{v}' = -\mathbf{v} \qquad \mathbf{w}' = -\mathbf{w} \qquad \text{but} \qquad \mathbf{u}' = +\mathbf{u} \qquad (1\text{-}2\text{-}19)$$

We shall shortly discuss more complex mathematical entities with more than three components. For the moment though, let us pause and see if we can understand the physical importance of what we have done.

One of our underlying physical principles is that there is no preferred direction or handedness[1] to the universe. This means that the basic laws of nature cannot depend on the coordinate system we choose to use for their formulation. Now basic physical laws are written down as equations. If the two sides of an equation do not have the same transformation properties then the form of the equation *will* depend on the coordinate system we chose, thereby violating our principle. For example, suppose our equation sets a scalar quantity equal to a pseudoscalar quantity. If we reverse the handedness of our system, one side of the equation would change sign and the other would not, leading to an obvious change in the appearance of the physical law. To avoid these problems we shall agree never to write down a basic equation in classical physics where the two sides do not behave identically under transformation. That means that we will always equate scalars with scalars, polar vectors with polar vectors, axial vectors with axial vectors, and so forth. In this way, if an equation is true in one coordinate system then the identical equation will be true in any system related to it by rotations or inversions.

We should emphasize that in all the above we are only talking about those equations which describe the *fundamental* physical laws. In applying the physical laws to any specific situation, we will usually find that there is a preferred coordinate system to use and hence what we have said above would not necessarily hold true. For example, if we were studying the trajectory followed by a baseball near the earth's surface we would naturally choose one of our coordinate axes in the upward direction. On the other hand, when we write down a general set of laws governing the behavior of magnetic and electric fields (Maxwell's equations), we will certainly insist that no preference be given to any coordinate system or to a particular handedness of our system.

The type of reasoning we have just described plays a particularly important role in electromagnetism, and we would like to take the liberty of drawing on some illustrations here even though we have not developed

---

[1] This belief has been shaken in recent years by the discovery of parity violation and the violation of time-reversal invariance in the weak interactions. Nevertheless it is still true to the best of our knowledge in classical electromagnetic theory.

the subject yet. (Hopefully the student taking this course has already studied some elementary physics before.) The first discovery we will make in electrostatics is Coulomb's law, where we will find that the force by one charge on another is proportional to the product of the charges and inversely proportional to the distance between them. Looking ahead to Eq. (2-1-1), we shall write

$$\mathbf{F}_{12} = \text{force by charge 1 on charge 2}$$

$$= \frac{q_1 q_2 (\mathbf{r}_2 - \mathbf{r}_1)}{|\mathbf{r}_2 - \mathbf{r}_1|^3}$$

Now, at the moment we do not know whether charge is a scalar or a pseudoscalar. Regardless of what it is though, the product of two charges $q_1 q_2$ is a scalar. Hence the right side of the equation is a polar vector. This in turn means that *force* is a polar vector. Since *force* is equal to *mass* times *acceleration*, we next conclude that *mass* is a scalar (under these three-dimensional transformations).

What about charge itself though? Whenever we see an elementary force in electromagnetism it is always proportional to the product of charges. Hence, there is no way of determining whether charge is a scalar or pseudoscalar quantity. Since it makes no difference, we will assume it to be a scalar. This implies immediately that electric field **E** is a polar vector since it is just equal to force per unit charge.

How about magnetic field? As we shall learn, a magnetic field **B** exerts a force on a charge q given by

$$\mathbf{F} = q \frac{\mathbf{v}}{c} \times \mathbf{B}$$

where c is the velocity of light. This tells us that **B** is an axial vector.

In Chap. 3 we will discuss an experiment to search for magnetic monopoles. These, if they exist, are elementary magnetic charges which are acted upon by a magnetic field in the same way as ordinary electric charges are acted upon by an electric field. We would have then, for a *magnetic charge* q_m, a force given by

$$\mathbf{F}_m = q_m \mathbf{B}$$

Since **B** is an axial vector, we conclude that q_m is a pseudoscalar quantity. Obviously, if we had begun by choosing electric charge to be a pseudoscalar, then we would now come out with magnetic charge as a scalar.

We can apply the principles we have discussed above in another manner to help us in determining the physical laws themselves. Suppose we have determined that a part of our physical law states that a particular component of a given polar (or axial) vector **v** is equal to the same component of another polar (or axial) vector **w**. We can choose our x axis along the direction in which the components are known to be equal and then summarize our knowledge by the statement that $v_x = w_x$ in this coordinate system. If we further know that there is *no* preferred system for the laws we are uncovering, then we can deduce that $v_y = w_y$ and $v_z = w_z$ also. In other words, **v** = **w** in all systems. We shall make extensive use of this procedure when we introduce magnetism in Chap. 3. The only difference will be that we will be working in the four-dimensional world of special relativity. It is at that point that the full beauty of the notions we have developed here will become apparent.

1-3 DIFFERENTIATION OF VECTORS WITH RESPECT TO TIME AND POSITION; THE "DEL" OPERATOR (∇) AS A VECTOR

Since we know how to add and subtract vectors, we know how to differentiate them. If $\mathbf{r}(t) = (x(t), y(t), z(t))$ is a position vector of a moving particle, then we have

$$\frac{d\mathbf{r}(t)}{dt} = \lim_{\Delta t \to 0} \frac{\mathbf{r}(t + \Delta t) - \mathbf{r}(t)}{\Delta t} \tag{1-3-1}$$

But

$$\mathbf{r}(t + \Delta t) - \mathbf{r}(t) = [x(t + \Delta t) - x(t),$$
$$y(t + \Delta t) - y(t), z(t + \Delta t) - z(t)]$$

and hence

$$\frac{d\mathbf{r}(t)}{dt} = \left(\frac{dx}{dt}, \frac{dy}{dt}, \frac{dz}{dt}\right) \tag{1-3-2}$$

We define $d\mathbf{r}/dt$ to be the velocity vector $\mathbf{v}(t)$ of our particle. Similarly we can define acceleration by

$$\mathbf{a} = \frac{d\mathbf{v}}{dt} = \left(\frac{d^2 x}{dt^2}, \frac{d^2 y}{dt^2}, \frac{d^2 z}{dt^2}\right) \tag{1-3-3}$$

We note immediately that both $\mathbf{v}(t)$ and $\mathbf{a}(t)$ are vectors.[1] This is so because the variable t is independent of the orientation of our coordinate system and can be kept constant as we carry out a rotation. If instead of t we had used x, the triplet created would not have been a vector at all.

Before we can treat derivatives with respect to the coordinates x, y, z, we must learn about partial differentiation, a trivial extension of ordinary differentiation. Let f be a function of the three variables x, y, z. The partial

---

[1] In the future we shall not differentiate between polar and axial vectors unless it is specifically important to do so. We shall call them both vectors.

derivative of f with respect to any of these variables is defined to be the ordinary derivative if we keep the other two variables fixed. Thus we have, for example,

$$\frac{\partial f(x,y,z)}{\partial x} = \lim_{\Delta x = 0} \frac{f(x + dx, y, z) - f(x,y,z)}{\Delta x} \tag{1-3-4}$$

Suppose now we start from some point (x,y,z) and move to a new, nearby position $(x + \Delta x, y + \Delta y, z + \Delta z)$. How much does f change? The answer is simple if we carry out our motion in three steps. First, we move from x to $x + dx$ and the function changes by an amount $(\partial f/\partial x)\,\Delta x$. Then we move in the y direction by an amount Δy. The function now changes by an amount $(\partial f/\partial y)\,\Delta y$. Last, we move in the z direction by an amount Δz. The function changes by an additional amount $(\partial f/\partial z)\,\Delta z$. (It is assumed in all this that Δx, Δy, and Δz are sufficiently small so that the derivatives do not vary significantly enough over this interval to affect these approximations.)

Before we add these changes together let us, in the customary manner, replace the Δ's by d's, signifying that we are dealing with infinitesimal displacements. We have then for the total change in f:

$$df = \frac{\partial f}{\partial x}\,dx + \frac{\partial f}{\partial y}\,dy + \frac{\partial f}{\partial z}\,dz \tag{1-3-5}$$

Now the triplet (dx,dy,dz) is surely a vector, and the change in f as we go from one place to another is surely a scalar (it does not depend on the orientation of our coordinate system). The right side of Eq. (1-3-5) looks like the scalar product of the triplet $(\partial f/\partial z, \partial f/\partial y, \partial f/\partial z)$ and the vector (dx,dy,dz). We expect then that the triplet $(\partial f/\partial x, \partial f/\partial y, \partial f/\partial z)$ is a vector, and we can demonstrate this explicitly. [This triplet is called the **gradient of f** or ∇f ("del f").]

To do so we must learn how to transform from one set of variables (x,y,z) to another (x',y',z'). Let us for convenience choose the displacement which led to df in such a way that only x' is changed and not y' or z'. If we divide Eq. (1-3-5) by dx' on both sides, we have

$$\frac{\partial f}{\partial x'} = \frac{\partial f}{\partial x}\frac{\partial x}{\partial x'} + \frac{\partial f}{\partial y}\frac{\partial y}{\partial x'} + \frac{\partial f}{\partial z}\frac{\partial z}{\partial x'} \tag{1-3-6}$$

Similarly, if we allow our displacements to be dy' and dz', we would obtain the equations

$$\frac{\partial f}{\partial y'} = \frac{\partial f}{\partial x}\frac{\partial x}{\partial y'} + \frac{\partial f}{\partial y}\frac{\partial y}{\partial y'} + \frac{\partial f}{\partial z}\frac{\partial z}{\partial y'} \tag{1-3-7}$$

$$\frac{\partial f}{\partial z'} = \frac{\partial f}{\partial x}\frac{\partial x}{\partial z'} + \frac{\partial f}{\partial y}\frac{\partial y}{\partial z'} + \frac{\partial f}{\partial z}\frac{\partial z}{\partial z'} \tag{1-3-8}$$

Going back to Eqs. (1-2-4), we find

$$\frac{\partial x}{\partial x'} = a_{11} \qquad \frac{\partial y}{\partial x'} = a_{12} \qquad \frac{\partial z}{\partial x'} = a_{13}$$

$$\frac{\partial x}{\partial y'} = a_{21} \qquad \frac{\partial y}{\partial y'} = a_{22} \qquad \frac{\partial z}{\partial y'} = a_{23} \qquad (1\text{-}3\text{-}9)$$

$$\frac{\partial x}{\partial z'} = a_{31} \qquad \frac{\partial y}{\partial z'} = a_{32} \qquad \frac{\partial z}{\partial z'} = a_{33}$$

Substituting back into (1-3-6) to (1-3-8), we obtain appropriate transformation equations for ∇f:

$$\frac{\partial f}{\partial x'} = a_{11} \frac{\partial f}{\partial x} + a_{12} \frac{\partial f}{\partial y} + a_{13} \frac{\partial f}{\partial z}$$

$$\frac{\partial f}{\partial y'} = a_{21} \frac{\partial f}{\partial x} + a_{22} \frac{\partial f}{\partial y} + a_{23} \frac{\partial f}{\partial z} \qquad (1\text{-}3\text{-}10)$$

$$\frac{\partial f}{\partial z'} = a_{31} \frac{\partial f}{\partial x} + a_{32} \frac{\partial f}{\partial y} = a_{33} \frac{\partial f}{\partial z}$$

Hence the gradient of f does indeed transform as a vector under rotation. The interesting thing, though, is that the transformation properties we are interested in do not concern f at all. They only concern the triplet of *differential operators* $\nabla = (\partial/\partial x, \partial/\partial y, \partial/\partial z)$. Although ∇ really means nothing unless something appears on its right, it nevertheless has all the characteristics of a polar vector under both rotation or inversion.

In any case let us see what the gradient is like "physically." We see from Eq. (1-3-5) that the function f changes by an amount $\nabla f \cdot d\mathbf{r}$ if we undergo a displacement $d\mathbf{r}$. Thus ∇f points along the direction in which f changes most rapidly. Its magnitude is the rate of change of f with respect to distance along that direction.

We shall have much more to do with our "del" operator as we proceed to develop a number of other mathematical and physical concepts.

1-4   THE NOTION OF FLUX; DIVERGENCE
OF A VECTOR FIELD; GAUSS' THEOREM

As we examine the world around us we find numerous instances where we need to speak of vector fields, that is, vector functions of position defined for all points within a given volume. The velocity or momentum density of fluids, electric and magnetic fields, and gravitational forces are examples that we come across quite readily. In each of these cases we will often make use of a very simple notion—the flux of the vector field $\mathbf{v}(x,y,z)$ through an imaginary infinitesimal bit of area dA with normal vector **n**. We first define it mathematically.

Flux through dA in direction of $\hat{n}$ = **v** · $\hat{n}$ dA

The reason for calling this quantity *flux* is reasonably clear. As we see from Fig. 1-3, if **v** is the velocity of a fluid of uniform density, then the flux is just the volume of fluid that passes through the surface ΔA in 1 sec.

Needless to say, we can now proceed to integrate the flux over a finite area A. We have then

$$\text{Flux of } \mathbf{v} \text{ through } A \text{ in direction of } \hat{n} = \int_A \mathbf{v} \cdot \hat{n} \, dA \qquad (1\text{-}4\text{-}1)$$

We now prove a simple but powerful theorem which relates the net outward-going flux of **v** through a closed surface to the space derivatives of **v** within the surface. This theorem will be of particular value to us in our study of electrostatics. We begin by taking the infinitesimal volume bounded by the range of coordinates x to $x + dx$, y to $y + dy$, z to $z + dz$, and shown in Fig. 1-4.

Let wall 1 be the wall parallel to the xy plane through the point (x,y,z).

Let wall 2 be the wall parallel to the xy plane through the point $(x + dx, y + dy, z + dz)$.

Let wall 3 be the wall parallel to the yz plane through the point (x,y,z).

Fig. 1-3 The flux through dA of the vector field **v** is just **v** · $\hat{n}$ dA. If **v** is the velocity of a fluid of uniform density, then **v** · $\hat{n}$ dA is just the volume of fluid passing through ΔA in 1 sec.

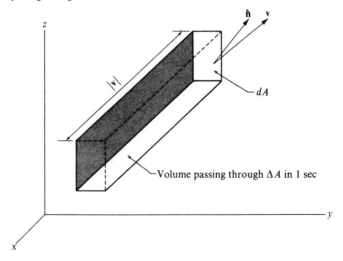

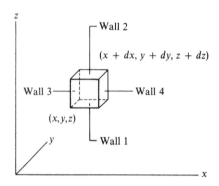

Fig. 1-4 We calculate the outward-going flux of the vector field **v** from the infinitesimal volume shown here.

Let wall 4 be the wall parallel to the yz plane through the point $(x + dx, y + dy, z + dz)$.

Let wall 5 be the wall parallel to the xz plane through the point (x,y,z).

Let wall 6 be the wall parallel to the xz plane through the point $(x + dx, y + dy, z + dz)$.

The area of wall 1 is $dx\,dy$. The outward-going normal to wall 1 is $-\hat{\mathbf{k}}$.

The area of wall 2 is $dx\,dy$. The outward-going normal to wall 2 is $+\hat{\mathbf{k}}$.

The area of wall 3 is $dy\,dz$. The outward-going normal to wall 3 is $-\hat{\mathbf{i}}$.

The area of wall 4 is $dy\,dz$. The outward-going normal to wall 4 is $+\hat{\mathbf{i}}$.

The area of wall 5 is $dx\,dz$. The outward-going normal to wall 5 is $-\hat{\mathbf{j}}$.

The area of wall 6 is $dx\,dz$. The outward-going normal to wall 6 is $+\hat{\mathbf{j}}$.

The outward-going flux through wall 1 is $-\mathbf{v}(1) \cdot \hat{\mathbf{k}}\, dx\, dy$.

The outward-going flux through wall 2 is $\mathbf{v}(2) \cdot \hat{\mathbf{k}}\, dx\, dy$.

The net outward-going flux through walls 1 and 2 is then

$$(\text{Net flux})_{1+2} = [\mathbf{v}(2) - \mathbf{v}(1)] \cdot \hat{\mathbf{k}}\, dx\, dy$$

$$= [v_z(2) - v_z(1)]\, dx\, dy$$

$$= \left(\frac{\partial v_z}{\partial z}\right) dz\, dx\, dy$$

Similarly

$$(\text{Net flux})_{2+3} = \left(\frac{\partial v_x}{\partial x}\right) dx\, dy\, dz$$

and

$$(\text{Net flux})_{5+6} = \left(\frac{\partial v_y}{\partial y}\right) dy\ dx\ dz$$

Noting that the volume of dV is $dx\ dy\ dz$, we have

$$\text{Net flux out} = \left(\frac{\partial v_x}{\partial x} + \frac{\partial v_y}{\partial y} + \frac{\partial v_z}{\partial z}\right) dV \tag{1-4-2}$$

We note now that the term $(\partial v_x/\partial x + \partial v_y/\partial y + \partial v_z/\partial z)$ is just $\nabla \cdot \mathbf{v}$; that is, it is the scalar product of the "del" operator and the vector function $\mathbf{v}$. Just like any scalar product of two vectors it must be invariant under rotation. This scalar product $\nabla \cdot \mathbf{v}$ is generally called the **divergence of v**.

As we see from Eq. (1-4-2) the divergence of a vector function of space is a measure of the extent to which there are local sources (or sinks) present, that is, the extent to which there is a net flux out of (or into) a region.

We can extend this very simply to a finite volume V enclosed by a surface S. The volume can be broken into infinitesimal bits dV. Let us sum together the flux leaving all bits. The flux leaving one volume bit will either pass through S or enter another volume bit. Hence we have the result

Sum of all flux leaving all volume bits = flux leaving through S

In mathematical terms we can write

$$\int_{\substack{\text{volume } V}} (\nabla \cdot \mathbf{v})\, dV = \int_{\substack{\text{surface } S \\ \text{bounding } V}} (\mathbf{v} \cdot \hat{\mathbf{n}})\, dA \tag{1-4-3}$$

This result, known as **Gauss' theorem,** will turn out to be tremendously useful as we proceed with our development of electromagnetic theory.

Returning to Eq. (1-4-2), let us make use of it to derive a differential equation for the flow of a fluid. In this case, $\mathbf{v}$ is the velocity of the fluid at a given point. Let ρ be the density of fluid. The flux of mass out of the infinitesimal volume dV is just $\nabla \cdot \rho\mathbf{v}\, dV$. The amount of mass in the volume is just $\rho\, dV$. Conservation of mass then tells us that

$$\nabla \cdot \rho\mathbf{v} + \frac{\partial \rho}{\partial t} = 0 \tag{1-4-4}$$

This equation will also be encountered later when we discuss charge conservation and will play a rather important role in our relativistic development of electromagnetism.

1-5 THE CURL OF A VECTOR FUNCTION
OF SPACE; STOKES' THEOREM

We have come across the notion of a line integral in the past, when we covered work and energy in our elementary mechanics course. At that time we learned that the bit of work dW done by a force $\mathbf{F}$ on a given point object undergoing a displacement $d\mathbf{l}$ is just $\mathbf{F} \cdot d\mathbf{l}$. If we wish to determine the work done in going from point A to point B along a given path, then we divide our path into infinitesimal bits and add together the contributions from each of the bits. That is,

$$W = \int_A^B \mathbf{F} \cdot d\mathbf{l}$$

Sometimes the path from A to B would be irrelevant; for any A or B the work done would be independent of the path taken between them. In that case we called the force conservative and observed that the *line integral* of $\mathbf{F} \cdot d\mathbf{l}$ around any closed loop would then be zero; that is, $\oint \mathbf{F} \cdot d\mathbf{l} = 0$ for a conservative force.

We now broaden our horizons a bit and consider the integral around a closed loop of the function $\mathbf{v} \cdot d\mathbf{l}$ where $\mathbf{v}(x,y,z)$ is a vector function of space and $d\mathbf{l}$ is an element of the loop. We propose to prove a very important theorem about this integral, relating it mathematically to a surface integral over any surface bounded by our closed loop. To begin with, though, we examine a simple rectangular path in the xy plane with sides that are of length dx and dy, respectively. We choose the direction of the path as shown in Fig. 1-5 and number the legs of the rectangle as indicated. Along leg (1),

$$\mathbf{v} \cdot d\mathbf{l} = [\mathbf{v}(1) \cdot \hat{\mathbf{i}}]\, dx$$
$$= v_x(1)\, dx$$

Along leg (2),
$$\mathbf{v} \cdot d\mathbf{l} = v_y(2)\, dy$$
Along leg (3),
$$\mathbf{v} \cdot d\mathbf{l} = -v_x(3)\, dx$$
Along leg (4),
$$\mathbf{v} \cdot d\mathbf{l} = -v_y(4)\, dy$$
Adding these together, we have
$$\oint \mathbf{v} \cdot d\mathbf{l} = [v_x(1) - v_x(3)]\, dx + [v_y(2) - v_y(4)]\, dy$$

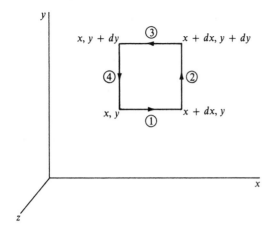

Fig. 1-5 We carry out the "line integral" of $\mathbf{v} \cdot d\mathbf{l}$ around an infinitesimal loop in the xy plane.

but

$$v_x(3) - v_x(1) = \frac{\partial v_x}{\partial y} dy$$

$$v_y(2) - v_y(4) = \frac{\partial v_y}{\partial x} dx$$

Hence

$$\oint_{\substack{\text{infinitesimal} \\ \text{loop}}} \mathbf{v} \cdot d\mathbf{l} = \left(\frac{\partial v_y}{\partial x} - \frac{\partial v_x}{\partial y}\right) dy\, dx \tag{1-5-1}$$

Now the left side of the equation is an invariant under rotation. (Is it a scalar or a pseudoscalar?) Hence we must rewrite the right side so that it also is an invariant under rotation.

$$\left(\frac{\partial v_y}{\partial x} - \frac{\partial v_x}{\partial y}\right) = (\nabla \times \mathbf{v})_z$$

$$dx\, dy = dA$$

Let $\hat{n}$ be a unit vector normal to dA in the direction given by a right-hand rule. That is, $\hat{n}$ points in the direction of the thumb if the fingers are along the direction of the path of integration. Then

$$\oint_{\substack{\text{infinitesimal} \\ \text{loop}}} \mathbf{v} \cdot d\mathbf{l} = (\nabla \times \mathbf{v}) \cdot \hat{n}\, dA \tag{1-5-2}$$

The right-hand side is now in the form where it clearly exhibits the same invariance properties as the left-hand side. Hence we can now rotate our coordinate system arbitrarily, and the equation will still be true. It is thus true in general regardless of the orientation of the bit of area relative to the x, y, z axes.

We now examine the situation of a finite loop C shown in Fig. 1-6. We begin by choosing a direction in which we are to go around the loop. We then cover the loop with a surface S and establish a normal to each point on S in accordance with the same right-hand rule we used earlier. We now break up S into infinitesimal loops as shown. Clearly

$$\oint_C \mathbf{v} \cdot d\mathbf{l} = \sum_i \oint_{\substack{\text{infinitesimal} \\ \text{loop } C_i}} \mathbf{v} \cdot d\mathbf{l}$$

$$= \sum_i (\nabla \times \mathbf{v}_i) \cdot \hat{\mathbf{n}}_i \, dA_i$$

The sum at the right is converted to an integral, and we have **Stokes' theorem**

$$\oint_C \mathbf{v} \cdot d\mathbf{l} = \int_{\substack{\text{surface bounded} \\ \text{by } C}} (\nabla \times \mathbf{v}) \cdot \hat{\mathbf{n}} \, dA \qquad (1\text{-}5\text{-}3)$$

We observe then that $\nabla \times \mathbf{v}$ is a measure of the extent to which $\mathbf{v}$ "curls" about, giving a nonzero line integral around a loop. If $\mathbf{F}$ is a conservative

Fig. 1-6   An arbitrary loop C is broken into a set of infinitesimal rectangular loops C_i.

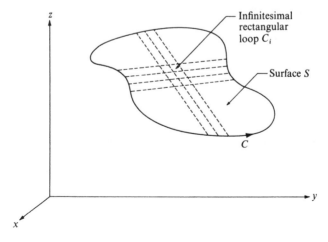

force, then we must have

$$\oint \mathbf{F} \cdot d\mathbf{l} = 0 \qquad \text{for any closed path}$$

Hence

$$\nabla \times \mathbf{F} = 0 \qquad \text{for a conservative force}$$

It is important to note here that the choice of surface was arbitrary. Let S_1 and S_2 be two surfaces, each bounded by C. Since either surface is satisfactory, we can write

$$\oint_{S_1} \nabla \times \mathbf{v} \cdot \hat{\mathbf{n}} \, dA = \oint_{S_2} \nabla \times \mathbf{v} \cdot \hat{\mathbf{n}} \, dA$$

If we let V be the volume enclosed by the two surfaces, we conclude from Gauss' theorem that

$$\int_V (\nabla \cdot \nabla \times \mathbf{v}) \, dV = 0$$

Hence we may expect that $\nabla \cdot \nabla \times \mathbf{v} = 0$ for any vector function $\mathbf{v}$. The reader may verify this result explicitly.

1-6 TENSORS OF THE SECOND RANK

We now go a bit beyond the notion of a vector to introduce a new type of mathematical object, a tensor of the second rank. We have seen that a vector is best defined in terms of its transformation properties under a rotation. The same type of definition will be given here. However, before that, let us play about a bit with a simple example.

Consider the set of nine numbers that are obtained by multiplying the three components of the vector (v_1, v_2, v_3) by three components of the vector (w_1, w_2, w_3). We can list these nine numbers conveniently in the form of a matrix which we call $\mathbf{T}$.

$$\mathbf{T} = \begin{bmatrix} v_1 w_1 & v_1 w_2 & v_1 w_3 \\ v_2 w_1 & v_2 w_2 & v_2 w_3 \\ v_3 w_1 & v_3 w_2 & v_3 w_3 \end{bmatrix}$$

If we had used a different coordinate system, our two vectors would, of course, have had components (v_1', v_2', v_3') and (w_1', w_2', w_3') and we would have come up with a listing

$$\mathbf{T}' = \begin{bmatrix} v_1' w_1' & v_1' w_2' & v_1' w_3' \\ v_2' w_1' & v_2' w_2' & v_2' w_3' \\ v_3' w_1' & v_3' w_2' & v_3' w_3' \end{bmatrix}$$

Now, we might ask, how is an element of $\mathbf{T}'$ related to an element of $\mathbf{T}$? The answer is trivial since we already know precisely how to transform the v's and w's into v''s and w''s [see Eq. (1-2-4)].

$$v_i' = \sum_{k=1}^{3} a_{ik} v_k$$

$$w_j' = \sum_{m=1}^{3} a_{jm} w_m$$

Hence we can obtain the element $v_i' w_j'$ by simple multiplication of the two relevant sums.

$$v_i' w_j' = \sum_{k=1}^{3} \sum_{m=1}^{3} (a_{ik} a_{jm})(v_k w_m) \tag{1-6-1}$$

Alternatively, we can go the other way by the transformation

$$v_i w_j = \sum_{m=1}^{3} \sum_{k=1}^{3} (a_{ki} a_{mj})(v_k' w_m') \tag{1-6-2}$$

We have then a clear-cut way in which the nine elements in the matrix $\mathbf{T}'$ can be written in terms of the nine elements in the matrix $\mathbf{T}$ and vice versa. Each element in one is a sum of coefficients times the elements of the other. The coefficients themselves only depend upon the rotation matrix $\mathbf{a}$ between the two coordinate systems.

We now define a **tensor of the second rank** as an array of nine objects which transforms in the same manner as the elements $v_i w_j$ under rotation. Thus we have, for any tensor $\mathbf{T}$,

$$T_{kl}' = \sum_{i,j} a_{ki} a_{lj} T_{ij} \tag{1-6-3}$$

Incidentally, having learned to multiply matrices [see Eq. (1-2-7)], we can rewrite Eq. (1-6-3) in a simple form.

$$\mathbf{T}' = \mathbf{a T a}^{-1} \tag{1-6-4}$$

where $\mathbf{a}^{-1}$ is the inverse rotation matrix. We remember that

$$a_{ij}^{-1} = a_{ji} \tag{1-6-5}$$

To get a better feeling for what a tensor actually is in physical terms, let us prove a simple mathematical theorem. We will show that the product of a second-rank tensor $\mathbf{T}$ and a vector $\mathbf{v}$ transforms as a vector. To do this, let us first write down the product $\mathbf{Tv}$ as it appears in both the primed and the unprimed coordinate systems. We will call the components of these products w_j and w_j', respectively.

$$w_j = \sum_{i} T_{ji} v_i \tag{1-6-6}$$

$$w'_j = \sum T'_{ji} v'_i \qquad (1\text{-}6\text{-}7)$$

What we would like to show, of course, is that the three numbers w_j in Eq. (1-6-6) transform into the three numbers in Eq. (1-6-7) precisely as would the three components of a vector.

To prove this theorem, let us just make use of the fundamental transformation properties of **T** and **v**.

$$T_{ji} = \sum_{kl} a_{kj} a_{li} T'_{kl}$$
$$v_i = \sum_m a_{mi} v'_m \qquad (1\text{-}6\text{-}8)$$

We substitute these expressions back into Eq. (1-6-6), obtaining thereby

$$w_j = \sum_{i,k,l,m} a_{kj} a_{li} a_{mi} T'_{kl} v'_m \qquad (1\text{-}6\text{-}9)$$

To evaluate this expression, we sum over i first. We recall Eq. (1-2-3), which can be written in simplified notation as

$$\sum_i a_{li} a_{mi} = \delta_{l,m} \qquad (1\text{-}6\text{-}10)$$

where $\delta_{l,m}$, called the **Kronecker** δ, is zero when m is unequal to l and 1 when m is equal to l. Substituting Eq. (1-6-10) back into Eq. (1-6-9), we obtain

$$w_j = \sum_{k,l,m} \delta_{lm} a_{kj} T'_{kl} v'_m$$

We next sum over m and observe that only the terms with $m = l$ remain.

$$w_j = \sum_{k,l} a_{kj} T'_{kl} v'_l$$

Finally, summing over l and making use of Eq. (1-6-7), we have

$$w_j = \sum_k a_{kj} w'_k \qquad (1\text{-}6\text{-}11)$$

This is exactly the transformation property we require of a vector. Hence we have proven our theorem.

We now have some insight into what a tensor really is. It is a *linear relationship* between two vectors of the form given by Eq. (1-6-6). Whenever we encounter a situation in physics where two vectors depend linearly on one another but point in different directions, the relationship between them will be tensorial in nature. For example, the stress and strain in material are linearly related when the material behaves elastically, yet they do not necessarily point in the same direction. The relationship between them is called the **stress tensor.**

To add two tensors we just add the individual components. This permits us to make a very simple decomposition of any tensor $\mathbf{T}$ into symmetric and antisymmetric parts. Let

$$\mathbf{T} = \mathbf{A} + \mathbf{S} \tag{1-6-12}$$

where

$$A_{ij} = \frac{T_{ij} - T_{ji}}{2} = -A_{ji} \tag{1-6-13}$$

$$S_{ij} = \frac{T_{ij} + T_{ji}}{2} = +S_{ji} \tag{1-6-14}$$

The tensor $\mathbf{A}$ has only three components, and the tensor $\mathbf{S}$ has six components. This decomposition would be pointless were it not for the fact that $\mathbf{A}$ *remains* antisymmetric and $\mathbf{S}$ *remains* symmetric as we rotate.

$$A'_{ij} = \sum_{kl} a_{ik}a_{jl}A_{kl} = -\sum_{kl} a_{ik}a_{jl}A_{lk} = -A'_{ji} \tag{1-6-15}$$

$$S'_{ij} = \sum_{kl} a_{ik}a_{jl}S_{kl} = +\sum_{kl} a_{ik}a_{jl}S_{lk} = +S'_{ji} \tag{1-6-16}$$

Let us take a look at the three components of A_{ij} in terms of their transformation properties under rotation.

A_{12} transforms as $(v_1 w_2 - v_2 w_1)$
A_{13} transforms as $(v_1 w_3 - v_3 w_1)$
A_{23} transforms as $(v_2 w_3 - v_3 w_2)$

Thus the triplet $(A_{23}, -A_{13}, A_{12})$ transforms like the vector product of two vectors. A vector product is simply an antisymmetric tensor of the second rank.

1-7 DIAGONALIZING A SECOND-RANK SYMMETRIC TENSOR

An interesting and important property of a symmetric tensor $\mathbf{S}$ is the fact that it can be "diagonalized." This means that we can rotate to a coordinate system given by a triplet of orthogonal unit vectors $\hat{\mathbf{i}}'$, $\hat{\mathbf{j}}'$, and $\hat{\mathbf{k}}'$ within which $\mathbf{S}'$ has only three diagonal elements, namely, those S'_{ij} for which $i = j$. All off-diagonal elements are equal to zero in this coordinate system. We will show explicitly that this is so by finding the vectors $\hat{\mathbf{i}}'$, $\hat{\mathbf{j}}'$, and $\hat{\mathbf{k}}'$ in terms of the initial base vectors $\hat{\mathbf{i}}$, $\hat{\mathbf{j}}$, and $\hat{\mathbf{k}}$ and then rotating to the new system.

Before doing this, though, let us assume the result to be true and write out the tensor S'_{ij} as we expect it to appear:

$$\mathbf{S}' = \begin{bmatrix} \lambda_1 & 0 & 0 \\ 0 & \lambda_2 & 0 \\ 0 & 0 & \lambda_3 \end{bmatrix} \tag{1-7-1}$$

We notice immediately that $\mathbf{S}'\hat{\mathbf{i}}' = \lambda_1\hat{\mathbf{i}}'$, $\mathbf{S}'\hat{\mathbf{j}}' = \lambda_2\hat{\mathbf{j}}'$, and $\mathbf{S}'\hat{\mathbf{k}}' = \lambda_3\hat{\mathbf{k}}'$. This gives us a clue as to how to proceed, starting out with the tensor $\mathbf{S}$ and its six components. We will search for unit vectors which satisfy the equation

$$\mathbf{S}\hat{\mathbf{n}} = \lambda\hat{\mathbf{n}} \qquad (1\text{-}7\text{-}2)$$

Hopefully we will find three such vectors which will be orthogonal and which we can then identify as $\hat{\mathbf{i}}'$, $\hat{\mathbf{j}}'$, and $\hat{\mathbf{k}}'$.

Let us write out Eq. (1-7-2) explicitly:

$$\begin{bmatrix} S_{11} & S_{12} & S_{13} \\ S_{12} & S_{22} & S_{23} \\ S_{13} & S_{23} & S_{33} \end{bmatrix} \begin{bmatrix} n_x \\ n_y \\ n_z \end{bmatrix} = \lambda \begin{bmatrix} n_x \\ n_y \\ n_z \end{bmatrix} \qquad (1\text{-}7\text{-}3)$$

Expanding the above equation, we find three equations for n_x, n_y, and n_z:

$$\begin{aligned} (S_{11} - \lambda)n_x + S_{12}n_y + S_{13}n_z &= 0 \\ S_{12}n_x + (S_{22} - \lambda)n_y + S_{23}n_z &= 0 \\ S_{13}n_x + S_{23}n_y + (S_{33} - \lambda)n_z &= 0 \end{aligned} \qquad (1\text{-}7\text{-}4)$$

In order that we have a nonzero solution for n_x, n_y, and n_z, we must set the determinant of these equations equal to zero. That is, we must have

$$\begin{Vmatrix} S_{11} - \lambda & S_{12} & S_{13} \\ S_{12} & S_{22} - \lambda & S_{23} \\ S_{13} & S_{23} & S_{33} - \lambda \end{Vmatrix} = 0 \qquad (1\text{-}7\text{-}5)$$

Clearly this yields a cubic equation for λ with three solutions which we can call λ_1, λ_2, and λ_3. We can now throw out one of Eqs. (1-7-4) because they are no longer all independent and use the other two to determine n_y/n_x and n_z/n_x for each choice of λ. Normalizing so that $n_x{}^2 + n_y{}^2 + n_z{}^2 = 1$ completes our job of determining $\hat{\mathbf{n}}_1$, $\hat{\mathbf{n}}_2$, and $\hat{\mathbf{n}}_3$.

Now, before we can identify $\hat{\mathbf{n}}_1$, $\hat{\mathbf{n}}_2$, and $\hat{\mathbf{n}}_3$ with $\hat{\mathbf{i}}'$, $\hat{\mathbf{j}}'$, and $\hat{\mathbf{k}}'$, we must show that they are orthogonal. Let us assume first that λ_1, λ_2, and λ_3 are all different. Then we have from Eq. (1-7-2)

$$\begin{aligned} \mathbf{S}\hat{\mathbf{n}}_1 &= \lambda_1\hat{\mathbf{n}}_1 \\ \mathbf{S}\hat{\mathbf{n}}_2 &= \lambda_2\hat{\mathbf{n}}_2 \\ \mathbf{S}\hat{\mathbf{n}}_3 &= \lambda_3\hat{\mathbf{n}}_3 \end{aligned} \qquad (1\text{-}7\text{-}6)$$

We take the scalar product of the first equation with $\hat{\mathbf{n}}_2$ and of the second equation with $\hat{\mathbf{n}}_1$ and then subtract the second equation from the first. This leads to

$$\hat{\mathbf{n}}_2 \cdot \mathbf{S}\hat{\mathbf{n}}_1 - \hat{\mathbf{n}}_1 \cdot \mathbf{S}\hat{\mathbf{n}}_2 = (\lambda_1 - \lambda_2)\hat{\mathbf{n}}_1 \cdot \hat{\mathbf{n}}_2 \qquad (1\text{-}7\text{-}7)$$

Because of the symmetry of $\mathbf{S}$ we see immediately that $\hat{\mathbf{n}}_2 \cdot \mathbf{S}\hat{\mathbf{n}}_1 = \hat{\mathbf{n}}_1 \cdot \mathbf{S}\hat{\mathbf{n}}_2$. Hence we conclude that

$$(\lambda_1 - \lambda_2)\hat{\mathbf{n}}_1 \cdot \hat{\mathbf{n}}_2 = 0 \tag{1-7-8}$$

Since $\lambda_1 \neq \lambda_2$, this equation implies that $\hat{\mathbf{n}}_1$ and $\hat{\mathbf{n}}_2$ are orthogonal. The same proof now demonstrates that $\hat{\mathbf{n}}_3$ is also orthogonal to both $\hat{\mathbf{n}}_1$ and $\hat{\mathbf{n}}_2$. Hence if λ_1, λ_2, and λ_3 are all different, we can identify them with $\hat{\mathbf{i}}'$, $\hat{\mathbf{j}}'$, and $\hat{\mathbf{k}}'$ and arrange the correspondence so as to make up a right-handed coordinate system.

The three constants λ_1, λ_2, and λ_3 are called the **eigenvalues** of Eq. (1-7-2) and the corresponding solutions are the **eigenvectors** $\hat{\mathbf{n}}_1$, $\hat{\mathbf{n}}_2$, and $\hat{\mathbf{n}}_3$. We might ask what happens, for example, if $\lambda_1 = \lambda_2$. It is easy to demonstrate then (see Prob. 1-10) that *any* vector normal to $\hat{\mathbf{n}}_3$ is an eigenvector [solution to Eq. (1-7-2)] with eigenvalue λ_1. Thus we can always choose two orthogonal ones which we can call $\hat{\mathbf{n}}_1$ and $\hat{\mathbf{n}}_2$, respectively, thereby completing our coordinate system.

We shall come back to the subject of tensors when we find them useful in electromagnetism. Suffice it to say at this point that the electric and magnetic fields which we have always thought of as vectors will turn out to be parts of a four-dimensional tensor of the second rank.

PROBLEMS

1-1. Demonstrate the following vector identities.
(a) $\mathbf{A} \times (\mathbf{B} \times \mathbf{C}) = (\mathbf{A} \cdot \mathbf{C})\mathbf{B} - (\mathbf{A} \cdot \mathbf{B})\mathbf{C}$
(b) $\mathbf{A} \cdot (\mathbf{B} \times \mathbf{C}) = (\mathbf{A} \times \mathbf{B}) \cdot \mathbf{C}$
(c) $\nabla uv = u\nabla v + v\nabla u$
(d) $\nabla \cdot u\mathbf{A} = u\nabla \cdot \mathbf{A} + \mathbf{A} \cdot \nabla u$
(e) $\nabla \times u\mathbf{A} = u\nabla \times \mathbf{A} + \nabla u \times \mathbf{A}$
(f) $\nabla \cdot (\mathbf{A} \times \mathbf{B}) = \mathbf{B} \cdot \nabla \times \mathbf{A} - \mathbf{A} \cdot \nabla \times \mathbf{B}$
(g) $\nabla \times (\nabla \times \mathbf{A}) = \nabla(\nabla \cdot \mathbf{A}) - \nabla^2\mathbf{A}$

where $\nabla^2\mathbf{A} = \dfrac{\partial^2 \mathbf{A}}{\partial x^2} + \dfrac{\partial^2 \mathbf{A}}{\partial y^2} + \dfrac{\partial^2 \mathbf{A}}{\partial z^2}$

(h) $\nabla(\mathbf{A} \cdot \mathbf{B}) = \mathbf{A} \times (\nabla \times \mathbf{B}) + (\mathbf{A} \cdot \nabla)\mathbf{B} + \mathbf{B} \times (\nabla \times \mathbf{A}) + (\mathbf{B} \cdot \nabla)\mathbf{A}$
(i) $\nabla \times (\mathbf{A} \times \mathbf{B}) = (\mathbf{B} \cdot \nabla)\mathbf{A} - (\mathbf{A} \cdot \nabla)\mathbf{B} + \mathbf{A}(\nabla \cdot \mathbf{B}) - \mathbf{B}(\nabla \cdot \mathbf{A})$

1-2. Show explicitly that
$\nabla \cdot (\nabla \times \mathbf{v}) = 0$ for any vector function $\mathbf{v}$
$\nabla \times \nabla f = 0$ for any function f

1-3. Prove that if C is a closed curve, S is the surface bounded by C, and φ is any function of space, then

$$\int_C \varphi \, d\mathbf{l} = \int_S (\hat{\mathbf{n}} \times \nabla\varphi) \, dA$$

1-4. Show that the magnitude of $\mathbf{a} \cdot (\mathbf{b} \times \mathbf{c})$ is equal to the volume of the parallelopiped determined by $\mathbf{a}$, $\mathbf{b}$, and $\mathbf{c}$.

1-5. Let $\mathbf{r}$ be a vector from the origin to the point (x,y,z), and let $\mathbf{r}'$ be a vector from the origin to the point (x',y',z'). Evaluate the following expressions in terms of $\mathbf{r}$ and $\mathbf{r}'$.

(a) $\nabla \dfrac{1}{|\mathbf{r} - \mathbf{r}'|}$

(b) $\nabla \dfrac{1}{|\mathbf{r} - \mathbf{r}'|^2}$

(c) $\nabla^2 \dfrac{1}{|\mathbf{r} - \mathbf{r}'|}$ for $\mathbf{r} \neq \mathbf{r}'$

(d) $\nabla \times \dfrac{\mathbf{r} - \mathbf{r}'}{|\mathbf{r} - \mathbf{r}'|^3}$ for $\mathbf{r} \neq \mathbf{r}'$

1-6. Prove the following vector identity for a volume V enclosed by a surface S.

$$\int_V \nabla \times \mathbf{v} \, dV = \int_S \hat{\mathbf{n}} \times \mathbf{v} \, dA$$

where $\mathbf{v}$ is any vector function of space. (Hint: Examine the components of this equation and use Gauss' theorem.)

1-7. The trace of a second-rank tensor is the sum of all of its diagonal elements. That is,

$$\text{Tr } \mathbf{T} = \sum_{k=1}^{3} T_{kk}$$

Show that $\text{Tr } \mathbf{T}$ is invariant under rotation.

1-8. Show that the sum of squares of all the elements of a second-rank tensor is invariant under rotation.

1-9. If $\mathbf{S}$ and $\mathbf{T}$ are two second-rank tensors, show that $\sum_{i,j} S_{ij} T_{ij}$ is an invariant under rotation.

1-10. In diagonalizing a symmetric tensor $\mathbf{S}$, we find that two of the eigenvalues (λ_1 and λ_2) are equal but the third (λ_3) is different. Show that *any* vector which is normal to $\hat{\mathbf{n}}_3$ is then an eigenvector of $\mathbf{S}$ with eigenvalue equal to λ_1.

2
Principles of Electrostatics

2-1 INTRODUCTION; COULOMB'S LAW

Our study of electromagnetism begins very simply with Coulomb's law. In the chapters ahead as we examine the beautiful structure of Maxwell's equations, we must try to remember that it all began as a law of force governing the interaction of two charged particles. It will be amazing to see how much ground we can cover on this one tank of fuel.

As we all know, there is an attribute which exists in matter and which we have come to call **electric charge.** Some objects may have positive charge and some negative. Once a standard charge sign has been chosen, all other charges can be classified in sign by whether they are attracted to or repelled by the standard charge. Having done this, we observe that all pairs of charges with like sign repel one another and all pairs with unlike sign attract one another. This pseudosexual rule can be codified quantitatively by observing that the force of one charge on another is proportional in magnitude to the product of the charges and inversely proportional to the square of their separation. We then write down **Coulomb's law.**

$\mathbf{F}_{12}$ = force by charge 1 on charge 2

$$= \frac{q_1 q_2 (\mathbf{r}_2 - \mathbf{r}_1)}{|\mathbf{r}_2 - \mathbf{r}_1|^3} \tag{2-1-1}$$

Of course we must set up a system of units, and we do so in the most natural way possible. We let distance be measured in centimeters and force in dynes. Equation (2-1-1) then serves to define the unit of charge, which we call the **esu** (electrostatic unit). Thus two equal charges, each of one electrostatic unit, will pull (or push) on each other with a force of one dyne if they are set one centimeter apart.

If we have a number of charges around, then the force on each by all the others can be obtained by simply adding the individual forces vectorially. If $q_1, q_2, \ldots, q_n$ are a set of charges at positions $\mathbf{r}_1, \mathbf{r}_2, \ldots, \mathbf{r}_n$, respectively, then the force on q_i is

$$\mathbf{F}_i = \sum_{j \neq i} \mathbf{F}_{ji} = \sum_{j \neq i} \frac{q_i q_j (\mathbf{r}_i - \mathbf{r}_j)}{|\mathbf{r}_i - \mathbf{r}_j|^3} \tag{2-1-2}$$

We now introduce the notion of **electric field.** Having set our charges $q_1, q_2, \ldots, q_n$ into their positions $\mathbf{r}_1, \ldots, \mathbf{r}_n$, we can place a very small charge at position $\mathbf{r} = (x,y,z)$. We then measure the force per unit charge on q in the limit as q approaches zero. (The limiting procedure is carried out to avoid disturbing the other charges.) The result is the electric field $\mathbf{E}$ at (x,y,z).

$$\mathbf{E}(x,y,z) = \lim_{q \to 0} \frac{\text{force on } q \text{ at } (x,y,z)}{q} \tag{2-1-3}$$

We have considered Coulomb's law and defined $\mathbf{E}$ for a system of point charges. There are no point charges in nature; we shall have to deal with charge distributions over finite volumes. Hence we introduce the notion of **charge density.** We define the function $\rho(x,y,z)$ to be the charge per unit volume at any point (x,y,z). Changing the sum in Eq. (2-1-2) to an integral and making use of our definition of electric field, we have then

$$\mathbf{E}(x,y,z) = \mathbf{E}(\mathbf{r}) = \int_{\substack{\text{all} \\ \text{space}}} \frac{\rho(\mathbf{r}')(\mathbf{r} - \mathbf{r}')}{|\mathbf{r} - \mathbf{r}'|^3} \, dV' \tag{2-1-4}$$

To develop a feeling for what we are doing in our integral, it is convenient to refer to Fig. 2-1. We consider a charged object located in space and a point given by the vector $\mathbf{r}$ at which we wish to evaluate the electric field.

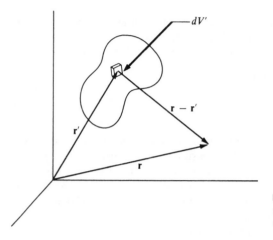

Fig. 2-1   We obtain the field at r by summing the contributions of each element dV'.

We break our volume up into little pieces dV' and let the vector $\mathbf{r}'$ go from the origin to dV'. The charge of dV' is just $\rho(\mathbf{r}')\,dV$. The field at $\mathbf{r}$ due to this bit of charge is just

$$\frac{\rho(\mathbf{r}')(\mathbf{r} - \mathbf{r}')}{|\mathbf{r} - \mathbf{r}'|^3}\,dV'$$

Equation (2-1-4) is obtained by adding all the contributions together. As we perform the integration, $\mathbf{r}$ is, of course, kept fixed.

2-2   THE DIVERGENCE OF E; GAUSS' LAW

The inverse square law governing the rate at which electric field drops off as a function of distance from its source has an immediate and important consequence. In a charge-free region the divergence of $\mathbf{E}$ is zero, and hence within such a region the flux of $\mathbf{E}$ into (or out of) any enclosed volume is zero. As we shall see, the flux of $\mathbf{E}$ out of any enclosed volume is just proportional to the total charge contained within the volume. This result, which is called **Gauss' law,** provides an exceedingly powerful tool for the calculation of electric field in a case of high symmetry. We shall make considerable use of it.

We wish then to evaluate $\nabla \cdot \mathbf{E}$ explicitly. Nothing could be simpler. Using Eq. (2-1-4), we obtain

$$\nabla \cdot \mathbf{E}(\mathbf{r}) = \nabla \cdot \int_{\substack{\text{all} \\ \text{space}}} \frac{\rho(\mathbf{r}')(\mathbf{r} - \mathbf{r}')}{|\mathbf{r} - \mathbf{r}'|^3}\,dV'$$

We can next interchange the order of differentiation and integration. To convince himself that this is a legitimate procedure, the student should carry out the same procedure with differences and sums. The difference between two sums of a set of terms is equal to the sum of the differences. We have then

$$\nabla \cdot \mathbf{E}(\mathbf{r}) = \int_{\substack{\text{all} \\ \text{space}}} \nabla \cdot \frac{\rho(\mathbf{r}')(\mathbf{r} - \mathbf{r}')}{|\mathbf{r} - \mathbf{r}'|^3} \, dV' \tag{2-2-1}$$

Since ρ is not a function of the variables x, y, z with respect to which we are differentiating, we can rewrite this as

$$\nabla \cdot \mathbf{E}(\mathbf{r}) = \int_{\substack{\text{all} \\ \text{space}}} \rho(\mathbf{r}') \, \nabla \cdot \frac{\mathbf{r} - \mathbf{r}'}{|\mathbf{r} - \mathbf{r}'|^3} \, dV' \tag{2-2-2}$$

But

$$\nabla \cdot \frac{\mathbf{r} - \mathbf{r}'}{|\mathbf{r} - \mathbf{r}'|^3} = \left(\nabla \frac{1}{|\mathbf{r} - \mathbf{r}'|^3} \right) \cdot (\mathbf{r} - \mathbf{r}') + \frac{1}{|\mathbf{r} - \mathbf{r}'|^3} \nabla \cdot (\mathbf{r} - \mathbf{r}')$$

Evaluating the various terms in the above expression, we have

$$\nabla \frac{1}{|\mathbf{r} - \mathbf{r}'|^3} = - \frac{3(\mathbf{r} - \mathbf{r}')}{|\mathbf{r} - \mathbf{r}'|^5}$$

$$\nabla \cdot (\mathbf{r} - \mathbf{r}') = \nabla \cdot \mathbf{r} = 3$$

and hence

$$\nabla \cdot \frac{\mathbf{r} - \mathbf{r}'}{|\mathbf{r} - \mathbf{r}'|^3} = \frac{-3}{|\mathbf{r} - \mathbf{r}'|^3} + \frac{3}{|\mathbf{r} - \mathbf{r}'|^3} = 0$$

We thus conclude that $\nabla \cdot \mathbf{E}$ is zero in any charge-free region, as long as $\mathbf{r}'$ can never equal $\mathbf{r}$. We need only consider then the integral over an infinitesimal sphere about the point $\mathbf{r} = \mathbf{r}'$:

$$\nabla \cdot \mathbf{E}(\mathbf{r}) = \int_{\substack{\text{sphere} \\ \text{about } \mathbf{r}}} \rho(\mathbf{r}') \, \nabla \cdot \frac{\mathbf{r} - \mathbf{r}'}{|\mathbf{r} - \mathbf{r}'|^3} \, dV'$$

We can remove $\rho(\mathbf{r}')$ from the integral since it does not vary much over the immediate neighborhood of $\mathbf{r}$. We evaluate it at $\mathbf{r}$.

$$\nabla \cdot \mathbf{E}(\mathbf{r}) = \rho(\mathbf{r}) \int_{\substack{\text{small sphere} \\ \text{about } \mathbf{r}}} \nabla \cdot \frac{\mathbf{r} - \mathbf{r}'}{|\mathbf{r} - \mathbf{r}'|^3} \, dV'$$

Now differentiating $(\mathbf{r} - \mathbf{r}')/|\mathbf{r} - \mathbf{r}'|^3$ with respect to x, y, or z is equivalent to differentiating with respect to $-x'$, $-y'$, and $-z'$. That is to say,

$$\nabla \cdot \frac{\mathbf{r} - \mathbf{r}'}{|\mathbf{r} - \mathbf{r}'|^3} = -\nabla' \cdot \frac{\mathbf{r} - \mathbf{r}'}{|\mathbf{r} - \mathbf{r}'|^3}$$

where

$$\nabla' = \left(\frac{\partial}{\partial x'}, \frac{\partial}{\partial y'}, \frac{\partial}{\partial z'} \right)$$

This gives us the result

$$\nabla \cdot \mathbf{E}(\mathbf{r}) = -\rho(\mathbf{r}) \int_{\substack{\text{sphere} \\ \text{about } \mathbf{r}}} \nabla' \cdot \frac{\mathbf{r} - \mathbf{r}'}{|\mathbf{r} - \mathbf{r}'|^3} \, dV'$$

By Gauss' theorem [Eq. (1-4-3)] we convert this to a surface integral.

$$\nabla \cdot \mathbf{E}(\mathbf{r}) = +\rho(\mathbf{r}) \int_{\substack{\text{surface} \\ \text{of sphere} \\ \text{about } \mathbf{r}}} \frac{\mathbf{r} - \mathbf{r}'}{|\mathbf{r} - \mathbf{r}'|^3} \cdot \frac{\mathbf{r} - \mathbf{r}'}{|\mathbf{r} - \mathbf{r}'|} \, dA'$$

$$= \rho(\mathbf{r}) \int_{\substack{\text{surface} \\ \text{of sphere} \\ \text{about } \mathbf{r}}} \frac{dA'}{|\mathbf{r} - \mathbf{r}'|^2}$$

The integral on the right is just equal to 4π. Thus we have

$$\nabla \cdot \mathbf{E}(\mathbf{r}) = 4\pi\rho(\mathbf{r}) \tag{2-2-3}$$

This result is one of the four basic equations of Maxwell and lies at the foundation of electrostatics. It says, in effect, that electric field can only have a net flux into or out of a region if there is charge within the region. We use this to derive Gauss' law by integrating over a volume V.

$$\int_V \nabla \cdot \mathbf{E}(\mathbf{r}) \, dV = 4\pi \int_V \rho(\mathbf{r}) \, dV \tag{2-2-4}$$

The left side of Eq. (2-2-4) can be transformed into a surface integral by means of Gauss' theorem.

$$\int_V \nabla \cdot \mathbf{E} \, dV = \int_S \mathbf{E} \cdot \hat{\mathbf{n}} \, dA$$

The right side of Eq. (2-2-4) is just 4π times the total charge Q within the volume. We have then

$$\int_S \mathbf{E} \cdot \hat{\mathbf{n}} \, dA = 4\pi Q \tag{2-2-5}$$

In other words, the total flux of electric field out of any given volume is just equal to 4π times the charge within the volume.

Gauss' law greatly facilitates the determination of electric field in situations characterized by high symmetry. We illustrate its application by calculating the field everywhere due to a spherically symmetrical, uniform charge distribution of radius R and total charge Q. For convenience we set the origin of our coordinate system at the center of the sphere, as shown in Fig. 2-2, and let r be the distance from the origin to the point at which we wish to determine the field. Obviously, by symmetry, the field must be radial in direction with magnitude only dependent upon r. We set an imaginary spherical surface at radius r and observe that the total flux of E out of the volume enclosed by this surface is just $4\pi r^2 E(r)$. If $r < R$ then the charge within the surface is just Qr^3/R^3. We have then

$$E(r < R) = \frac{Qr}{R^3}$$
$$E(r \geqq R) = \frac{Q}{r^2}$$

(2-2-6)

2-3 A FEW WORDS ABOUT MATERIALS; CONDUCTORS

At this point we must say a few words about the electrical nature of materials. Later, in Chap. 10, we will try to give a more detailed and extensive picture of what actually goes on at the microscopic level. For now we will be rather brief and somewhat incomplete.

As we all know, matter is constituted of positively charged heavy nuclei surrounded by negatively charged light electrons. The scale of physical dimensions of macroscopic bodies is determined by the natural radius of an electron cloud about the nucleus, about 10^{-8} cm. Hence about 10^{24} electrons are typically packed into a cubic centimeter of solid material and provide the bonds which keep it together. The outermost

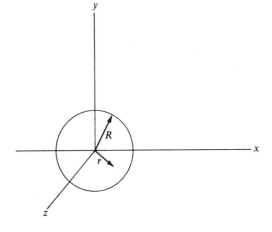

Fig. 2-2 We use Gauss' law to determine the electric field everywhere due to a uniform spherical charge distribution of radius R.

electrons in each atom are relatively easy to move about and can act as charge carriers in conductors.

Now, to get some physical feeling for the strength of electrostatics, we should point out that if every electron were removed from each of two 1-gram pith balls and they were placed 1 cm apart, the force of repulsion between them would be about 10^{29} dynes or about 10^{20} tons! A more relevant illustration (suggested to the author by Prof. D. A. Bromley) is the fact that if all the electrons were removed from one-tenth of a cubic millimeter of the material on the nose cone of an Apollo rocket in some unspecified manner and brought down to the pad, the attraction between these electrons and the remaining positive charges on the nose cone would be sufficient to anchor the rocket firmly in place despite full thrust from the first stage. Remarkable indeed! We see thus that the fraction of available electrons which are involved when we charge or discharge an object is infinitesimal.

Some materials have the property that all their electron clouds are quite strongly bound to the individual positive nuclei. In this case the application of a reasonable local electric field within the material serves only to stretch the bonds between electrons and individual nuclei and not to break them. As we shall learn shortly, the material is then polarized but no real current flows. Such materials are called **dielectrics.**

Within some materials, called **conductors,** the application of an electric field causes the outermost electrons about each nucleus to move relatively freely from atom to atom and to continue moving as long as any field remains within the conductor. If our conductor is isolated, then the moving charge will pile up somewhere until it just neutralizes the applied field within the conductor. At that point no further charge will flow. (There is a tacit assumption here that the momentum picked up by the electrons as they accelerate can be ignored. This assumption is valid because each electron can only accelerate for a very short time before colliding with an atom.)

The fact that the electric field within a conductor is zero in static equilibrium permits us to deduce that the charge density is zero within the conductor [see Eq. (2-2-3)]. Hence whatever charge has piled up must be on the surface of the conductor.

Let us apply some of what we have just learned to the solution of a simple problem. We have two concentric conducting spheres, as shown in Fig. 2-3. The inner sphere has inner radius a and outer radius b. The outer sphere has inner radius c and outer radius d. We place a charge Q_1 on the inner sphere and a charge Q_2 on the outer sphere. We would like to know how the charge is distributed and what the value of electric field is at every point in space. We let σ_a, σ_b, σ_c, and σ_d be the unknown surface charge densities per unit area on the surfaces with radii a, b, c, and d, respectively.

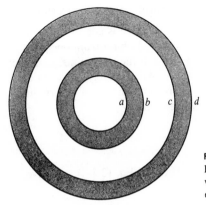

Fig. 2-3 Two concentric conducting spheres have charges Q_1 and Q_2 placed on them. We wish to evaluate the surface charge densities and electric fields.

Again symmetry tells us that the electric field must be in a radial direction and its magnitude can only depend on the radius from the center, r. The imaginary gaussian surfaces through which we evaluate the flux of $\mathbf{E}$ are, of course, spheres at various radii. We see immediately that $\mathbf{E} = 0$ for $0 \leqq r \leqq a$ since there is no charge within the inner sphere. For $a \leqq r \leqq b$, we are within the conducting material, and so $\mathbf{E} = 0$. Hence there can be no charge density on the inner surface of the inner conductor ($\sigma_a = 0$). All of Q_1 must appear on surface b, and we conclude that

$$\sigma_b = \frac{Q_1}{4\pi b^2}$$

Between the two spheres we have

$$E(b \leqq r \leqq c) = \frac{4\pi Q_1}{4\pi r^2} = \frac{Q_1}{r^2}$$

Inasmuch as there is no field within the second conductor ($c \leqq r \leqq d$) the total charge on the inner surface of the outer conductor must be $-Q_1$. Hence

$$\sigma_c = \frac{-Q_1}{4\pi c^2}$$

Finally the remaining charge on the outer conductor appears on its outer surface, leading to the result that

$$\sigma_d = \frac{Q_2 + Q_1}{4\pi d^2}$$

The field outside the second conductor is thus

$$E(r \geqq d) = \frac{Q_1 + Q_2}{r^2}$$

2-4   THE CONSERVATIVE NATURE OF ELECTROSTATICS; POTENTIAL

We will now demonstrate that the electrostatic force is conservative. That is to say, the work done by it as we take a charged object from one point to another is independent of the path taken between the two points. (Another example of a conservative force is gravitation.) Alternatively, the work done by this force as we go around a closed loop is zero. The proof is very simple and rests upon an application of Stokes' theorem, viz.,

$$\oint_c \mathbf{E} \cdot d\mathbf{l} = \int_s \nabla \times \mathbf{E} \cdot \hat{\mathbf{n}} \, dA$$

Thus we need only show that $\nabla \times \mathbf{E} = 0$. Using Eq. (2-1-4), we have

$$\nabla \times \mathbf{E} = \int_{\substack{\text{all} \\ \text{space}}} \rho(\mathbf{r}') \nabla \times \frac{\mathbf{r} - \mathbf{r}'}{|\mathbf{r} - \mathbf{r}'|^3} \, dV' \tag{2-4-1}$$

But

$$\nabla \times \frac{\mathbf{r} - \mathbf{r}'}{|\mathbf{r} - \mathbf{r}'|^3} = \nabla \frac{1}{|\mathbf{r} - \mathbf{r}'|^3} \times (\mathbf{r} - \mathbf{r}') + \frac{\nabla \times (\mathbf{r} - \mathbf{r}')}{|\mathbf{r} - \mathbf{r}'|^3}$$

Now $\nabla \times (\mathbf{r} - \mathbf{r}') = \nabla \times \mathbf{r} = 0$. Also, $\nabla(1/|\mathbf{r} - \mathbf{r}'|^3)$ is the same direction as $\mathbf{r} - \mathbf{r}'$, and hence

$$\nabla \frac{1}{|\mathbf{r} - \mathbf{r}'|^3} \times (\mathbf{r} - \mathbf{r}') = 0$$

We conclude that $\nabla \times \mathbf{E} = 0$, at least as long as $\mathbf{r}$ is not equal to $\mathbf{r}'$ at any point in our integral. To cope with the contribution from the neighborhood where $\mathbf{r} = \mathbf{r}'$, we again convert our volume integral into a surface integral.

$$\nabla \times \mathbf{E} = \int_{\substack{\text{small} \\ \text{sphere} \\ \text{about } \mathbf{r}}} \rho(\mathbf{r}') \nabla \times \frac{\mathbf{r} - \mathbf{r}'}{|\mathbf{r} - \mathbf{r}'|^3} \, dV'$$

We can take $\rho(\mathbf{r}')$ out of the integral and evaluate it at $\mathbf{r}$, leaving us with the equation

$$\nabla \times \mathbf{E} = \rho(\mathbf{r}) \int_{\substack{\text{sphere} \\ \text{about } \mathbf{r}}} \nabla \times \frac{\mathbf{r} - \mathbf{r}'}{|\mathbf{r} - \mathbf{r}'|^3} \, dV'$$

Now, for any vector function $\mathbf{v}$, we have the general identity

$$\int_{\text{vol}} \nabla \times \mathbf{v} \, dV = \int_{\text{surface}} \hat{\mathbf{n}} \times \mathbf{v} \, dA \tag{2-4-2}$$

(This identity is easily proven by examining its components and using

Gauss' theorem.) Changing ∇ to $-\nabla'$ and using (2-4-2), we convert our expression for $\nabla \times \mathbf{E}$ to a surface integral.

$$\nabla \times \mathbf{E} = -\rho(\mathbf{r}) \int_{\substack{\text{surface} \\ \text{of sphere} \\ \text{about } \mathbf{r}}} \hat{\mathbf{n}}' \times \frac{\mathbf{r} - \mathbf{r}'}{|\mathbf{r} - \mathbf{r}'|^3} \, dA' \qquad (2\text{-}4\text{-}3)$$

The right side of (2-4-3) is zero since the normal $\hat{\mathbf{n}}'$ is just parallel to $\mathbf{r} - \mathbf{r}'$. Hence, for the case of *electrostatics*,

$$\nabla \times \mathbf{E} = 0 \qquad (2\text{-}4\text{-}4)$$

We conclude that $\mathbf{E}$ is conservative. The integral $\int \mathbf{E} \cdot d\mathbf{l}$ between two points is independent of path.

We can now choose an arbitrary reference point and evaluate $\int \mathbf{E} \cdot d\mathbf{l}$ from any point $\mathbf{r}$ to this reference point. It is conventional to take the reference point at infinity in the case of electrostatics. We define the function $\varphi(\mathbf{r})$, called the **potential function,** by the equation

$$\varphi(\mathbf{r}) \equiv \int_{\mathbf{r} \text{ to } \infty} \mathbf{E} \cdot d\mathbf{l} \qquad (2\text{-}4\text{-}5)$$

If we evaluate the potential function φ at two points A and B, then the work done by $\mathbf{E}$ per unit charge moved from A to B (if all other charges are kept fixed) is just $\varphi(A) - \varphi(B)$. Letting point A be given by $\mathbf{r}$ and point B by $\mathbf{r} + \Delta\mathbf{r}$, we have

$$\varphi(A) - \varphi(B) = \varphi(\mathbf{r}) - \varphi(\mathbf{r} + \Delta\mathbf{r}) = \mathbf{E} \cdot \Delta\mathbf{r}$$

But

$$\varphi(\mathbf{r} + \Delta\mathbf{r}) - \varphi(\mathbf{r}) = \nabla\varphi \cdot \Delta\mathbf{r}$$

Since $\Delta\mathbf{r}$ is arbitrary, we conclude that

$$\nabla\varphi = -\mathbf{E} \qquad (2\text{-}4\text{-}6)$$

To summarize, the conservative nature of electrostatics permits us to set up a potential function $\varphi(\mathbf{r})$ defined for every point in space such that the gradient of φ is equal to the negative of the electric field. The power of this observation will become apparent when we discover that it is often easier to determine φ than it is to determine $\mathbf{E}$ directly.

Let us evaluate the potential function at a distance r from a point charge q.

$$\varphi(r) = \int_r^\infty \frac{q}{r^2} \, dr = \frac{q}{r} \qquad (2\text{-}4\text{-}7)$$

If we have a distribution of charge $\rho(\mathbf{r}')$, we sum the contributions to the potential and obtain

$$\int_{\substack{\text{all} \\ \text{space}}} \frac{\rho(\mathbf{r}')}{|\mathbf{r} - \mathbf{r}'|} \, dV' \tag{2-4-8}$$

In the case of a conductor the absence of any internal electric field ensures that it is an equipotential. We might return then to the problem of the two concentric spheres with charges Q_1 and Q_2, respectively (see Fig. 2-3), and evaluate the potentials on these spheres. The potential on the inner sphere is obtained by integrating from radius b outward.

$$\varphi(b) = \int_b^c E(r) \, dr + \int_d^\infty E(r) \, dr$$

$$= \int_b^c \frac{Q_1}{r^2} + \int_d^\infty \frac{Q_1 + Q_2}{r^2} \, dr$$

$$= Q_1 \left(\frac{1}{b} - \frac{1}{c} \right) + \frac{Q_1 + Q_2}{d}$$

The potential on the outer sphere is just $\varphi(d) = (Q_1 + Q_2)/d$.

We now make a very simple application of what we have just learned to demonstrate that the electric field at the surface of a conductor is perpendicular to the surface. Figure 2-4 shows a portion of that surface and the path for which we wish to evaluate $\oint \mathbf{E} \cdot d\mathbf{l}$. The legs of the path which cross the surface are assumed to be of infinitesimal and negligible length.

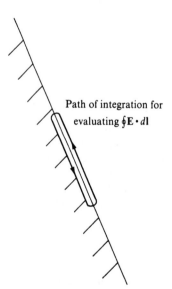

Path of integration for evaluating $\oint \mathbf{E} \cdot d\mathbf{l}$

Fig. 2-4   We evaluate $\oint \mathbf{E} \cdot d\mathbf{l}$ for the path shown at the surface of a conductor. In order that the integral be equal to zero we must have no tangential electric field just outside the conducting surface.

Since there is no field inside the conductor, we conclude that the contribution to the leg outside the conductor must be zero. Hence there is no tangential component of electric field at the conducting surface.

We can use Gauss' law to obtain the magnitude of **E** at the surface of a conductor with surface charge density $= \sigma$. We enclose the surface, as shown in Fig. 2-5a, within a flat thin box of negligible thickness and area ΔA and evaluate the flux out of the box.

$$\text{Flux} = (E)(\Delta A) = 4\pi\sigma\Delta A$$

Hence the field at the surface of a conductor is

$$E = 4\pi\sigma \tag{2-4-9}$$

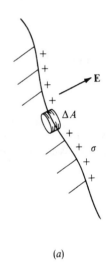

(a)

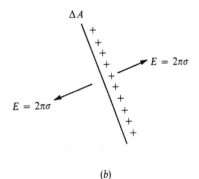

(b)

Fig. 2-5 (a) We evaluate the field at a conducting surface in terms of σ, the charge density per unit area. (b) A little section of the surface charge considered all by itself would give rise to a field of $2\pi\sigma$ on either side of it.

If we now have a good look at our little area ΔA, we see that the surface charge on it (σ), if considered all by itself, would give rise to a field of $2\pi\sigma$ pointing in on one side of it and $2\pi\sigma$ pointing out on the other side (see Fig. 2-5b). That means that all the rest of the conductor must give rise to a field $2\pi\sigma$ pointing out, just at the surface. On the inside of the surface this field from the rest of the conductor cancels the ingoing $2\pi\sigma$ from the little segment ΔA. On the outside of the surface the field from the rest of the conductor augments the field from the segment ΔA to give us the result of Eq. (2-4-9).

Knowing how much of the field is contributed by the local surface charge σ and how much is contributed by the rest of the conductor permits us to calculate the force on the little bit of surface area ΔA. We just multiply the local charge ($\sigma\Delta A$) by the field due to the remaining charges ($2\pi\sigma$) and find

$$F = 2\pi\sigma^2\Delta A \tag{2-4-10}$$

In other words, the surface of a conductor feels an outward-going pressure given by

$$P = 2\pi\sigma^2 \tag{2-4-11}$$

Let us apply what we have just learned to a very simple problem. We will calculate the force between two charged conducting plates, each of area A and separated by a distance d (see Fig. 2-6). We will assume the

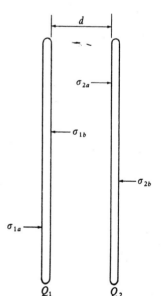

Fig. 2-6 Two conducting plates of area A are separated by a distance d as shown. A charge Q_1 is placed on one and a charge Q_2 on the other. We would like to find the total force that one plate exerts on the other.

lateral dimensions of each plate to be large compared with d and the thickness to be small compared with d. We will also assume a charge Q_1 on the first plate and a charge Q_2 on the second plate.

Referring to Fig. 2-6, we have taken the charge densities on the four relevant surfaces to be σ_{1a}, σ_{1b}, σ_{2a}, and σ_{2b}, respectively. Of course, their values are subject to the constraint that

$$\sigma_{1a} + \sigma_{1b} = \frac{Q_1}{A}$$

$$\sigma_{2a} + \sigma_{2b} = \frac{Q_2}{A}$$

(2-4-12)

We can next calculate the fields in each conductor in terms of σ_{1a}, σ_{1b}, σ_{2a}, σ_{2b} and then set these fields equal to zero. Each sheet of charge makes its contribution of $2\pi\sigma$ with a sign which depends on whether it is to the left or right of the region of interest:

$$E \text{ in conductor } 1 = 2\pi\sigma_{1a} - 2\pi\sigma_{1b} - 2\pi\sigma_{2a} - 2\pi\sigma_{2b} = 0$$

$$E \text{ in conductor } 2 = 2\pi\sigma_{1a} + 2\pi\sigma_{1b} + 2\pi\sigma_{2a} - 2\pi\sigma_{2b} = 0$$

(2-4-13)

We can solve the four equations above for the four unknowns, and we obtain

$$\sigma_{1a} = \sigma_{2b} = \frac{Q_1 + Q_2}{2A}$$

$$\sigma_{1b} = -\sigma_{2a} = \frac{Q_1 - Q_2}{2A}$$

(2-4-14)

To evaluate the force on each plate we make use of Eq. (2-4-11). We will take as positive a force to the right and as negative a force to the left.

$$F_1 = \text{force on plate } 1 = -2\pi\sigma_{1a}^2 A + 2\pi\sigma_{1b}^2 A$$

$$= \frac{-2\pi Q_1 Q_2}{A}$$

(2-4-15)

$$F_2 = \text{force on plate } 2 = -2\pi\sigma_{2a}^2 A + 2\pi\sigma_{2b}^2 A$$

$$= \frac{2\pi Q_1 Q_2}{A}$$

(2-4-16)

Just as we expected, we find the forces to be equal and opposite. The magnitude of the force, to the approximation used here, does not depend on d. Obviously, when d becomes large and of the same order as the transverse dimensions of the plates, this approximation will no longer hold.

Incidentally, the electric field between the plates is given by $4\pi\sigma_{1b}$ or $-4\pi\sigma_{2a}$.

$$E_{\text{between plates}} = \frac{2\pi(Q_1 - Q_2)}{A} \qquad (2\text{-}4\text{-}17)$$

We will return to this rather simple example again as we develop more sophisticated techniques for evaluating the forces on collections of charges. In the meantime, we go on to a detailed study of techniques for determining the potential function φ.

2-5 SOME IMPORTANT THEOREMS ABOUT POTENTIAL FUNCTIONS; BOUNDARY CONDITIONS AND UNIQUENESS

As we have already pointed out, finding the potential $\varphi(\mathbf{r})$ is a convenient way of determining the field. The fundamental differential equation obeyed by the function φ can be obtained from Eq. (2-2-3) by substituting $-\nabla\varphi$ for $\mathbf{E}$.

$$\nabla^2\varphi \equiv \frac{\partial^2\varphi}{\partial x^2} + \frac{\partial^2\varphi}{\partial y^2} + \frac{\partial^2\varphi}{\partial z^2} = -4\pi\rho \qquad (2\text{-}5\text{-}1)$$

(The operator $\nabla^2 = \partial^2/\partial x^2 + \partial^2/\partial y^2 + \partial^2/\partial z^2$ is called the **Laplacian.**) We will now demonstrate a remarkable set of theorems which relate to the uniqueness of our solutions for φ subject to Eq. (2-5-1) and appropriate boundary conditions.

As the first step along the way, we prove a simple lemma called the **mean value theorem.** The theorem states that in a charge-free region the average value of potential on the surface of any hypothetical sphere is equal to the value of the potential at the center of the sphere.

Let our spherical surface have radius a. Let $\bar{\varphi}$ be the average potential on the surface. Then we have

$$\bar{\varphi} = \int_{\text{surface}} \varphi\, \frac{dA}{4\pi a^2} = \frac{1}{4\pi}\int_{\text{surface}} \varphi\, d\Omega \qquad (2\text{-}5\text{-}2)$$

where $d\Omega$ is the element of solid angle subtended by dA. Differentiating $\bar{\varphi}$ with respect to a, we obtain

$$\frac{d\bar{\varphi}}{da} = \frac{1}{4\pi}\int_{\text{surface}} \frac{\partial\varphi}{\partial a}\, d\Omega = \frac{1}{4\pi}\int_{\text{surface}} \nabla\varphi\cdot\hat{\mathbf{n}}\, d\Omega \qquad (2\text{-}5\text{-}3)$$

Substituting back again for $d\Omega$, we rewrite Eq. (2-5-3) as

$$\frac{d\bar{\varphi}}{da} = \frac{1}{4\pi a^2}\int_{\text{surface}} \nabla\varphi\cdot\hat{\mathbf{n}}\, dA = \frac{1}{4\pi a^2}\int_{\substack{\text{vol} \\ \text{of sphere}}} \nabla^2\varphi\, dV \qquad (2\text{-}5\text{-}4)$$

Since $\nabla^2\varphi = 0$ in a charge-free region, we have $d\bar{\varphi}/da = 0$, and hence $\bar{\varphi}$

does not depend on the radius of the sphere. Thus we have proven our theorem.

$$\varphi_{center} = \bar{\varphi}(a = 0) = \bar{\varphi}(a) \tag{2-5-5}$$

There is an obvious byproduct of this theorem, called **Earnshaw's theorem,** which states that φ cannot have a local maximum or minimum in a charge-free region. The proof of this theorem requires only that we take a small sphere about our presumed maximum or minimum. The potential on the surface of the sphere will presumably always be less than (or more than) its value at the center, violating the mean value theorem.

We shall note here that the two theorems we have just proved relate to any function $f(x,y,z)$ which satisfies the equation $\nabla^2 f = 0$. We will make use of this fact in our further work.

We next examine a situation in which we have a set of N conductors held at potentials $\varphi_1, \varphi_2, \ldots, \varphi_N$. In addition we have a charge density $\rho(\mathbf{r})$ specified for all points outside the conductors. We will show that there is *one* and only *one* solution to the differential equation (2-5-1) which goes to zero at infinity and which is equal to the appropriate potential on each of the N conductors.

One solution to the problem is, of course, given by Eq. (2-4-8), where we take care to integrate over the surface charge distributions on all the conductors. That this is the only solution is not so obvious.

Let $\varphi_a(\mathbf{r})$ and $\varphi_b(\mathbf{r})$ be two solutions which satisfy the boundary conditions and the differential equation (2-5-1). Then, letting $f = \varphi_a - \varphi_b$, we have

$$\nabla^2 f = \nabla^2 \varphi_a - \nabla^2 \varphi_b = -4\pi\rho + 4\pi\rho = 0$$

Therefore f has no minima or maxima anywhere. Since f is zero on the surface of every conductor and also at ∞, it must be zero everywhere. Thus $\varphi_a = \varphi_b$ and our theorem is proved.

Instead of fixing the potentials at each conductor, we might set each total charge to a definite value. Thus the N conductors would have charges $Q_1, Q_2, \ldots, Q_N$, respectively. In addition the charge density outside the conductors is again taken as $\rho(\mathbf{r})$. We will demonstrate that the solution for φ which satisfies Eq. (2-5-1) and leads to the appropriate values of Q_i is unique. This is a very powerful result since it implies that the charge Q_i will arrange itself in a unique way on the ith conductor. Again the proof is carried forth by first assuming that φ_a and φ_b are both solutions to our problem and then examining $f = \varphi_a - \varphi_b$.

We remember that the field at the surface of a conductor is just equal to $4\pi\sigma\hat{n}$. Hence we have

$$Q_i = \int_{s_i} \sigma_i \, dA_i = \frac{-1}{4\pi} \int_{s_i} \nabla\varphi \cdot \hat{n} \, dA_i \tag{2-5-6}$$

If both φ_a and φ_b satisfy Eq. (2-5-6), then f satisfies the equation

$$\int_{S_i} \nabla f \cdot \hat{n} \, dA_i = 0 \tag{2-5-7}$$

Just as before, $\nabla^2 f = 0$ outside the conductor and $f = 0$ at ∞. Again Earnshaw's theorem requires that f have no maximum or minimum outside the conductor.

Suppose for a moment that f is not zero on every conductor. Then there is one conductor for which f is either the most negative or the most positive. Since no minimum or maximum can exist outside the conductors, this particular conductor must be an *absolute* extremum. Hence $\nabla f \cdot \hat{n}$ must have the same positive or negative sign at all points on the surface and Eq. (2-5-7) would be contradicted. We conclude that f must be zero on all the conductors. Again this leads us to the result that $f = 0$ everywhere.

Now, of what use are these uniqueness theorems anyway? The answer is very simple. There are many general techniques for solving Eq. (2-5-1) subject to specific and complete boundary conditions. Once a solution has been found which satisfies the complete set of boundary conditions, our job is done. We need to look no further; we have the only possible solution to the physical problem at hand. Furthermore, if by a bit of cleverness we can see the answer quickly, all the more power to us.

To see the possibilities inherent in this approach let us solve a problem which would seem well nigh impossible if handled in a routine way. We place a large sheet of conductor on the yz plane and ground it ($\varphi = 0$ for the conductor). We then bring a charge q to the point $x = a$, along the x axis (see Fig. 2-7). We would like to find the field at all points with $x \geq 0$, and we would like to know how much force is exerted on the charge.

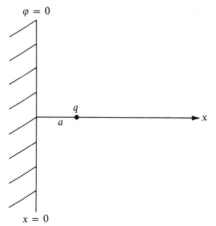

Fig. 2-7 A charge q is placed at a distance a from a grounded plane.

What exactly are the requirements that must be met by φ in the region $x \geqq 0$? First, $\nabla^2\varphi$ must be zero everywhere except at the charge itself. Second, $\varphi = 0$ at $x = 0$ and at $x = \infty$.

Suppose we were to examine an entirely different problem, where the conducting sheet was removed and replaced by a charge $-q$ on the x axis at the point $x = -a$. This problem would be simple to solve. But the most obvious fact about the new problem is that the potential φ is zero at $x = 0$ and corresponds to the same charge distribution for $x > 0$ as we had previously. Hence it is *exactly* the potential we seek. The field in the region $x \geqq 0$ is then also the same for the two problems. Finally we conclude that the force on a charge q at a distance a from a grounded plane is just $q^2/4a^2$ toward the plane. This technique for finding φ is called the **method of images.**

We will very shortly come back to potential theory and develop some techniques for determining potentials. In the meantime, though, we go back to charge distributions and the nature of matter to investigate the notion of electric dipole moment and polarization.

2-6 ELECTRIC DIPOLE MOMENT; POLARIZATION; DISPLACEMENT FIELD

Most matter is electrically neutral but with the property that an applied electric field can cause some polarization. This can happen in one of two ways. If the atoms are basically symmetrical, then they can be unbalanced or polarized to the extent that the applied electric field can compete with the atomic fields. Alternatively, the molecules themselves may be constructed in such a way that they are not electrically symmetrical. In that case, if the molecules are free to rotate, as in a liquid, they can be "lined up" by the applied field. The extent of alignment will then depend upon the temperature because the molecules are continually being depolarized by collisions. Figure 2-8 illustrates these two types of behavior.

In any case, we deal with the field-distribution characteristic of the electric dipole. Let us begin by examining the field due to two equal and opposite point charges set a small distance l apart. We draw a vector $\mathbf{l}$ from the negative to the positive charge, as shown in Fig. 2-9, and let $\mathbf{r}$ be the vector from the negative charge to the point P at which we wish to evaluate the field. We let $\mathbf{r}'$ be the vector from $+q$ to P. We will assume throughout that $|l| \ll r$ and hence that terms of the order of l^2/r^2 can be ignored. Clearly, we can now write

$$\varphi(\mathbf{r}) = q\left(\frac{1}{r'} - \frac{1}{r}\right) \tag{2-6-1}$$

Now

$$\mathbf{r}' = \mathbf{r} - \mathbf{l} \tag{2-6-2}$$

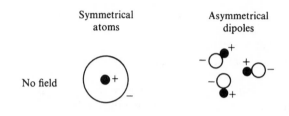

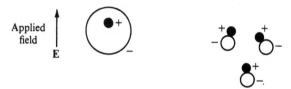

Fig. 2-8 The application of an electric field will polarize matter by either polarizing the molecular charge distributions or aligning the permanent molecular dipoles.

and hence

$$r' = \sqrt{r^2 + l^2 - 2\mathbf{r} \cdot \mathbf{l}} \tag{2-6-3}$$

Ignoring terms of order l^2/r^2, we rewrite Eq. (2-6-3) as

$$r' \cong r \sqrt{1 - \frac{2\mathbf{r} \cdot \mathbf{l}}{r^2}}$$

$$\cong r\left(1 - \frac{\mathbf{r} \cdot \mathbf{l}}{r^2}\right) \tag{2-6-4}$$

Substituting back into Eq. (2-6-1) and remembering that

$$1 - \frac{\mathbf{r} \cdot \mathbf{l}}{r^2} \cong \frac{1}{1 + (\mathbf{r} \cdot \mathbf{l})/r^2}$$

we find

$$\varphi(\mathbf{r}) = \frac{+q\mathbf{l} \cdot \mathbf{r}}{r^3} \tag{2-6-5}$$

We now define a new vector **p** which we call the **dipole moment** of this system.

$$\mathbf{p} = q\mathbf{l} \tag{2-6-6}$$

We finally have then

$$\varphi(\mathbf{r}) = -\mathbf{p} \cdot \nabla \frac{1}{r} \tag{2-6-7}$$

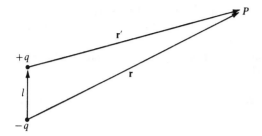

Fig. 2-9 We wish to evaluate
the field due to two closely spaced,
equal, and opposite charges.

The important thing to note here is that the potential drops off as $1/r^2$, and hence the field drops off as $1/r^3$. It is interesting to examine the field distributions qualitatively. We note first that $\varphi = 0$ for all points in the equatorial plane about **p** (such that $\mathbf{p} \cdot \mathbf{r} = 0$). Above this plane, in the direction of **p**, φ is positive. Below this plane φ is negative. Hence, in the equatorial plane the gradient of φ must be in the direction of **p** and the direction of **E** must be opposite to **p**. Just along the line containing **p** we find the field in the same direction as **p**. A picture of this field distribution is shown in Fig. 2-10.

We next examine the more realistic problem of a spatially confined charge distribution given by $\rho(\mathbf{r}')$ such that the net charge of the distribution is zero. An example of this is, of course, the atomic system. We are

Fig. 2-10   The field distribution due to an electric dipole **p**.

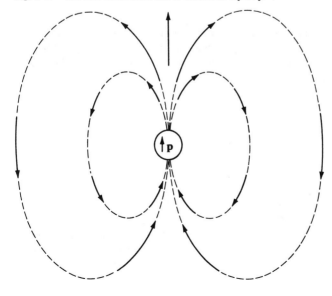

again interested in the field at a distance from the charge distribution which is large compared with its size. We shall show that to first approximation this field looks like a dipole field.

The charge distribution is shown in Fig. 2-11. The potential $\varphi(\mathbf{r})$ at position $\mathbf{r}$ is given by Eq. (2-4-8):

$$\varphi(\mathbf{r}) = \int_{\substack{\text{charge}\\\text{distribution}}} \frac{\rho(\mathbf{r}')}{|\mathbf{r} - \mathbf{r}'|} \, dV' \tag{2-6-8}$$

To first approximation, noting Eq. (2-6-4), we have

$$\frac{1}{|\mathbf{r} - \mathbf{r}'|} = \frac{1}{r}\left(1 + \frac{\mathbf{r}' \cdot \mathbf{r}}{r^2}\right) \tag{2-6-9}$$

Substituting back in Eq. (2-6-8), we have

$$\varphi(\mathbf{r}) = \frac{1}{r}\int_{\substack{\text{charge}\\\text{distribution}}} \rho(\mathbf{r}')\, dV' + \frac{\mathbf{r}}{r^3} \cdot \int_{\substack{\text{charge}\\\text{distribution}}} \rho(\mathbf{r}')\mathbf{r}'\, dV' \tag{2-6-10}$$

Remembering that the net charge is zero, we remove the first term on the right of Eq. (2-6-10), leaving us with the old result for a simple dipole:

$$\varphi(\mathbf{r}) = -\mathbf{p} \cdot \nabla \frac{1}{r}$$

where

$$\mathbf{p} = \int_{\substack{\text{charge}\\\text{distribution}}} \rho(\mathbf{r}')\mathbf{r}'\, dV' \tag{2-6-11}$$

We see that any "neutral" charge distribution can be represented, to first approximation, by a dipole whose dipole moment is given by Eq. (2-6-11). That this integral reduces to Eq. (2-6-6) in the case of two equal and opposite point charges is obvious.

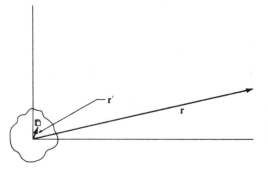

Fig. 2-11 We integrate over a charge distribution with no net charge to find the dipole field at large distances.

Now, if we look into a bit of matter, we see immense numbers of these little dipoles, and it becomes necessary to talk about **dipole moment per unit volume.** We define the vector field $\mathbf{P}(\mathbf{r})$ such that for a little bit of volume dV at position $\mathbf{r}$ the dipole moment $d\mathbf{p}$ will be given by

$$d\mathbf{p} = \mathbf{P} \, dV \qquad (2\text{-}6\text{-}12)$$

We now wish to find the potential $\varphi(\mathbf{r})$ which results from a distribution of *free charge* $\rho_F(\mathbf{r})$ and dipole moment per unit volume $\mathbf{P}(\mathbf{r})$. By *free charge* we mean the charge which is not paired off as part of a dipole. Using Eqs. (2-4-8) and (2-6-7), we write

$$\varphi(\mathbf{r}) = \int_{\substack{\text{all} \\ \text{space}}} \frac{\rho_F(\mathbf{r}')\, dV'}{|\mathbf{r} - \mathbf{r}'|} - \int_{\substack{\text{all} \\ \text{space}}} \mathbf{P}(\mathbf{r}') \cdot \nabla \frac{1}{|\mathbf{r} - \mathbf{r}'|}\, dV' \qquad (2\text{-}6\text{-}13)$$

Remembering again that

$$\nabla \frac{1}{|\mathbf{r} - \mathbf{r}'|} = -\nabla' \frac{1}{|\mathbf{r} - \mathbf{r}'|}$$

we have

$$\varphi(\mathbf{r}) = \int_{\substack{\text{all} \\ \text{space}}} \frac{\rho_F(\mathbf{r}')\, dV'}{|\mathbf{r} - \mathbf{r}'|} + \int_{\substack{\text{all} \\ \text{space}}} \nabla' \cdot \frac{\mathbf{P}(\mathbf{r}')}{|\mathbf{r} - \mathbf{r}'|}\, dV'$$

$$- \int_{\substack{\text{all} \\ \text{space}}} \frac{\nabla' \cdot \mathbf{P}(\mathbf{r}')}{|\mathbf{r} - \mathbf{r}'|}\, dV' \qquad (2\text{-}6\text{-}14)$$

The second term on the right-hand side of Eq. (2-6-14) can be transformed into a surface integral at infinity. Since $\mathbf{P} = 0$ at infinity, we can write

$$\int_{\substack{\text{all} \\ \text{space}}} \nabla' \cdot \frac{\mathbf{P}(\mathbf{r}')}{|\mathbf{r} - \mathbf{r}'|}\, dV' = \int_{\substack{\text{surface} \\ \text{at } \infty}} \frac{\mathbf{P}(\mathbf{r}') \cdot \hat{\mathbf{n}}'\, dA'}{|\mathbf{r} - \mathbf{r}'|} = 0 \qquad (2\text{-}6\text{-}15)$$

Hence we write

$$\varphi(\mathbf{r}) = \int_{\substack{\text{all} \\ \text{space}}} \frac{\rho_F(\mathbf{r}')}{|\mathbf{r} - \mathbf{r}'|}\, dV' + \int_{\substack{\text{all} \\ \text{space}}} \frac{\rho_P(\mathbf{r}')}{|\mathbf{r} - \mathbf{r}'|}\, dV' \qquad (2\text{-}6\text{-}16)$$

where

$$\rho_P \equiv -\nabla \cdot \mathbf{P} \qquad (2\text{-}6\text{-}17)$$

We conclude that the entire effect of polarization can be taken into account by introducing a *polarization charge* equal to $-\nabla \cdot \mathbf{P}$ and calculating as though this were a real charge.

We can now go back to our basic differential equation for **E**, Eq. (2-2-3), and separate ρ into its two parts.

$$\nabla \cdot \mathbf{E} = 4\pi\rho$$

$$= 4\pi(\rho_F + \rho_P)$$

$$= 4\pi(\rho_F - \nabla \cdot \mathbf{P}) \tag{2-6-18}$$

Rearranging terms, we have

$$\nabla \cdot (\mathbf{E} + 4\pi\mathbf{P}) = 4\pi\rho_F \tag{2-6-19}$$

We define a new vector field **D** everywhere in space by the equation

$$\mathbf{D} = \mathbf{E} + 4\pi\mathbf{P} \tag{2-6-20}$$

Substituting back into Eq. (2-6-19), we obtain the appropriate Maxwell equation in the presence of matter.

$$\nabla \cdot \mathbf{D} = 4\pi\rho_F \tag{2-6-21}$$

We conclude then that the free charge density ρ_F can be regarded as a local source of **D** in the same way as the total charge density ρ can be regarded as a source of **E**. That is where the similarity ends, however. Specifying the total charge density everywhere determines **E** completely by means of Eq. (2-1-4). Specifying ρ_F does not determine **D** completely. For example, a block of permanently polarized material with no free charge can certainly give rise to an electric field near it. Hence we would have a **D** field in the complete absence of any ρ_F. The magnitude of **D** would depend on the magnitude and distribution of **P**.

Returning to Eq. (2-6-20), we note that there is often a linear relationship between the electric field at a given point and the polarization at that point. We can then write

$$\mathbf{P} = \chi\mathbf{E} \tag{2-6-22}$$

where χ, called the **electric susceptibility,** is in general a second-rank tensor. In the simple-minded case where χ is just a number, it is usual to write

$$\mathbf{D} = \varepsilon\mathbf{E} \tag{2-6-23}$$

where ε, called the **dielectric constant,** is given by

$$\varepsilon = 1 + 4\pi\chi \tag{2-6-24}$$

We can establish some very simple rules governing the behavior of the **E** and **D** fields at the boundary between two regions of differing dielectric constant. We assume that no free charge is present at the boundary. Figure 2-12 shows a portion of such a boundary between regions I and II, respectively.

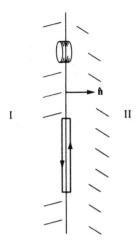

Fig. 2-12 We determine the behavior of **E** and **D** at the interface between two dielectrics.

We first place a flat thin volume across the boundary and apply Gauss' theorem. Since there is no free charge within the volume, the flux of **D** out of the volume is zero. The normal components of **D** on either side of the boundary are thus equal.

$$\mathbf{D_I} \cdot \hat{\mathbf{n}} = \mathbf{D_{II}} \cdot \hat{\mathbf{n}} \tag{2-6-25}$$

We then take a path of integration shown and evaluate the line integral $\int \mathbf{E} \cdot d\mathbf{l}$ around the path. Since it must be equal to zero, we conclude that the tangential components of **E** must be the same on either side of the boundary.

$$E^I_{\text{tang}} = E^{II}_{\text{tang}} \tag{2-6-26}$$

These rules will make it possible to calculate the direction and magnitudes of $\mathbf{E_{II}}$ and $\mathbf{D_{II}}$ from the direction and magnitudes of $\mathbf{E_I}$ and $\mathbf{D_I}$ if we are given the two dielectric constants.

2-7 THE ENERGY OF A CHARGE DISTRIBUTION

We will now determine the energy of an electrostatic charge distribution. In doing so, we will at first ignore the energy corresponding to the formation of the dielectrics themselves but evaluate only the energy which is added to the system through the introduction of the free charge. The energy of the system will be considered to be zero when all free charge has been removed to infinity and only the dielectrics and conductors are left in position. All polarizations will be assumed proportional to the applied fields.

Let us take ρ_F and φ to be the *final* distributions of free charge and potential, respectively, after our system has been completely set up. If we had somehow decided to place a factor of α as much charge everywhere, then our potential would be changed by the same factor α. That is to say, if the charge distribution ρ_F leads to the potential distribution φ, then the charge distribution $\alpha\rho_F$ would lead to the potential distribution $\alpha\varphi$. (We must repeat that this depends upon the assumed *linearity* of the dielectrics.)

Now let us bring in the charge from infinity, a little bit at a time. That is, let α go from 0 to 1, in infinitesimal steps. As we go from α to $\alpha + d\alpha$, we have

$$dU = \text{increase in energy}$$

$$= \int_{\substack{\text{all} \\ \text{space}}} (\alpha\varphi)(\rho_F \, d\alpha) \, dV \tag{2-7-1}$$

since $\alpha\varphi$ is the potential to which the additional bit of charge $\rho_F \, d\alpha \, dV$ is brought. The final energy is then

$$U = \int_0^1 \alpha \, d\alpha \int_{\substack{\text{all} \\ \text{space}}} \varphi\rho_F \, dV$$

$$= \frac{1}{2} \int_{\substack{\text{all} \\ \text{space}}} \rho_F\varphi \, dV \tag{2-7-2}$$

If we have conductors present, then φ is a constant within their boundaries and we have

$$U = \frac{1}{2} \int_{\substack{\text{outside} \\ \text{cond}}} \rho_F\varphi \, dV + \frac{1}{2} \sum_{\text{cond}} Q_i\varphi_i \tag{2-7-3}$$

where Q_i is the charge of the ith conductor.

We can rewrite U in a somewhat different form. Let us remember that $\nabla \cdot \mathbf{D} = 4\pi\rho_F$. Hence, going back to Eq. (2-7-2), we can write

$$U = \frac{1}{2} \int_{\substack{\text{all} \\ \text{space}}} \rho_F\varphi \, dV = \frac{1}{8\pi} \int_{\substack{\text{all} \\ \text{space}}} (\nabla \cdot \mathbf{D})\varphi \, dV$$

$$= \frac{1}{8\pi} \int_{\substack{\text{all} \\ \text{space}}} [\nabla \cdot (\varphi\mathbf{D}) - \nabla\varphi \cdot \mathbf{D}] \, dV$$

$$= \frac{1}{8\pi} \int_{S_\infty} \varphi\mathbf{D} \cdot \hat{\mathbf{n}} \, dA + \frac{1}{8\pi} \int_{\substack{\text{all} \\ \text{space}}} \mathbf{D} \cdot \mathbf{E} \, dV \tag{2-7-4}$$

where S_∞ is a surface at infinity. But **D** goes down as $1/r^2$ and φ goes down as $1/r$. Hence

$$\int_{S_\infty} \varphi \mathbf{D} \cdot \hat{\mathbf{n}} \, dA \to 0 \qquad \text{as } r \to \infty$$

We are then left with

$$U = \frac{1}{8\pi} \int_{\substack{\text{all} \\ \text{space}}} \mathbf{D} \cdot \mathbf{E} \, dV \qquad (2\text{-}7\text{-}5)$$

At this point we might ask what the expression for energy would look like if we really knew what was going on in the material and chose to include the energy of formation of the dielectric into our tally. We would then have no need to introduce **D** at all but would use only the total charge density ρ and the electric field **E**. The energy of the system would then be written simply as

$$U_{\text{total}} = \frac{1}{8\pi} \int_{\substack{\text{all} \\ \text{space}}} E^2 \, dV \qquad (2\text{-}7\text{-}6)$$

or

$$U_{\text{total}} = \frac{1}{2} \int_{\substack{\text{all} \\ \text{space}}} \rho\varphi \, dV \qquad (2\text{-}7\text{-}7)$$

In general, Eq. (2-7-6) will give a smaller value for the energy than Eq. (2-7-5) because **D** is usually larger than **E**. This is because the dielectric itself has a net negative energy corresponding to the binding of negative and positive charges.

We now apply these results to a very simple example. We will calculate the energy stored in a parallel-plate capacitor, with no dielectric present, and then make use of this to determine the force between the plates (see Fig. 2-13). We assume a charge of $+Q$ on one plate and $-Q$ on the other.

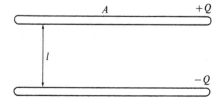

Fig. 2-13 We calculate the energy stored in a parallel-plate capacitor. We again find the force between capacitor plates but this time by considering the change in energy as we reduce l while keeping Q constant.

The area of the plates is A and their separation is l. By our first method we have

$$U = \frac{1}{2} \sum Q_i \varphi_i = \frac{1}{2} Q \, \Delta\varphi$$

where $\Delta\varphi$ is the potential difference between the plates.

$$\Delta\varphi = El = 4\pi\sigma l = \frac{4\pi Q l}{A}$$

Hence

$$U = \frac{2\pi Q^2 l}{A} \tag{2-7-8}$$

By our second method

$$U = \frac{1}{8\pi} E^2 \text{ (volume between plates)} = \frac{1}{8\pi} \left(\frac{4\pi Q}{A}\right)^2 lA = \frac{2\pi Q^2 l}{A}$$

We note that the stored energy decreases as the separation between plates decreases. Since *all* the stored energy is in the gap, this tells us that the force between plates is

$$F = -\frac{dU}{dl} = \frac{-2\pi Q^2}{A} \tag{2-7-9}$$

with the minus sign indicating that the force tends to decrease l. This is in agreement with the result obtained earlier [see Eq. (2-4-15)]. If our plates were attached to a battery ($\Delta\varphi$ constant), the situation would be somewhat different. The energy between the plates would be

$$U = \frac{1}{8\pi} \left(\frac{\Delta\varphi}{l}\right)^2 lA = \frac{(\Delta\varphi)^2 A}{8\pi l}$$

Here the energy stored between the plates *increases* as the gap decreases. Indeed, the change in electrostatic energy for a given small displacement is equal and opposite to that given by Eq. (2-7-9). Obviously the total energy stored in the system must still decrease as before, when l is decreased (since the force is just as attractive as before). Thus the battery runs down by an amount equal to twice the mechanical work done by the fields.

2-8 THE GENERAL THEORY OF CAPACITANCE

We will investigate the notion of capacitance in a rather general way, beginning with one or two conductors but then extending ourselves to an indefinite number of conductors. This will give us a tremendous amount of insight into the linear nature of electromagnetism in addition to providing powerful techniques for the solution of specific problems.

We start with one of the most important of all principles in physics and one which we will encounter many times when we deal with quantum-mechanical systems. This is the **superposition principle.** Within the present context the principle states that if a charge distribution ρ_A leads to a potential distribution φ_A and a charge distribution ρ_B leads to a potential distribution φ_B, then the charge distribution $\rho_A + \rho_B$ will lead to the potential distribution $\varphi_A + \varphi_B$. (We must be careful here to use the *entire* charge density if we want to be rigorously correct. Otherwise nonlinearities in the dielectric will lead us into difficulty.) We need only go back to Eq. (2-4-8) to see that this principle is obviously true. The value of this principle will become apparent in cases where the solutions to $\nabla^2\varphi_A = -4\pi\rho_A$ and $\nabla^2\varphi_B = -4\pi\rho_B$ are easier to come by than the solution to $\nabla^2\varphi = -4\pi(\rho_A + \rho_B)$. We will then solve for φ_A and φ_B and add these solutions together to obtain φ.

Let us consider first the case of only one conductor in space and nothing else (no other charge). If we place an amount of charge Q_0 on the conductor, its potential will rise to some value φ_0. An amount of charge αQ_0 would lead to a potential $\alpha\varphi_0$. Clearly then there is a proportionality between the charge on the conductor and its potential which we can express as

$$Q = C\varphi_{\text{cond}} \tag{2-8-1}$$

The proportionality constant C is defined to be the **capacitance** of a single conductor. In the case of a conducting sphere of radius R, where the potential is just given by

$$\varphi_{\text{sphere}} = \frac{Q}{R} \tag{2-8-2}$$

we see that the capacitance is just R.

It is amusing at this point to calculate the capacitance of the earth. Its radius is roughly 6500 km or 6.5×10^8 cm. This is not a very large number. Indeed, converting to "practical" units, we would find it to be equal to only about 700 microfarads. Alternatively, we can observe that if we removed the electrons from only a gram or so of material, we could raise the potential of the earth by 140 million volts. This in turn would lead to an electric field at the surface of about 0.2 volt/cm. Clearly then the positive and negative charges on the earth are very well balanced indeed, and it takes rather little to disturb this balance. That is why the flood of charged particles coming from a solar flare plays such havoc with radio communications on earth.

We next consider the case of two conductors where one has charge

$+Q$ and the other has charge $-Q$ (see Fig. 2-14). Let φ_1 be the potential on conductor 1 and φ_2 be the potential on conductor 2. If we multiply Q by a constant factor, then both φ_1 and φ_2 will be multiplied by the same factor. Hence $\varphi_1 - \varphi_2$ is proportional to Q; that is, $Q = C(\varphi_1 - \varphi_2)$. The proportionality constant C is the capacitance of a two-conductor system.

It is clear that the linearity of electrostatics permits us to specify important electrostatic properties of conductors which are quite independent of their charges or potentials but depend only on the geometry of the system. We generalize now to a system of N conductors and no extraneous charge. Let the charges on the conductors be $Q_1, Q_2, Q_3, \ldots, Q_i, \ldots,$ Q_N. The potentials will then be $\varphi_1, \varphi_2, \ldots, \varphi_i, \ldots, \varphi_N$. We seek a way of specifying $Q_1, \ldots, Q_N$ in terms of $\varphi_1, \ldots, \varphi_N$, dependent only upon the geometry of the situation.

We begin by solving a subsidiary problem. Let us ground all conductors except the first one. That is, let $\varphi_i = 0$ for $i \neq 1$. For this first subsidiary problem we will have a charge $Q_i(1)$ on the ith conductor. If every charge were to be multiplied by a factor α, then every potential would be multiplied by the same factor α. All those for $i \neq 1$ would, of course, remain zero. The potential on the first conductor would become equal to $\alpha\varphi_1$. Hence we can write

$$Q_1(1) = C_{11}\varphi_1$$
$$Q_2(1) = C_{21}\varphi_1$$
$$Q_3(1) = C_{31}\varphi_1$$
$$\cdots\cdots\cdots\cdots$$
$$Q_N(1) = C_{N1}\varphi_1 \qquad\qquad (2\text{-}8\text{-}3)$$

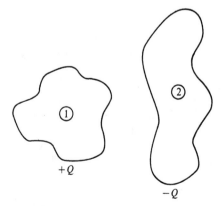

$+Q$

$-Q$

Fig. 2-14 Two conductors, one with charge $+Q$ and one with charge $-Q$.

We now solve a *second* subsidiary problem. We ground all conductors but the second and set the second to φ_2. We have then

$$Q_1(2) = C_{12}\varphi_2$$
$$Q_2(2) = C_{22}\varphi_2$$
$$\cdots\cdots\cdots\cdots$$
$$Q_N(2) = C_{N2}\varphi_2 \tag{2-8-4}$$

It is clear that we can go on like this and solve a total of N such subsidiary problems.

We now recall the superposition principle. If we add together the solutions to subsidiary problems 1 to N, we will have the solution to the problem where each conductor i is raised to its appropriate potential φ_i. Thus

$$Q_1 = C_{11}\varphi_1 + C_{12}\varphi_2 + \cdots + C_{1N}\varphi_N$$

or

$$Q_i = \sum_{j=1}^{N} C_{ij}\varphi_j \tag{2-8-5}$$

The numbers C_{ij} (called **coefficients of capacitance**) tell us what we want to know about the geometry of the situation.

In physical terms then, if we can determine the full set of C_{ij} for any given conductor configuration, we will be prepared to determine immediately the full set of charges $Q_1, \ldots, Q_N$ from the given set of potentials $\varphi_1, \ldots, \varphi_N$. Of course, if we are given the charges $Q_1, \ldots, Q_N$, we can determine the potentials by means of N simultaneous equations. We now wish to prove that $C_{ij} = C_{ji}$. To do this, we will first prove a more powerful theorem.

Green's Reciprocity Theorem. Let charges $Q_1, Q_2, \ldots, Q_N$ correspond to potentials $\varphi_1, \varphi_2, \ldots, \varphi_N$. Let charges $Q_1^*, Q_2^*, \ldots, Q_N^*$ correspond to potentials $\varphi_1^*, \varphi_2^*, \ldots, \varphi_N^*$. Then

$$\sum_i \varphi_i Q_i^* = \sum_i \varphi_i^* Q_i \tag{2-8-6}$$

Proof: We first note that

$$Q_i = \int_{S_i} \sigma_i(\mathbf{r}_i)\, dA_i \tag{2-8-7}$$

$$Q_i^* = \int_{S_i} \sigma_i^*(\mathbf{r}_i)\, dA_i \tag{2-8-8}$$

Using Eq. (2-4-8), we derive

$$\varphi_i = \varphi_i(\mathbf{r}_i) = \sum_{j=1}^{N} \int_{S_j} \frac{\sigma_j(\mathbf{r}_j')}{|\mathbf{r}_i - \mathbf{r}_j'|} \, dA_j'$$

$$= \text{constant depending only on } i \tag{2-8-9}$$

$$\varphi_i^* = \varphi_i^*(\mathbf{r}_i) = \sum_{j=1}^{N} \int_{S_j} \frac{\sigma_j^*(\mathbf{r}_j')}{|\mathbf{r}_i - \mathbf{r}_j'|} \, dA_j' \tag{2-8-10}$$

We have then

$$Q_i \varphi_i^* = \left[\int_{S_i} \sigma_i(\mathbf{r}_i) \, dA_i \right] \varphi_i^* \tag{2-8-11}$$

But φ_i^* is independent of $\mathbf{r}_i$ as long as $\mathbf{r}_i$ is at a point on the conductor. Hence we can bring it into the integral

$$Q_i \varphi_i^* = \int_{S_i} \sigma_i(\mathbf{r}_i) \varphi_i^*(\mathbf{r}_i) \, dA_i \tag{2-8-12}$$

and

$$\sum_i \varphi_i^* Q_i = \sum_{i=1}^{N} \sum_{j=1}^{N} \int_{S_j} \int_{S_i} \frac{\sigma_i(\mathbf{r}_i) \sigma_j^*(\mathbf{r}_j')}{|\mathbf{r}_i - \mathbf{r}_j'|} \, dA_j' \, dA_i \tag{2-8-13}$$

Also, similarly

$$\sum_i \varphi_i Q_i^* = \sum_{i=1}^{N} \sum_{j=1}^{N} \int_{S_j} \int_{S_i} \frac{\sigma_i^*(\mathbf{r}_i) \sigma_j(\mathbf{r}_j')}{|\mathbf{r}_i - \mathbf{r}_j'|} \, dA_j' \, dA_i \tag{2-8-14}$$

We can now interchange the dummy indices i and j as well as the primed and unprimed variables. We see then that the two expressions are identical. Thus

$$\sum_{i=1}^{N} \varphi_i Q_i^* = \sum_{i=1}^{N} \varphi_i^* Q_i \tag{2-8-15}$$

To prove that $C_{ij} = C_{ji}$, we take as our two cases:

1. Starred case: Let

    $$\varphi_j^* = \varphi_0 \qquad \text{all other } \varphi_i = 0$$

    Then

    $$Q_i^* = C_{ij} \varphi_0$$

2. Unstarred case: Let

    $$\varphi_i = \varphi_0 \qquad \text{all other } \varphi_j = 0$$

Then

$$Q_j = C_{ji}\varphi_0$$

Green's reciprocity theorem then tells us

$$\sum_k Q_k\varphi_k^* = Q_j\varphi_j^* = Q_j\varphi_0 = C_{ji}\varphi_0{}^2 = \sum Q_k^*\varphi_k = Q_i^*\varphi_i = C_{ij}\varphi_0{}^2$$

Hence

$$C_{ij} = C_{ji} \tag{2-8-16}$$

We observe one of the physical consequences of this statement in the case of two conductors. If conductor 1 is placed at potential φ_0 and conductor 2 is grounded, a certain charge Q_0 will be induced on the latter. If, on the other hand, conductor 2 is placed at potential φ_0 and conductor 1 is grounded, the same charge Q_0 will now appear on conductor 1.

It is instructive to calculate explicitly that this is so in the case of two concentric conducting spherical shells, one of radius a and the other of radius b $(a < b)$.

Situation 1. Inner sphere a is at potential φ_0. Outer sphere b is at potential 0. Let Q_a = charge on inner conductor. Then we calculate the potential on the inner conductor in the usual manner:

$$\varphi_0 = \int_a^b \frac{Q_a}{r^2}\,dr = \frac{Q_a}{a} - \frac{Q_a}{b}$$

Solving for Q_a, we have

$$Q_a = \frac{ab}{b - a}\varphi_0$$

Since $E = 0$ outside the outer sphere, we must have $Q_b = -Q_a$ and thus

$$Q_b = \frac{ab}{a - b}\varphi_0 \tag{2-8-17}$$

Situation 2. Inner sphere a is grounded. Outer sphere b is at potential φ_0. The potential at $r = a$ is equal to 0. This means that we can write

$$\int_a^b \frac{Q_a}{r^2}\,dr + \varphi_0 = 0$$

$$Q_a\frac{b - a}{ab} + \varphi_0 = 0$$

$$Q_a = \frac{ab}{a - b}\varphi_0 \tag{2-8-18}$$

Thus Q_b (situation 1) $= Q_a$ (situation 2).

As we mentioned earlier, we could solve the set of N equations (2-8-5) and obtain the φ's in terms of the Q's. We would then obtain the set of equations

$$\varphi_i = \sum_{j=1}^{N} P_{ij} Q_j \qquad (2\text{-}8\text{-}19)$$

where the P_{ij}'s are called the **potential coefficients.** We can again make use of Green's reciprocity theorem to prove that $P_{ij} = P_{ji}$.

1. Unstarred case: Let

$$Q_j = Q_0 \qquad \text{all other } Q_i = 0$$

Then

$$\varphi_i = P_{ij} Q_0$$

2. Starred case: Let

$$Q_i^* = Q_0 \qquad \text{all other } Q_j^* = 0$$

Then

$$\varphi_j^* = P_{ji} Q_0$$

Substituting into (2-8-15), we have

$$\sum_k Q_k \varphi_k^* = Q_j \varphi_j^* = P_{ji} Q_0^{\,2}$$

$$\sum_k Q_k^* \varphi_k = Q_i^* \varphi_i = P_{ij} Q_0^{\,2}$$

and hence

$$P_{ij} = P_{ji} \qquad (2\text{-}8\text{-}20)$$

The physical content of this statement is quite remarkable. Imagine that we have a set of conductors. If we put a charge Q_A on the fourth one, for example, and no charge on any of the others, we will produce a certain potential φ_B on the seventh one (for example). If the same charge Q_A were put on the seventh one instead (all others uncharged), the potential on the fourth one would now be φ_B. Thus there is a profound, reciprocal relationship among the charges and potentials on all the conductors.

Making use of Eq. (2-7-3), we have for the energy of a set of charged conductors

$$U = \frac{1}{2} \sum \varphi_i Q_i$$

$$= \frac{1}{2} \sum_{i,j} C_{ij} \varphi_i \varphi_j \qquad (2\text{-}8\text{-}21)$$

$$U = \frac{1}{2} \sum_{i,j} P_{ij} Q_i Q_j \qquad (2\text{-}8\text{-}22)$$

We can prove a rather nice theorem now. Let δU_Q be the change in electrostatic energy of the system if the conductors are slightly displaced while the charges are held constant. Let δU_φ be the change in electrostatic energy of the system of conductors if their displacement is the same as before, while the potentials are held constant. Then we will prove that $\delta U_Q = -\delta U_\varphi$.

First we treat the constant-potential case:

$$\delta U_\varphi = \tfrac{1}{2}\delta \left(\sum C_{ij}\varphi_i\varphi_j \right)$$

$$= \frac{1}{2} \sum (\delta C_{ij})\varphi_i\varphi_j \qquad (2\text{-}8\text{-}23)$$

since φ_i, φ_j remained fixed. To treat the case of fixed charge, we note that in this case

$$\delta Q_i = 0 = \delta \left(\sum_j C_{ij}\varphi_j \right)$$

$$= \sum_j (\delta C_{ij})\varphi_j + \sum_j C_{ij}\delta\varphi_j$$

Hence

$$\sum_j C_{ij}\delta\varphi_j = -\sum_j (\delta C_{ij})\varphi_j \qquad (2\text{-}8\text{-}24)$$

Now for the fixed-charge case

$$\delta U_Q = \frac{1}{2} \sum_{ij} (\delta C_{ij})\varphi_i\varphi_j + \frac{1}{2} \sum_{ij} C_{ij}(\delta\varphi_i)\varphi_j$$

$$+ \frac{1}{2} \sum_{ij} C_{ij}\varphi_i(\delta\varphi_j) \qquad (2\text{-}8\text{-}25)$$

We first interchange the dummy indices i and j in the last term on the right side of Eq. (2-8-25). We then remember that $C_{ij} = C_{ji}$. Thus the last two terms on the right side of Eq. (2-8-25) are equal, and we can write

$$\delta U_Q = \frac{1}{2} \sum_{i,j} (\delta C_{ij})\varphi_i\varphi_j + \sum_{ij} C_{ij}(\delta\varphi_j)\varphi_i \qquad (2\text{-}8\text{-}26)$$

Using Eq. (2-8-24), we have

$$\delta U_Q = \frac{1}{2} \sum_{ij} (\delta C_{ij})\varphi_i\varphi_j - \sum_{i,j} (\delta C_{ij})\varphi_j\varphi_i \qquad (2\text{-}8\text{-}27)$$

$$\delta U_Q = \delta U_\varphi - 2\delta U_\varphi = -\delta U_\varphi \qquad (2\text{-}8\text{-}28)$$

Our theorem is proved.

The physical consequences of this theorem are quite interesting. Imagine that we have a given arrangement of conductors with charges and potential on them. If the conductors are isolated and unconnected to the outside world, then the charges will remain fixed in any displacement. If δU_Q is *positive*, then we must do mechanical work of magnitude δU_Q against the electrostatic forces to perform the displacement. Now if, on the other hand, the conductors were connected to a set of batteries which maintained the potentials fixed, the change in electrostatic energy for the same displacement would be δU_φ. Since the forces are the same as before, depending only upon the charges and positions of the conductors, the mechanical work that we do to carry out the displacement is exactly the same as before (δU_Q). Since a positive δU_Q implies an equal negative δU_φ, we must be putting an amount of energy $2\delta U_Q$ into the battery when we perform the displacement.

An example of this was discussed earlier in the case of a parallel-plate capacitor (see page 54).

2-9 CYLINDRICAL AND SPHERICAL COORDINATES

As we have already surmised, much of our study of electrostatics must be devoted to the development of techniques for the solution of Laplace's equation (2-5-1) subject to appropriate boundary conditions. In general these boundary conditions will include the specification of either the total charge or the potential on each conductor. The potential at infinity will be assumed to be zero.

We will soon see that only a limited number of problems can be solved in closed form or in terms of a simple power series. These are, in general, problems that exhibit a great deal of symmetry about a point or a line and hence call naturally for the use of cylindrical or spherical coordinates. Accordingly we begin by developing the important differential forms (gradient, curl, and divergence) in these coordinate systems.

The cylindrical coordinate system, illustrated in Fig. 2-15, makes use of the three variables r, θ, z to specify the location of a point in space. The coordinate r is just the distance from the point to the z axis. The angle θ is the projected azimuthal angle as measured from the x axis and z is, of course, identical to the usual z coordinate. The coordinates (r,θ,z) and the old coordinates (x,y,z) are related through the transformation equations

$$
\begin{aligned}
r &= \sqrt{x^2 + y^2} \quad &\text{or} \quad & x = r\cos\theta \\
\tan\theta &= y/x \quad &\text{or} \quad & y = r\sin\theta \\
& & & z = z
\end{aligned}
\qquad (2\text{-}9\text{-}1)
$$

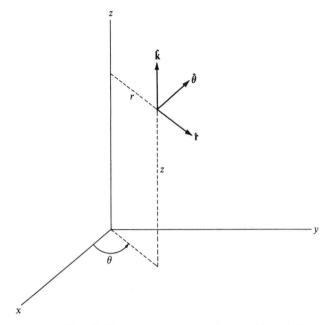

Fig. 2-15 The cylindrical coordinate system makes use of the variables r, θ, z to specify the location of a point in space.

It is convenient to set up three unit vectors $\hat{r}$, $\hat{\theta}$, and $\hat{k}$ at a point of interest. The unit vector $\hat{r}$ is drawn in the direction of increasing r, and the unit vector $\hat{\theta}$ is drawn in the direction of increasing θ, and naturally the vector $\hat{k}$ is drawn in the direction of increasing z. The three form an orthogonal coordinate system with $\hat{r} \times \hat{\theta} = \hat{k}$.

We note immediately that the new triad of unit vectors can be related to the old triad $\hat{i}$, $\hat{j}$, and $\hat{k}$ through the transformations

$$\hat{r} = \cos\theta\,\hat{i} + \sin\theta\,\hat{j} \qquad \text{or} \qquad \hat{i} = \cos\theta\,\hat{r} - \sin\theta\,\hat{\theta}$$

$$\hat{\theta} = -\sin\theta\,\hat{i} + \cos\theta\,\hat{j} \qquad \text{or} \qquad \hat{j} = \sin\theta\,\hat{r} + \cos\theta\,\hat{\theta}$$

$$\hat{k} = \hat{k} \qquad\qquad (2\text{-}9\text{-}2)$$

Proceeding first to find an expression for $\nabla\varphi$ in this coordinate system, we write, using Eq. (2-9-2),

$$\nabla\varphi = \frac{\partial\varphi}{\partial x}\,\hat{i} + \frac{\partial\varphi}{\partial y}\,\hat{j} + \frac{\partial\varphi}{\partial z}\,\hat{k}$$

$$= \left(\frac{\partial \varphi}{\partial x} \cos \theta + \frac{\partial \varphi}{\partial y} \sin \theta \right) \hat{\mathbf{r}} + \left(\frac{\partial \varphi}{\partial y} \cos \theta - \frac{\partial \varphi}{\partial x} \sin \theta \right) \hat{\boldsymbol{\theta}}$$

$$+ \frac{\partial \varphi}{\partial z} \hat{\mathbf{k}} \quad (2\text{-}9\text{-}3)$$

We remember though that $\partial \varphi / \partial x$ and $\partial \varphi / \partial y$ can be rewritten as

$$\frac{\partial \varphi}{\partial x} = \frac{\partial \varphi}{\partial r} \frac{\partial r}{\partial x} + \frac{\partial \varphi}{\partial \theta} \frac{\partial \theta}{\partial x} + \frac{\partial \varphi}{\partial z} \frac{\partial z}{\partial x}$$

$$\frac{\partial \varphi}{\partial y} = \frac{\partial \varphi}{\partial r} \frac{\partial r}{\partial y} + \frac{\partial \varphi}{\partial \theta} \frac{\partial \theta}{\partial y} + \frac{\partial \varphi}{\partial z} \frac{\partial z}{\partial y}$$

$$(2\text{-}9\text{-}4)$$

Using Eq. (2-9-1), we find that

$$\frac{\partial r}{\partial x} = \cos \theta \qquad \frac{\partial \theta}{\partial x} = \frac{-\sin \theta}{r}$$

$$\frac{\partial r}{\partial y} = \sin \theta \qquad \frac{\partial \theta}{\partial y} = \frac{\cos \theta}{r}$$

$$(2\text{-}9\text{-}5)$$

$$\frac{\partial z}{\partial x} = \frac{\partial z}{\partial y} = 0$$

If we substitute these results back into Eq. (2-9-3), we finally have

$$\nabla \varphi = \left(\frac{\partial \varphi}{\partial r} \right) \hat{\mathbf{r}} + \frac{1}{r} \left(\frac{\partial \varphi}{\partial \theta} \right) \hat{\boldsymbol{\theta}} + \left(\frac{\partial \varphi}{\partial z} \right) \hat{\mathbf{k}} \quad (2\text{-}9\text{-}6)$$

The identical result can be obtained immediately if we think of the meaning of $\nabla \varphi$ in physical terms. The three components of $\nabla \varphi$ are the rates of change of φ with respect to distance in the three mutually perpendicular directions given by $\hat{\mathbf{r}}$, $\hat{\boldsymbol{\theta}}$, and $\hat{\mathbf{k}}$. Going along $\hat{\mathbf{r}}$ first, we noted that the rate of change of φ with respect to distance in that direction is just $\partial \varphi / \partial r$. Going along $\hat{\boldsymbol{\theta}}$, we see that a differential distance in this direction is just $r\, d\theta$. Hence the rate of change of φ with respect to distance in the θ direction is given by $\dfrac{1}{r} \dfrac{\partial \varphi}{\partial \theta}$. Finally, the rate of change of φ with respect to distance in the z direction is $\partial \varphi / \partial z$. Equation (2-9-6) is then confirmed directly.

We next consider the divergence of a vector function $\mathbf{F}$ as expressed in cylindrical coordinates. We could begin with cartesian coordinates and convert all derivatives appropriately. It is more elegant, however, to just calculate the flux out of a small volume element dV and set the result equal to $\nabla \cdot \mathbf{F}\, dV$. As our volume element dV, we choose that which is bounded by the coordinates r and $r + dr$, θ and $\theta + d\theta$, z and $z + dz$, as shown in

Fig. 2-16. We indicate the bounding surfaces of dV by the numbers ① to ⑥.

Surface ① is at radius r and is bounded by θ, $\theta + d\theta$, z, and $z + dz$.

Surface ② is at radius $r + dr$ and is bounded by θ, $\theta + d\theta$, z, and $z + dz$.

Surface ③ is at height z and is bounded by r, $r + dr$, θ, and $\theta + d\theta$.

Surface ④ is at height $z + dz$ and is bounded by r, $r + dr$, θ, and $\theta + d\theta$.

Surface ⑤ is at angle θ and is bounded by z, $z + dz$, r, and $r + dr$.

Surface ⑥ is at angle $\theta + d\theta$ and is bounded by z, $z + dz$, r, and $r + dr$.

The areas of the various surfaces are

$$dA_① = r\,d\theta\,dz \qquad dA_③ = dA_④ = r\,dr\,d\theta$$

$$dA_② = (r + dr)d\theta\,dz \qquad dA_⑤ = dA_⑥ = dr\,dz \qquad (2\text{-}9\text{-}7)$$

The fluxes out of the volume are then

Flux out through ① $= -F_r(1)r\,d\theta\,dz$

Flux out through ② $= +F_r(2)(r + dr)d\theta\,dz$

Flux out through ③ $= -F_z(3)r\,dr\,d\theta$

Flux out through ④ $= +F_z(4)r\,dr\,d\theta \qquad (2\text{-}9\text{-}8)$

Flux out through ⑤ $= -F_\theta(5)dr\,dz$

Flux out through ⑥ $= +F_\theta(6)dr\,dz$

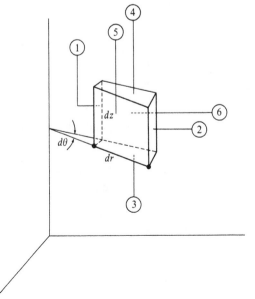

Fig. 2-16 We wish to find the flux out of an infinitesimal volume dV shown above.

Remembering that

$$F_r(2) - F_r(1) = \frac{\partial F_r}{\partial r} dr$$

$$F_z(4) - F_z(3) = \frac{\partial F_z}{\partial z} dz$$

$$F_\theta(6) - F_\theta(5) = \frac{\partial F_\theta}{\partial \theta} d\theta$$

we have (ignoring all terms which are the product of four differentials)

$$\text{Net flux out} = \frac{\partial F_r}{\partial r} r \, dr \, d\theta \, dz + \frac{F_r}{r} r \, dr \, d\theta \, dz + \frac{\partial F_z}{\partial z} r \, dr \, d\theta \, dz$$

$$+ \left(\frac{1}{r} \frac{\partial F_\theta}{\partial \theta} \right) r \, dr \, d\theta \, dz \quad (2\text{-}9\text{-}9)$$

The volume dV is just $r \, dr \, d\theta \, dz$. We can then rewrite Eq. (2-9-9) into the form

$$\text{Net flux out} = \left[\frac{1}{r} \frac{\partial}{\partial r} (rF_r) + \frac{1}{r} \frac{\partial F_\theta}{\partial \theta} + \frac{\partial F_z}{\partial z} \right] dV \quad (2\text{-}9\text{-}10)$$

The divergence of $\mathbf{F}$ is thus given by

$$\nabla \cdot \mathbf{F} = \frac{1}{r} \frac{\partial}{\partial r} (rF_r) + \frac{1}{r} \frac{\partial F_\theta}{\partial \theta} + \frac{\partial F_z}{\partial z} \quad (2\text{-}9\text{-}11)$$

Inserting our expression for $\nabla \varphi$ [Eq. (2-9-6)] into the above, we find that the laplacian is given by

$$\nabla^2 \varphi = \nabla \cdot \nabla \varphi = \frac{1}{r} \frac{\partial}{\partial r} \left(r \frac{\partial \varphi}{\partial r} \right) + \frac{1}{r^2} \frac{\partial^2 \varphi}{\partial \theta^2} + \frac{\partial^2 \varphi}{\partial z^2} \quad (2\text{-}9\text{-}12)$$

We next find the three components of $\nabla \times \mathbf{F}$ in cylindrical coordinates. To do so we make use of Stokes' theorem and evaluate the line integral of $\mathbf{F} \cdot d\mathbf{l}$ around three infinitesimal loops, one perpendicular to $\hat{\mathbf{r}}$, one perpendicular to $\hat{\boldsymbol{\theta}}$, and one perpendicular to $\hat{\mathbf{k}}$. We take a loop perpendicular to $\hat{\mathbf{r}}$ first, bounded by the coordinates θ, $\theta + d\theta$, z, and $z + dz$. We have then, applying Stokes' theorem,

$$\hat{\mathbf{r}} \cdot (\nabla \times \mathbf{F}) r \, d\theta \, dz = [F_\theta(z) - F_\theta(z + dz)] r \, d\theta$$

$$+ [F_z(\theta + d\theta) - F_z(\theta)] dz$$

$$= \left(\frac{1}{r} \frac{\partial F_z}{\partial \theta} - \frac{\partial F_\theta}{\partial z} \right) r \, d\theta \, dz$$

Hence

$$(\nabla \times \mathbf{F}) \cdot \hat{\mathbf{r}} = \frac{1}{r} \frac{\partial F_z}{\partial \theta} - \frac{\partial F_\theta}{\partial z} \tag{2-9-13}$$

For a loop perpendicular to θ and bounded by r, $r + dr$, z, and $z + dz$, we have

$$\hat{\boldsymbol{\theta}} \cdot (\nabla \times \mathbf{F}) \, dr \, dz = [F_z(r) - F_z(r + dr)] \, dz$$
$$+ [F_r(z + dz) - F_r(z)] \, dr$$
$$= \left(\frac{\partial F_r}{\partial z} - \frac{\partial F_z}{\partial r} \right) dr \, dz$$

Hence

$$(\nabla \times \mathbf{F}) \cdot \hat{\boldsymbol{\theta}} = \frac{\partial F_r}{\partial z} - \frac{\partial F_z}{\partial z} \tag{2-9-14}$$

Finally we take a loop perpendicular to $\hat{\mathbf{k}}$. We have then

$$(\nabla \times \mathbf{F} \cdot \hat{\mathbf{k}}) r \, dr \, d\theta = [F_r(\theta) - F_r(\theta + d\theta)] \, dr$$
$$+ F_\theta(r + dr)(r + dr) \, d\theta - F_\theta(r) r \, d\theta$$
$$= \left(\frac{\partial F_\theta}{\partial r} + \frac{F_\theta}{r} - \frac{1}{r} \frac{\partial F_r}{\partial \theta} \right) r \, dr \, d\theta$$

Thus

$$(\nabla \times \mathbf{F} \cdot \hat{\mathbf{k}}) = \frac{\partial F_\theta}{\partial r} + \frac{F_\theta}{r} - \frac{1}{r} \frac{\partial F_r}{\partial \theta} \tag{2-9-15}$$

Combining Eqs. (2-9-13) to (2-9-15), we have

$$\nabla \times \mathbf{F} = \left(\frac{1}{r} \frac{\partial F_z}{\partial \theta} - \frac{\partial F_\theta}{\partial z} \right) \hat{\mathbf{r}} + \left(\frac{\partial F_r}{\partial z} - \frac{\partial F_z}{\partial r} \right) \hat{\boldsymbol{\theta}}$$
$$+ \left(\frac{\partial F_\theta}{\partial r} + \frac{F_\theta}{r} - \frac{1}{r} \frac{\partial F_r}{\partial \theta} \right) \hat{\mathbf{k}} \tag{2-9-16}$$

We shall return very shortly and make use of cylindrical coordinates for the solution of specific problems. In the meantime we turn to the spherical coordinate system shown in Fig. 2-17.

The spherical coordinates of a point are its distance from the origin r and its polar and azimuthal angles θ and ψ. (Note: We use ψ to denote azimuthal angle to avoid confusion with the potential function φ.) The transformation equations relating x, y, and z to r, θ, and ψ are given by

$$\begin{aligned} x &= r \sin \theta \cos \psi \\ y &= r \sin \theta \sin \psi \\ z &= r \cos \theta \end{aligned} \tag{2-9-17}$$

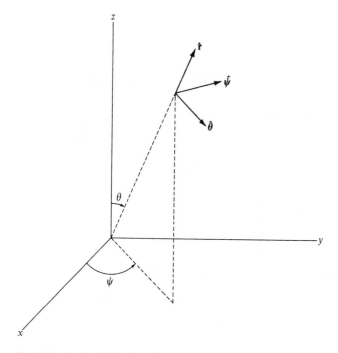

Fig. 2-17 In the spherical coordinate system a point is designated by its distance from the origin r, its polar angle θ, and its azimuthal angle ψ.

or

$$r = \sqrt{x^2 + y^2 + z^2}$$

$$\cos \theta = \frac{z}{\sqrt{x^2 + y^2 + z^2}}$$

(2-9-18)

$$\cos \psi = \frac{x}{\sqrt{x^2 + y^2}}$$

Again we set up an orthogonal coordinate system at the point of interest. The unit vector $\hat{r}$ points in the direction of increasing r. The unit vectors $\hat{\theta}$ and $\hat{\psi}$ point, respectively, in the directions of increasing θ and ψ. As can be seen from Fig. 2-17, $\hat{r} \times \hat{\theta} = \hat{\psi}$.

Using the same arguments as before, we evaluate $\nabla\varphi$ by determining the rate of change of φ with respect to distance along each of the three mutually perpendicular unit vectors. The rate of change of φ with respect to distance along $\hat{r}$ is just $\partial\varphi/\partial r$. The rate of change of φ with respect to distance along the $\hat{\theta}$ direction is $\dfrac{1}{r}\dfrac{\partial\varphi}{\partial\theta}$. The rate of change of φ with respect

to distance in the ψ direction is $\dfrac{1}{r\sin\theta}\dfrac{\partial\varphi}{\partial\psi}$. Hence we have

$$\nabla\varphi = \frac{\partial\varphi}{\partial r}\,\hat{\mathbf{r}} + \frac{1}{r}\frac{\partial\varphi}{\partial\theta}\,\hat{\boldsymbol{\theta}} + \frac{1}{r\sin\theta}\frac{\partial\varphi}{\partial\psi}\,\hat{\boldsymbol{\psi}} \tag{2-9-19}$$

Using precisely the same techniques as we used in dealing with cylindrical coordinates, we can evaluate $\nabla\cdot\mathbf{F}$, $\nabla\times\mathbf{F}$, and $\nabla^2\varphi$. We obtain

$$\nabla\cdot\mathbf{F} = \frac{1}{r^2}\frac{\partial}{\partial r}(r^2 F_r) + \frac{1}{r\sin\theta}\frac{\partial}{\partial\theta}(\sin\theta\,F_\theta) + \frac{1}{r\sin\theta}\frac{\partial F_\psi}{\partial\psi} \tag{2-9-20}$$

$$\nabla\times\mathbf{F} = \frac{1}{r\sin\theta}\left[\frac{\partial}{\partial\theta}(\sin\theta\,F_\psi) - \frac{\partial F_\theta}{\partial\psi}\right]\hat{\mathbf{r}}$$

$$+ \left[\frac{1}{r\sin\theta}\frac{\partial F_r}{\partial\psi} - \frac{1}{r}\frac{\partial(rF_\psi)}{\partial r}\right]\hat{\boldsymbol{\theta}} + \frac{1}{r}\left[\frac{\partial(rF_\theta)}{\partial r} - \frac{\partial F_r}{\partial\theta}\right]\hat{\boldsymbol{\psi}} \tag{2-9-21}$$

$$\nabla^2\varphi = \frac{1}{r^2}\frac{\partial}{\partial r}\left(r^2\frac{\partial\varphi}{\partial r}\right) + \frac{1}{r^2\sin\theta}\frac{\partial}{\partial\theta}\left(\sin\theta\frac{\partial\varphi}{\partial\theta}\right)$$

$$+ \frac{1}{r^2\sin^2\theta}\frac{\partial^2\varphi}{\partial\psi^2} \tag{2-9-22}$$

We now proceed to make some use of all this newly acquired knowledge.

2-10 SOLVING LAPLACE'S EQUATION IN CARTESIAN COORDINATES

Now that we have learned to express Laplace's equation in three useful coordinate systems (cartesian, cylindrical, and spherical), we might give a bit of thought to the problem of solving the equation. In general we will find ourselves with an infinitude of possible solutions, and we will have to refer to the specific boundary conditions at hand to find the correct combination. What we hope to do is to make our procedure orderly, remembering always that only one solution of Laplace's equation can satisfy the complete set of boundary conditions.

We start with the simplest system, cartesian coordinates. The procedure we follow here will be a quite general one and will be justified more thoroughly after we have worked with it some. We will search for solutions to $\nabla^2\varphi = 0$ of the form

$$\varphi(x,y,z) = X(x)\,Y(y)\,Z(z) \tag{2-10-1}$$

Inserting this product form into Laplace's equation, we obtain, after dividing through by $\varphi(x,y,z)$,

$$\frac{1}{X}\frac{d^2X}{dx^2} + \frac{1}{Y}\frac{d^2Y}{dy^2} + \frac{1}{Z}\frac{d^2Z}{dz^2} = 0 \tag{2-10-2}$$

Since the three variables x, y, and z are all independent, the three parts to Eq. (2-10-2) must all be constants. That is,

$$\frac{d^2X}{dx^2} = \alpha^2 X$$

$$\frac{d^2Y}{dy^2} = \beta^2 Y \tag{2-10-3}$$

$$\frac{d^2Z}{dz^2} = \gamma^2 Z$$

provided that

$$\alpha^2 + \beta^2 + \gamma^2 = 0 \tag{2-10-4}$$

(Needless to say, α, β, and γ cannot all be real.) We have then

$$\begin{aligned} X &= A_1 e^{\alpha x} + B_1 e^{-\alpha x} \\ Y &= A_2 e^{\beta y} + B_2 e^{-\beta y} \\ Z &= A_3 e^{\gamma z} + B_3 e^{-\gamma z} \end{aligned} \tag{2-10-5}$$

The possible solutions of the form given by Eq. (2-10-1) are thus innumerable—every choice of A_1, B_1, A_2, B_2, A_3, and B_3 will do. In addition we can choose α, β, and γ as we like provided that they satisfy the simple constraint given by Eq. (2-10-4). What are we to do next? The answer is best given by solving a specific problem in which we apply a set of boundary conditions to determine φ completely.

We consider a box made up of six plane sheets as shown in Fig. 2-18. The planes at $x = 0$, $y = 0$, $z = 0$, $x = a$, and $y = b$ are grounded ($\varphi = 0$). The plane at $z = c$ has a potential distribution on it given by $\varphi_0(x,y)$. We wish to know the potential everywhere within the box.

We begin by inserting the boundary condition at $x = 0$ into Eq. (2-10-5). Since $\varphi(x = 0)$ must vanish for all y and z, we must have

$$X(0) = A_1 + B_1 = 0 \tag{2-10-6}$$

Similarly, making use of the boundary conditions at $y = z = 0$, we can write

$$A_2 + B_2 = 0 \tag{2-10-7}$$
$$A_3 + B_3 = 0 \tag{2-10-8}$$

If we now go to the wall at $x = a$, we find

$$A_1 e^{\alpha a} + B_1 e^{-\alpha a} = 0 \tag{2-10-9}$$

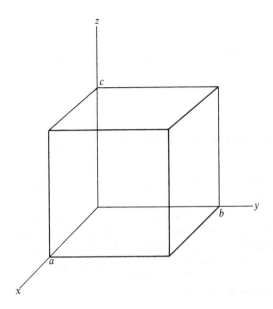

Fig. 2-18 We solve for the potential on the inside of a box, all the walls of which are grounded except for one. That wall at $z = c$ has a potential distribution on it given by $\varphi_0(x,y)$.

In order that Eqs. (2-10-9) and (2-10-6) both hold we must have

$$e^{\alpha a} = e^{-\alpha a}$$

or

$$e^{2\alpha a} = 1 \tag{2-10-10}$$

For Eq. (2-10-10) to be true, we must have

$$\alpha = \frac{ni\pi}{a} \tag{2-10-11}$$

where n is an integer. Combining what we learned and remembering that

$$\sin \theta = \frac{e^{i\theta} - e^{-i\theta}}{2i} \tag{2-10-12}$$

we can conclude that

$$X(x) \propto \sin \frac{n\pi x}{a} \tag{2-10-13}$$

Similarly

$$Y(y) \propto \sin \frac{m\pi y}{b} \tag{2-10-14}$$

where m is also an integer. For every choice of integers m and n we have then a possible solution

$$\varphi_{nm}(x,y,z) = A_{n,m} \sin \frac{n\pi x}{a} \sin \frac{m\pi y}{b} \sinh \gamma_{n,m} z \qquad (2\text{-}10\text{-}15)$$

where

$$\gamma_{n,m} = \sqrt{\frac{n^2\pi^2}{a^2} + \frac{m^2\pi^2}{b^2}} \qquad (2\text{-}10\text{-}16)$$

We are now ready to insert the final boundary condition at $z = c$. We have then

$$\varphi(x,y,c) = \varphi_0(x,y) = \sum_{n,m=1}^{\infty} (A_{n,m} \sinh \gamma_{nm} c) \sin \frac{n\pi}{a} x \ \sin \frac{m\pi}{b} y \qquad (2\text{-}10\text{-}17)$$

Our job now seems to be reduced to finding the set of coefficients $A_{n,m}$. But, do such a set of coefficients exist? At this point we must state without proof the fundamental theorem of Fourier analysis.[1] *Any* function of x defined over the interval from $x = 0$ to $x = a$ can be expressed as a linear combination of terms of the form $\sin(n\pi x/a)$. That is, for any $f(x)$ there exist a set of numbers B_n such that

$$f(x) = \sum_{n=1}^{\infty} B_n \sin \frac{n\pi}{a} x$$

over the interval from 0 to a. In the case of two dimensions x and y, any function $f(x,y)$ defined over the interval $0 \leq x \leq a$ and $0 \leq y \leq b$ can be represented as a double Fourier series:

$$f(x,y) = \sum_{n,m=1}^{\infty} B_{n,m} \sin \frac{n\pi}{a} x \sin \frac{m\pi}{b} y \qquad (2\text{-}10\text{-}18)$$

Returning to Eq. (2-10-17), we find the coefficients $A_{n,m}$ by a standard procedure. We multiply each side by $\sin \frac{k\pi}{a} x \sin \frac{l\pi}{b} y$ where both k and l are integers. We then integrate over both x and y. Remembering that

$$\int_0^a \sin \frac{n\pi}{a} x \sin \frac{k\pi}{a} x \, dx = \frac{a}{2} \delta_{nk}$$

$$\int_0^b \sin \frac{m\pi}{b} y \sin \frac{l\pi}{b} y \, dy = \frac{b}{2} \delta_{ml} \qquad (2\text{-}10\text{-}19)$$

we have

$$A_{n,m} = \frac{4}{ab \sinh \gamma_{n,m} c} \int_0^a dx \int_0^b dy \, \varphi_0(x,y) \sin \frac{n\pi}{a} x \sin \frac{m\pi}{b} y \qquad (2\text{-}10\text{-}20)$$

[1] For further explanation see Chap. 6.

We are thus in a position to evaluate the coefficients $A_{n,m}$ explicitly, and we have solved our problem.

Let us go back now and say a few words about the key step in our procedure for solving Laplace's equation. We searched for solutions in the form of products of functions of the independent variables. We then noted that at least in the solution of one simple boundary-value problem we could write our answer in terms of a sum over these solutions. The remarkable fact is that *any* solution to Laplace's equation can be expressed as a sum over such product functions and that is why we have lost no generality by proceeding in this manner. We will always look for these product-function solutions and then spend our time determining the coefficients from the boundary conditions.

2-11 SOLVING LAPLACE'S EQUATION IN CYLINDRICAL COORDINATES

As we pointed out earlier, most simple potential problems with a great deal of symmetry can be handled best by means of cylindrical or spherical coordinates. In this section we consider the solution of Laplace's equation in cylindrical coordinates subject to appropriate boundary conditions. Making use of Eq. (2-9-12), we write

$$\nabla^2\varphi = \frac{1}{r}\frac{\partial}{\partial r}\left(r\frac{\partial\varphi}{\partial r}\right) + \frac{1}{r^2}\frac{\partial^2\varphi}{\partial\theta^2} + \frac{\partial^2\varphi}{\partial z^2} = 0 \qquad (2\text{-}11\text{-}1)$$

For simplicity let us begin by limiting our discussion to cases where φ does not depend on z. That is to say, $\partial^2\varphi/\partial z^2 = 0$. We look again for product solutions. In this case they have the form

$$\varphi(r,\theta) = R(r)\Theta(\theta) \qquad (2\text{-}11\text{-}2)$$

Our equation becomes

$$\frac{\Theta}{r}\frac{d}{dr}\left(r\frac{dR}{dr}\right) + \frac{R}{r^2}\frac{d^2\Theta}{d\theta^2} = 0$$

Let us multiply by r^2 and divide by $R\Theta$. We have then

$$\frac{r}{R}\frac{d}{dr}\left(r\frac{dR}{dr}\right) = -\frac{1}{\Theta}\frac{d^2\Theta}{d\theta^2} \qquad (2\text{-}11\text{-}3)$$

Note that the right side of the equation depends only on θ and the left side only on r. Since the two are independent variables, the two sides must separately be equal to a constant, which we write as k^2. We have then

$$\frac{d^2\Theta}{d\theta^2} = -k^2\Theta \qquad (2\text{-}11\text{-}4)$$

and

$$r \frac{d}{dr}\left(r \frac{dR}{dr}\right) = k^2 R \qquad (2\text{-}11\text{-}5)$$

Consider first the case where $k \neq 0$. The first equation then has as a solution

$$\Theta = A_k \cos k\theta + B_k \sin k\theta \qquad (2\text{-}11\text{-}6)$$

In the event that $k = 0$, we have

$$\Theta = A_0 + \alpha\theta \qquad (2\text{-}11\text{-}7)$$

Solutions corresponding to imaginary k, that is, negative k^2, cannot be single valued as we increase θ by 2π. Hence they are excluded. In fact the requirement of single valuedness implies that $A_k = B_k = 0$ unless k is an integer. It also implies that $\alpha = 0$. Hence we can write in general

$$\Theta = A_n \cos n\theta + B_n \sin n\theta \qquad \text{where } n = 0, 1, 2, \ldots \qquad (2\text{-}11\text{-}8)$$

To find $R(r)$ we first expand in a power series:

$$R(r) = \sum_{m=-\infty}^{\infty} C_m r^m \qquad (2\text{-}11\text{-}9)$$

Substituting back into our differential equation (with $k = n \neq 0$), we find

$$r \frac{d}{dr}\left(r \frac{dR}{dr}\right) = \sum C_m m^2 r^m = n^2 \sum C_m r^m \qquad (2\text{-}11\text{-}10)$$

Hence

$$\sum C_m (m^2 - n^2) r^m = 0 \qquad (2\text{-}11\text{-}11)$$

In order that this be zero, we must have each and every term zero, since no power of r can be expressed in terms of any other powers of r. Hence either

$$C_m = 0$$

or

$$m^2 = n^2 \qquad \text{and} \qquad m = \pm n$$

In the event that $n = 0$, we have

$$\frac{d}{dr}\left(r \frac{dR}{dr}\right) = 0$$

$$r \frac{dR}{dr} = \beta$$

$$R = \beta \ln r + \gamma \qquad (2\text{-}11\text{-}12)$$

Finally then we can combine all our constants and write the general solution as

$$\varphi(r,\theta) = \sum_{n=1}^{\infty} A_n r^n \cos n\theta + \sum_{n=1}^{\infty} B_n r^n \sin n\theta + \sum_{n=1}^{\infty} C_n r^{-n} \cos n\theta$$
$$+ \sum_{n=1}^{\infty} D_n r^{-n} \sin n\theta + C_0 \ln r + D_0 \qquad (2\text{-}11\text{-}13)$$

where all the constants are arbitrary.

We now apply this to the solution of a simple problem. Let us examine the case of an infinitely long, conducting, uncharged cylinder in an electric field which approaches uniformity at large distances from the cylinder. The field is taken to be perpendicular to the axis of the cylinder. The radius of the cylinder is a (see Fig. 2-19). For large x we have $\varphi = -E_0 x$ (we obviously must remove the restriction $\varphi = 0$ at infinity if we want a constant electric field at infinity).

$$\varphi(\text{large } x) = -E_0 r \cos \theta \qquad (2\text{-}11\text{-}14)$$

The condition that the cylinder be uncharged yields

$$\lambda = \frac{\text{charge}}{\text{unit length}} = 0$$

$$= -\frac{1}{4\pi} \int_{S \text{ unit length}} \frac{\partial \varphi}{\partial r} dA = -\frac{a}{4\pi} \int_0^{2\pi} \frac{\partial \varphi}{\partial r} d\theta \qquad (2\text{-}11\text{-}15)$$

We expect, by the symmetry inherent in the situation, to find that $\varphi = 0$ on the cylinder. Let us assume this to be true and see where we get. Since at large r we must have $\varphi \to -E_0 r \cos \theta$, we can set [see Eq. (2-11-13)]

$$\begin{aligned} A_n &= 0 &&\text{for } n \neq 1 \\ B_n &= 0 &&\text{for all } n \\ A_1 &= -E_0 \end{aligned} \qquad (2\text{-}11\text{-}16)$$

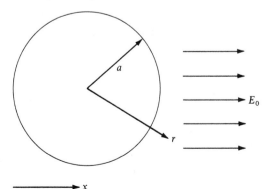

Fig. 2-19 An infinitely long, conducting cylinder is placed in a uniform field.

Now at $r = a$, $\varphi = 0$. Hence

$$-E_0 a \cos \theta + \sum_{n=1}^{\infty} C_n a^{-n} \cos n\theta$$

$$+ \sum_{n=1}^{\infty} D_n a^{-n} \sin n\theta + C_0 \ln a + D_0 = 0 \qquad \text{for all } \theta \quad (2\text{-}11\text{-}17)$$

Since all functions $\cos n\theta$ are linearly independent, we have immediately

$$\begin{aligned}
D_n &= 0 & &\text{for all } n \\
C_0 &= 0 & D_0 &= 0 \\
C_n &= 0 & &\text{for } n \neq 1 \\
-E_0 a &+ C_1 a^{-1} = 0
\end{aligned} \qquad (2\text{-}11\text{-}18)$$

Thus

$$C_1 = E_0 a^2 \qquad (2\text{-}11\text{-}19)$$

We have then

$$\varphi = E_0 \left(\frac{a^2}{r} - r \right) \cos \theta \qquad (2\text{-}11\text{-}20)$$

We can calculate the surface charge everywhere:

$$\sigma = -\frac{1}{4\pi} \left(\frac{\partial \varphi}{\partial r} \right)_a$$

$$= \frac{E_0}{4\pi} \left(\frac{a^2}{a^2} + 1 \right) \cos \theta = \frac{E_0}{2\pi} \cos \theta \qquad (2\text{-}11\text{-}21)$$

The charge per unit length is

$$\lambda = \frac{E_0 a}{2\pi} \int_0^{2\pi} \cos \theta \, d\theta = 0 \qquad (2\text{-}11\text{-}22)$$

This completes our solution.

We now return to Laplace's equation (2-11-1) and remove the restriction of z independence. We search for solutions of the form

$$\varphi(r,\theta,z) = R(r)\Theta(\theta)Z(z) \qquad (2\text{-}11\text{-}23)$$

Substituting, dividing through by φ, and multiplying by r^2, we rewrite Eq. (2-11-1) as

$$\frac{r}{R} \frac{d}{dr} \left(r \frac{dR}{dr} \right) + \frac{r^2}{Z} \frac{d^2 Z}{dz^2} = -\frac{1}{\Theta} \frac{d^2 \Theta}{d\theta^2} \qquad (2\text{-}11\text{-}24)$$

Again we realize that the two sides of Eq. (2-11-24) must be set equal to a constant. Referring back to Eqs. (2-11-4) to (2-11-8), we have once more

$$\Theta = A_n \cos n\theta + B_n \sin n\theta \tag{2-11-25}$$

where n is an integer. Inserting this back into Eq. (2-11-24), we obtain

$$\frac{1}{rR}\frac{d}{dr}\left(r\frac{dR}{dr}\right) - \frac{n^2}{r^2} = -\frac{1}{Z}\frac{d^2Z}{dz^2} \tag{2-11-26}$$

We next find the possible solutions for Z by requiring that both sides of Eq. (2-11-26) be equal to a constant which we set equal to $-\alpha^2$. We have then

$$\frac{d^2Z}{dz^2} = \alpha^2 Z \tag{2-11-27}$$

The solutions for nonzero values of α are just

$$Z_\alpha(z) = C_\alpha e^{\alpha z} + D_\alpha e^{-\alpha z} \tag{2-11-28}$$

In the event that $\alpha = 0$, the solution to Eq. (2-11-27) becomes

$$Z_0(z) = C_0 + D_0 z \tag{2-11-29}$$

Finally we must find the solution for $R_{\alpha,n}(r)$ corresponding to a particular choice of n and α. Our differential equation is

$$\frac{1}{r}\frac{d}{dr}\left(r\frac{dR_{\alpha,n}}{dr}\right) + \left(\alpha^2 - \frac{n^2}{r^2}\right)R_{\alpha,n} = 0 \tag{2-11-30}$$

It is convenient at this point to let $u = \alpha r$ for the case where $\alpha \neq 0$. We then divide Eq. (2-11-30) by α^2 and obtain the standard form for Bessel's equation:

$$\frac{1}{u}\frac{d}{du}\left(u\frac{dR_{\alpha,n}}{du}\right) + \left(1 - \frac{n^2}{u^2}\right)R_{\alpha,n} = 0$$

The solutions to this equation are best expressed as a power series in u. As usual there are two linearly independent solutions, $J_n(u)$ and $N_n(u)$. The first of these, $J_n(u)$, is called a **Bessel function of the first kind** and is defined as

$$J_n(u) = \left(\frac{u}{2}\right)^n \sum_{k=0}^{\infty} \frac{(-1)^k}{k!(n+k)!}\left(\frac{u}{2}\right)^{2k} \tag{2-11-31}$$

The function $N_n(u)$ is often called a **Neumann function** and is defined as

$$N_n(u) = \frac{2}{\pi}J_n(u)\ln\left(\frac{u}{2}\right) - \frac{1}{\pi}\left(\frac{u}{2}\right)^n \sum_{k=0}^{\infty} \frac{(-1)^k}{k!(n+k)!}\left(\frac{u}{2}\right)^{2k}$$

$$\cdot \left(\sum_{j=1}^{k}\frac{1}{j} + \sum_{j=1}^{n+k}\frac{1}{j}\right) - \frac{1}{\pi}\left(\frac{u}{2}\right)^{-n}\sum_{k=0}^{n-1}\frac{(n-k-1)!}{k!}\left(\frac{u}{2}\right)^{2k}$$

$$\tag{2-11-32}$$

Needless to say the use of a digital computer is strongly recommended when dealing with either of these functions. In any case, the general solution to the radial equation is

$$R_{\alpha,n}(r) = E_{\alpha,n} J_n(\alpha r) + F_{\alpha,n} N_n(\alpha r) \qquad (2\text{-}11\text{-}33)$$

where $E_{\alpha,n}$ and $F_{\alpha,n}$ are arbitrary constants.

In the event that $\alpha = 0$ the two solutions for $R_{0,n}(r)$ are the same as those for the differential equation (2-11-5).

$$R_{0,n}(r) = E_{0,n} r^n + E_{0,n} r^{-n} \qquad (2\text{-}11\text{-}34)$$

Finally, if n is also equal to zero, we recall that

$$R_{0,0}(r) = E_{0,0} + F_{0,0} \ln r \qquad (2\text{-}11\text{-}35)$$

Again the values of the various coefficients will be determined by boundary conditions. Quite often we will find that only specific and discrete values of α will permit us to fit our boundary conditions and the first part of our effort will be devoted to finding these values. We will then make use of completeness and orthogonality to determine the actual values of the coefficients themselves.

The detailed solution of problems of this sort is somewhat beyond the scope of this book. The reader is referred to innumerable treatises on classical potential theory or to more advanced texts in the field of electromagnetism.[1]

2-12   THE SOLUTION TO LAPLACE'S EQUATION IN SPHERICAL COORDINATES

Perhaps the most useful of all coordinate systems in the study of electrostatics, particularly at the microscopic level, is that of the spherical coordinates. We will limit ourselves to physical situations with complete rotational symmetry about the z axis (axial symmetry). The extension to nonsymmetric situations is straightforward and is left to the reader in consultation with more advanced textbooks.

Before proceeding with Laplace's equation, it is convenient to make a change of variable. We let $\mu = \cos \theta$ and have thus

$$\frac{\partial}{\partial \theta} = \frac{\partial \mu}{\partial \theta} \frac{\partial}{\partial \mu} = -\sin \theta \frac{\partial}{\partial \mu} \qquad (2\text{-}12\text{-}1)$$

Substituting back into Laplace's equation (remembering that there is no ψ dependence), we have

$$\nabla^2 \varphi = \frac{1}{r^2} \frac{\partial}{\partial r} \left(r^2 \frac{\partial \varphi}{\partial r} \right) + \frac{1}{r^2} \frac{\partial}{\partial \mu} (1 - \mu^2) \frac{\partial \varphi}{\partial \mu} = 0 \qquad (2\text{-}12\text{-}2)$$

We search for a product solution of the form

[1] See, for example, J. D. Jackson, "Classical Electrodynamics," p. 75, Wiley, New York, 1962.

$$\varphi = F(r)P(\mu) \tag{2-12-3}$$

Substituting Eq. (2-12-3) into Eq. (2-12-2), we obtain in the same manner as before

$$\frac{1}{F}\frac{d}{dr}\left(r^2\frac{dF}{dr}\right) = \frac{-1}{P}\frac{d}{d\mu}\left[(1 - \mu^2)\frac{dP}{d\mu}\right] = k \tag{2-12-4}$$

Let us begin by examining the radial equation. In most problems that we are likely to encounter φ will be finite at all points except perhaps at $r = 0$ and $r = \infty$. The reason for allowing φ to possibly diverge at $r = 0$ is that the region about $r = 0$ may be excluded from the problem. For example, if we are dealing with a conducting sphere of radius a, we are only interested in φ for $r > a$. The reason for allowing φ to possibly diverge as $r \to \infty$ is to permit the inclusion of cases like a constant electric field out to large distances. Since at all other points φ is finite, we can expand $F(r)$ in a power series, allowing all possible negative and positive powers of r.

$$F(r) = \sum_{m=-\infty}^{\infty} A_m r^m \tag{2-12-5}$$

If we now substitute this power series back into the right side of Eq. (2-12-4), we find that

$$\frac{d}{dr}\left(r^2\frac{dF}{dr}\right) = \sum_{m=-\infty}^{\infty} A_m m(m + 1)r^m = kF = \sum_{m=-\infty}^{\infty} A_m k r^m \tag{2-12-6}$$

In order that this equality holds for all r, it must hold for each and every coefficient of a given power of r. Hence

$$m(m + 1) = k \tag{2-12-7}$$

We conclude then that k can take on only specific and discrete values, namely, those which can be obtained through the product of m and $m + 1$ where m is an integer. The values of k which are possible are all positive, as can be determined from inspection.

For each value of k that permits a solution, there are two values of m which correspond to that k. We list some of these in order in Eq. (2-12-8).

k	Possible m	$F(r)$
0	0, −1	$F_0(r) = A_0 + B_0 r^{-1}$
2	1, −2	$F_1(r) = A_1 r + B_1 r^{-2}$
6	2, −3	$F_2(r) = A_2 r^2 + B_2 r^{-3}$
. . . . . .	. . . .	. . . .
$n(n + 1)$	$n, -(n + 1)$	$F_n(r) = A_n r^n + B_n r^{-(n+1)}$
. . . . . .	. . . .	. . . .

$$\tag{2-12-8}$$

Of course, for each value of k (or n) that is possible, we have a corresponding equation for $P(\mu)$. The solution to this equation will be called $P_n(\mu)$. Our final solution for φ will then be

$$\varphi(r,\mu) = \sum_{n=0}^{\infty} (A_n r^n + B_n r^{-(n+1)}) P_n(\mu) \tag{2-12-9}$$

We now proceed to find $P_n(\mu)$, by means of the differential equation

$$\frac{d}{d\mu}\left[(1 - \mu^2)\frac{dP_n}{d\mu}\right] + n(n + 1)P_n = 0 \tag{2-12-10}$$

As usual we will expand $P_n(\mu)$ in a power series. This time, since we want $P_n(\mu)$ to remain finite for all μ between 0 and 1, we take only positive integral powers of μ.

$$P_n(\mu) = \sum_{l=0}^{\infty} A_{n,l}\mu^l \tag{2-12-11}$$

We evaluate the first term on the left side of Eq. (2-12-10).

$$\frac{d}{d\mu}\left[(1 - \mu^2)\frac{dP_n}{d\mu}\right] = \sum_{l=0}^{\infty} A_{n,l}l(l - 1)\mu^{l-2} - \sum_{l=0}^{\infty} A_{n,l}l(l + 1)\mu^l \tag{2-12-12}$$

The first sum on the right side of Eq. (2-12-12) can be rewritten slightly if we note that the first two terms are zero.

$$\sum_{l=0}^{\infty} A_{n,l}l(l - 1)\mu^{l-2} = \sum_{l=2}^{\infty} A_{n,l}l(l - 1)\mu^{l-2}$$

$$= \sum_{l=0}^{\infty} A_{n,l+2}(l + 2)(l + 1)\mu^l \tag{2-12-13}$$

We can now enter all this information into Eq. (2-12-10), coming up with a relationship among the various coefficients:

$$\sum_{l=0}^{\infty} \left[A_{n,l+2}(l + 2)(l + 1) - A_{n,l}l(l + 1) + A_{n,l}n(n + 1)\right]\mu^l = 0 \tag{2-12-14}$$

As we have done many times before, we set each coefficient of a given power of μ equal to zero. We then have

$$\frac{A_{n,l+2}}{A_{n,l}} = \frac{l(l + 1) - n(n + 1)}{(l + 1)(l + 2)} \tag{2-12-15}$$

This equation, known as a **recursion relation,** is extremely powerful! It relates any given coefficient to the one two places further down the line.

Thus, if the first two are known, then the entire series is known. (This technique is a very commonly used one for the solution of differential equations in terms of power series. The reader might try the method on the harmonic equation $d^2x/dt^2 = -k^2x$.)

To develop our functions $P_n(\mu)$ we first consider the case where n is even. We notice immediately that if l is odd, there is no value of l for which $l(l + 1) - n(n + 1)$ is equal to zero. Hence, if one odd term exists, then *all* odd terms exist. As we take l larger and larger, the ratio $A_{n,l+2}/A_{n,l}$ approaches unity. We are clearly in for trouble then at $\mu = 1$ because our series diverges at that point. There is one and only one solution to our dilemma, to set $A_{n,1}$ equal to zero, in which case all odd terms disappear. As far as even l is concerned, there is no problem. When we come to the term where l is equal to n, the ratio $A_{n,l+2}/A_{n,l}$ becomes equal to zero. Thus $A_{n,n+2} = 0$ and the series terminates, *all further terms vanishing*.

We conclude then that if n is even, then $P_n(\mu)$ contains only even powers of μ. Only terms from μ^0 to μ^n are present.

On the other hand, if n is odd, we can apply the identical argument to exclude even powers of μ. Again the highest power of μ in the series is μ^n. The series runs from μ^1 to μ^n.

It is useful to find the first few solutions for $P_n(\mu)$. Since any solution can be multiplied by an arbitrary constant and still satisfy the same differential equation, we must choose some normalization. The usual convention requires that

$$P_n(1) = 1 \qquad (2\text{-}12\text{-}16)$$

Obviously, for this normalization,

$$P_n(-1) = (-1)^n \qquad (2\text{-}12\text{-}17)$$

The functions $P_n(\mu)$ are called **Legendre polynomials**. Since $P_0(\mu)$ is a constant, our normalization requires that

$$P_0(\mu) = 1 \qquad (2\text{-}12\text{-}18)$$

Next we note that $P_1(\mu)$ is proportional to μ. By our convention then, we have

$$P_1(\mu) = \mu \qquad (2\text{-}12\text{-}19)$$

To obtain $P_2(\mu)$ we must first make use of Eq. (2-12-15) to obtain $A_{2,2}/A_{2,0}$.

$$\frac{A_{2,2}}{A_{2,0}} = \frac{-6}{2} = -3 \qquad (2\text{-}12\text{-}20)$$

Choosing $A_{2,0}$ according to our normalization, we have

$$P_2(\mu) = \frac{3\mu^2 - 1}{2} \qquad\qquad (2\text{-}12\text{-}21)$$

Going on to $P_3(\mu)$, we first find that

$$\frac{A_{3,3}}{A_{3,1}} = \frac{-5}{3} \qquad\qquad (2\text{-}12\text{-}22)$$

Fixing $A_{3,1}$ by convention, we have

$$P_3(\mu) = \frac{5\mu^3 - 3\mu}{2} \qquad\qquad (2\text{-}12\text{-}23)$$

We could proceed in this manner ad infinitum to find all the functions $P_n(\mu)$. It is more convenient, however, to make use of Rodrigues' formula

$$P_n(\mu) = \frac{1}{2^n n!} \frac{d^n}{d\mu^n} (\mu^2 - 1)^n \qquad\qquad (2\text{-}12\text{-}24)$$

We then note that the most general solution to Laplace's equation in the case of axial symmetry can be written as

$$\varphi(r,\theta) = \sum_{n=0}^{\infty} \left(A_n r^n + \frac{B_n}{r^{n+1}} \right) P_n(\cos\theta) \qquad\qquad (2\text{-}12\text{-}25)$$

Again our basic problem will be the determination of the coefficients. This can be done through matching boundary conditions at a complete set of boundaries to a charge-free region. (These conditions can consist of the specification of either the potential at the boundary or of the normal component of electric field at the boundary.) Alternatively, we may be given a charge distribution and asked to find the potential outside of the distribution. In the next two sections we will discuss some specific methods of handling these problems.

Before we conclude this section we should note that one of the remarkable facts about the functions $P_n(\mu)$ is that they form a complete and orthogonal set over the range $-1 \leqq \mu \leqq 1$. That is to say, any function of μ can be expressed as a series in $P_n(\mu)$ and no one of the $P_n(\mu)$ can be expressed entirely in terms of the others.

We note first that

$$\int_{-1}^{1} P_n(\mu) P_m(\mu)\, d\mu = 0 \qquad \text{if } n \neq m \qquad\qquad (2\text{-}12\text{-}26)$$

We also note that

$$\int_{-1}^{1} [P_n(\mu)]^2\, d\mu = \frac{2}{2n+1} \qquad\qquad (2\text{-}12\text{-}27)$$

The fact that the Legendre polynomials form a complete set permits us to write, for any function $f(\mu)$ that is finite in the range $-1 \leq \mu \leq 1$,

$$f(\mu) = \sum_{n=0}^{\infty} A_n P_n(\mu) \tag{2-12-28}$$

We now multiply both sides of Eq. (2-12-28) by $P_m(\mu)$ and integrate from $\mu = -1$ to $\mu = +1$:

$$\int_{-1}^{1} f(\mu) P_m(\mu) \, d\mu = \sum_{n} \int_{-1}^{1} A_n P_n(\mu) P_m(\mu) \, d\mu$$

$$= \frac{2}{2m+1} A_m$$

Hence

$$A_m = \frac{2m+1}{2} \int_{-1}^{1} f(\mu) P_m(\mu) \, d\mu \tag{2-12-29}$$

Obviously, were we to choose $f(\mu)$ to be $P_n(\mu)$, then A_m would be zero unless $n = m$. Our $P_n(\mu)$ thus form an orthogonal set in the sense that no one can be expressed in terms of any of the others.

2-13 SOLVING BOUNDARY-VALUE PROBLEMS IN SPHERICAL COORDINATES WITH AZIMUTHAL SYMMETRY

We now develop some of the techniques for evaluating the coefficients in Eq. (2-12-25), the general solution to the azimuthally symmetrical Laplace equation. In particular, we will assume here that the potential or its derivative normal to the surface is specified at each and every boundary of our charge-free region, and we will then force the potential to fit. We will find that the set of coefficients will then be completely determined.

Our discussion is best done in terms of a set of specific problems, each of which has some widespread applicability. The extension to other classes of problems will be straightforward for the reader once he has grasped the general principles.

We begin by specifying the potential $\varphi(a, \theta)$ over a spherical surface of radius a (see Fig. 2-20). We will assume that no other charges are present anywhere and that the potential at infinity is zero. We are interested in knowing the potential everywhere, both inside and outside the sphere.

Since the potential at infinity is zero, the outside potential can be written as

$$\varphi_{\text{out}}(r, \theta) = \sum_{n=0}^{\infty} \frac{B_n^{\text{out}}}{r^{n+1}} P_n(\cos \theta) \tag{2-13-1}$$

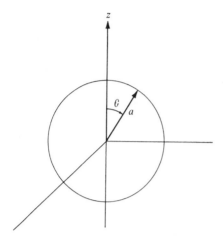

Fig. 2-20 The potential $\varphi(a,\theta)$ (or its derivative with respect to r) is specified at the spherical boundary $r = a$. It is assumed that no other charge is present and that $\varphi(\infty) = 0$. We would like to know the potential $\varphi(r,\theta)$ everywhere, both outside and within the sphere.

This potential must be equal to $\varphi(a,\theta)$ at $r = a$, leading to the observation that

$$\varphi(a,\theta) = \sum_{n=0}^{\infty} \frac{B_n^{\text{out}}}{a^{n+1}} P_n(\cos\theta) \tag{2-13-2}$$

Multiplying both sides by $P_m(\cos\theta)$ and integrating with respect to $\cos\theta$ from $\cos\theta = -1$ to $\cos\theta = +1$ [see Eq. (2-12-29)], we have

$$B_m^{\text{out}} = \frac{(2m+1)a^{m+1}}{2} \int_{-1}^{1} \varphi(a,\theta) P_m(\cos\theta) \, d\cos\theta \tag{2-13-3}$$

We have thus found the potential $\varphi(r,\theta)$ outside the sphere. Inside the sphere we must have $B_n = 0$ or else our solution will diverge at $r = 0$. We have then

$$\varphi_{\text{in}}(r,\theta) = \sum_{n=0}^{\infty} A_n^{\text{in}} r^n P_n(\cos\theta) \tag{2-13-4}$$

Again we are told what the potential $\varphi(r,\theta)$ is at $r = a$. Substitution of $r = a$ into Eq. (2-13-4) yields

$$\varphi(a,\theta) = \sum_{n=0}^{\infty} A_n^{\text{in}} a^n P_n(\cos\theta) \tag{2-13-5}$$

The coefficients are thus given by

$$A_m^{\text{in}} = \frac{2m+1}{2a^m} \int_{-1}^{1} \varphi(a,\theta) P_m(\cos\theta) \, d\cos\theta \tag{2-13-6}$$

Note incidentally that the continuity of φ across our surface at $r = a$ implies

$$\frac{B_m{}^{\text{out}}}{A_m{}^{\text{in}}} = a^{2m+1} \tag{2-13-7}$$

Now that we have such a powerful result, let us try it on the simplest physical situation and see if it works. We will take $\varphi(a,\theta)$ to be a constant, namely, φ_0. In that case we know from experience that the potential inside should be constant (this is like a conducting spherical shell) and the potential outside should be

$$\varphi(r) = \varphi_0 \frac{a}{r} \tag{2-13-8}$$

Let us see if we come up with the same answers with our general method. We have

$$\int_{-1}^{1} \varphi(a,\theta) P_m(\cos\theta)\, d\cos\theta = \varphi_0 \int_{-1}^{1} P_m(\cos\theta)\, d\cos\theta$$

$$= \varphi_0 \int_{-1}^{1} P_0(\cos\theta) P_m(\cos\theta)\, d\cos\theta$$

$$= \begin{cases} 2\varphi_0 & \text{if } m = 0 \\ 0 & \text{if } m \neq 0 \end{cases} \tag{2-13-9}$$

Thus

$$\begin{aligned}
A_0{}^{\text{in}} &= \varphi_0 \\
A_m{}^{\text{in}} &= 0 \qquad \text{for } m \neq 0 \\
B_0{}^{\text{out}} &= a\varphi_0 \\
B_m{}^{\text{out}} &= 0 \qquad \text{for } m \neq 0
\end{aligned} \tag{2-13-10}$$

We finally obtain, as expected,

$$\varphi_{\text{out}}(r,\theta) = \frac{a\varphi_0}{r} \tag{2-13-11}$$

$$\varphi_{\text{in}}(r,\theta) = \varphi_0$$

As we said earlier, we could specify the normal derivative of the potential at the surface ($r = a$) and use that to obtain the coefficients $B_n{}^{\text{out}}$ or $A_n{}^{\text{in}}$. Since in general the derivative of φ with respect to r is not continuous at the surface (there being some charge at the surface), we must use $(\partial\varphi_{\text{out}}/\partial r)_{r=a}$ to determine the $B_n{}^{\text{out}}$ coefficients *or* $(\partial\varphi_{\text{in}}/\partial r)_{r=a}$ to determine the $A_m{}^{\text{in}}$ coefficients. Once we have determined either of these sets, the other can be determined by requiring that φ be continuous. That is to say, $\varphi_{in}(a,\theta) = \varphi_{out}(a,\theta)$.

For example, we can write

$$\left(\frac{\partial\varphi_{out}}{\partial r}\right)_{r=a} = -\sum_{n=0}^{\infty} (n + 1) \frac{B_n^{out}}{a^{n+2}} P_n(\cos\theta) \tag{2-13-12}$$

This would yield the equation for B_m^{out}:

$$B_m^{out} = \frac{-(2m + 1)a^{m+2}}{2(m + 1)} \int_{-1}^{1} \left(\frac{\partial\varphi_{out}}{\partial r}\right)_{r=a} P_m(\cos\theta) \, d\cos\theta \tag{2-13-13}$$

The continuity condition expressed in Eq. (2-13-7) would then tells us A_m^{in}:

$$A_m^{in} = \frac{B_m^{out}}{a^{2m+1}} \tag{2-13-14}$$

Alternatively we might have specified $(\partial\varphi_{in}/\partial r)_{r=a}$, obtained the coefficients A_m^{in}, and then applied Eq. (2-13-14) to get B_m^{out}.

Let us now attempt a slightly more difficult problem. Let the potentials be specified on each of two concentric spheres at radii a and b, respectively. We would like to find the potential between the spheres. We begin by writing down the potentials at a and b:

$$\varphi(a,\theta) = \sum \left(A_n a^n + \frac{B_n}{a^{n+1}}\right) P_n(\cos\theta) \tag{2-13-15}$$

$$\varphi(b,\theta) = \sum \left(A_n b^n + \frac{B_n}{b^{n+1}}\right) P_n(\cos\theta) \tag{2-13-16}$$

From Eq. (2-13-15) we determine the linear combination $A_m a^m + B_m/a^{m+1}$ for arbitrary m:

$$A_m a^m + \frac{B_m}{a^{m+1}} = \frac{2m + 1}{2} \int_{-1}^{1} \varphi(a,\theta) P_m(\cos\theta) \, d\cos\theta \tag{2-13-17}$$

We obtain another equation for A_m and B_m from Eq. (2-13-16):

$$A_m b^m + \frac{B_m}{b^{m+1}} = \frac{2m + 1}{2} \int_{-1}^{1} \varphi(b,\theta) P_m(\cos\theta) \, d\cos\theta \tag{2-13-18}$$

Using both Eqs. (2-13-17) and (2-13-18), we can find the A_m and the B_m coefficients, completing the solution to our problem.

We next consider the problem of a spherical grounded conducting ball of radius a that is placed in a "constant" field. (Obviously, after the ball is introduced the field is no longer constant near the ball.) We assume that the potential is given by

$$\varphi(r,\theta) \cong -E_0 r \cos\theta \tag{2-13-19}$$

at large distances from the ball. At $r = a$, the potential is taken to be zero. Now referring back to our general solution, Eq. (2-12-25), we can immediately set all A_n except for A_1 equal to zero. A_1 is of course equal to $-E_0$.

$$A_1 = -E_0$$
$$A_n = 0 \quad \text{for } n \neq 1$$
(2-13-20)

At $r = a$, we have

$$\varphi(a,\theta) = 0 = -E_0 a \cos \theta + \sum_{n=0}^{\infty} \frac{B_n}{a^{n+1}} P_n(\cos \theta)$$
(2-13-21)

In order that this hold for all θ, we must set the coefficient of each and every Legendre polynomial equal to zero. Hence

$$B_1 = E_0 a^3$$
$$B_n = 0 \quad \text{for } n \neq 1$$
(2-13-22)

We finally have then

$$\varphi(r,\theta) = \left(-E_0 r + \frac{E_0 a^3}{r^2}\right)\cos \theta$$
(2-13-23)

The (normal) electric field at the surface is given by $-(\partial\varphi/\partial r)_{r=a}$.

$$E_n = -\left(\frac{\partial\varphi}{\partial r}\right)_{r=a} = 3E_0 \cos \theta$$
(2-13-24)

The surface charge density is

$$\sigma = \frac{E_n}{4\pi} = \frac{3E_0 \cos \theta}{4\pi}$$
(2-13-25)

The total charge is zero, as might be expected.

Finally, we consider what would happen if the conducting sphere we have just considered were to be replaced by a dielectric sphere of uniform dielectric constant ε. The absence of free charge ensures that $\nabla \cdot \mathbf{D} = 0$ within the sphere. But $\nabla \cdot \mathbf{D} = \varepsilon\nabla \cdot \mathbf{E} = -\varepsilon\nabla^2\varphi$ in the case where ε is a constant. Hence φ satisfies Laplace's equation within the dielectric as well as outside of it. Only at the surface of the dielectric is $\nabla^2\varphi$ not equal to zero.

We will allow $\varphi^{\mathrm{I}}(r,\theta)$ to be the solution within the dielectric and $\varphi^{\mathrm{II}}(r,\varphi)$ to be the solution outside. Our boundary conditions are as follows:

1. $\varphi^{\mathrm{II}} \to -E_0 r \cos \theta$ as $r \to \infty$.
2. The potential is continuous at the boundary of the dielectric. That is, $\varphi^{\mathrm{I}}(a,\theta) = \varphi^{\mathrm{II}}(a,\theta)$.
3. The potential is finite at $r = 0$.

4. The normal component of the displacement field **D** is continuous at the boundary. Thus

$$\varepsilon\left(\frac{\partial\varphi^{\mathrm{I}}}{\partial r}\right)_{r=a} = \left(\frac{\partial\varphi^{\mathrm{II}}}{\partial r}\right)_{r=a}$$

The continuity of the tangential component of **E** at the boundary is of course assured by boundary condition 2.

Expressing both $\varphi^{\mathrm{I}}(r,\theta)$ and $\varphi^{\mathrm{II}}(r,\theta)$ in conventional form, we have

$$\varphi^{\mathrm{I}}(r,\theta) = \sum_{n=0}^{\infty}\left(A_n^{\mathrm{I}}r^n + \frac{B_n^{\mathrm{I}}}{r^{n+1}}\right)P_n(\cos\theta) \qquad (2\text{-}13\text{-}26)$$

$$\varphi^{\mathrm{II}}(r,\theta) = \sum_{n=1}^{\infty}\left(A_n^{\mathrm{II}}r^n + \frac{B_n^{\mathrm{II}}}{r^{n+1}}\right)P_n(\cos\theta) \qquad (2\text{-}13\text{-}27)$$

Applying boundary condition 1 tells us that

$$A_1^{\mathrm{II}} = -E_0$$
$$A_n^{\mathrm{II}} = 0 \qquad \text{for } n \neq 1 \qquad (2\text{-}13\text{-}28)$$

Applying boundary condition 3 forces us to write

$$B_n^{\mathrm{I}} = 0 \qquad \text{for } all \ n \qquad (2\text{-}13\text{-}28')$$

Using the remaining coefficients, we next apply condition 2.

$$\sum_{n=0}^{\infty} A_n^{\mathrm{I}}a^n P_n(\cos\theta) = -E_0 a\cos\theta + \sum_{n=0}^{\infty}\frac{B_n^{\mathrm{II}}}{a^{n+1}}P_n(\cos\theta) \qquad (2\text{-}13\text{-}29)$$

We deduce from this that

$$A_1^{\mathrm{I}}a = -E_0 a + \frac{B_1^{\mathrm{II}}}{a^2} \qquad (2\text{-}13\text{-}30)$$

$$A_n^{\mathrm{I}}a^n = \frac{B_n^{\mathrm{II}}}{a^{n+1}} \qquad \text{for } n \neq 1 \qquad (2\text{-}13\text{-}31)$$

Finally, we apply condition 4.

$$\varepsilon\sum_{n=0}^{\infty} A_n^{\mathrm{I}}na^{n-1}P_n(\cos\theta) = -E_0\cos\theta$$

$$-\sum_{n=0}^{\infty}\frac{(n+1)B_n^{\mathrm{II}}}{a^{n+2}}P_n(\cos\theta) \qquad (2\text{-}13\text{-}32)$$

We deduce, as before, that

$$\varepsilon A_1^{\mathrm{I}} = -E_0 - \frac{2B_1^{\mathrm{II}}}{a^3} \qquad (2\text{-}13\text{-}33)$$

$$\varepsilon A_n^{\;I} n a^{n-1} = - \frac{(n+1)B_n^{\;II}}{a^{n+2}} \qquad \text{for } n \neq 1 \tag{2-13-34}$$

Equation (2-13-31) implies that $A_n^{\;I}$ and $B_n^{\;II}$ have the same sign, for $n \neq 1$. On the other hand, Eq. (2-13-34) implies that they have opposite signs for $n \neq 1$. We conclude that

$$A_n^{\;I} = B_n^{\;II} = 0 \qquad \text{for } n \neq 1 \tag{2-13-35}$$

Finally, combining Eqs. (2-13-30) and (2-13-33), we have

$$A_1^{\;I} = - \frac{3E_0}{\varepsilon + 2}$$
$$\tag{2-13-36}$$
$$B_1^{\;II} = \frac{\varepsilon - 1}{\varepsilon + 2} E_0 a^3$$

We can now write down the solution for φ^I and φ^{II} everywhere:

$$\varphi^I(r,\theta) = - \frac{3E_0 r \cos \theta}{\varepsilon + 2} \tag{2-13-37}$$

$$\varphi^{II}(r,\theta) = - E_0 r \cos \theta + \left(\frac{\varepsilon - 1}{\varepsilon + 2} \right) \frac{E_0 a^3}{r^2} \cos \theta \tag{2-13-38}$$

A glance at Eq. (2-13-37) tells us that the field within the dielectric is a constant in the z direction and is equal to

$$\mathbf{E}(\text{inside}) = \frac{3}{\varepsilon + 2} E_0 \hat{\mathbf{k}} \tag{3-13-39}$$

In the event that $\varepsilon = 1$ (no dielectric) this reduces as expected to $E_0 \hat{\mathbf{k}}$. The field outside the dielectric is clearly composed of the original constant field $E_0 \hat{\mathbf{k}}$ and a field which has a characteristic dipole distribution with dipole moment of

$$\mathbf{p} = \frac{\varepsilon - 1}{\varepsilon + 2} E_0 a^3 \hat{\mathbf{k}} \tag{2-13-40}$$

We might check to see if the dipole moment we obtain this way agrees with what we expect from integrating the dipole moment per unit volume $\mathbf{P}$ over the sphere. Inside the dielectric we have

$$\mathbf{P} = \frac{\mathbf{D} - \mathbf{E}}{4\pi}$$

$$= \frac{(\varepsilon - 1)\mathbf{E}}{4\pi}$$

$$= \frac{(\varepsilon - 1)}{4\pi} \left(\frac{3}{\varepsilon + 2} \right) E_0 \hat{\mathbf{k}} \tag{2-13-41}$$

Inasmuch as the dipole moment per unit volume is constant, we can obtain the total dipole moment through multiplying $\mathbf{P}$ by the volume of the sphere.

$$\mathbf{p} = \mathbf{P}(\tfrac{4}{3}\pi a^3)$$

$$= \frac{\varepsilon - 1}{\varepsilon + 2} E_0 a^3 \hat{\mathbf{k}} \qquad \text{as before}$$

This completes our study of spherical boundary conditions. We will next learn how to determine the all-important coefficients A_n and B_n in the event that the charge distribution is completely known.

2-14 THE MULTIPOLE EXPANSION OF AN AZIMUTHALLY SYMMETRICAL CHARGE DISTRIBUTION

Suppose a distribution of charge in space were limited in physical extent and azimuthally symmetrical and we wanted to find the potential it produced in some charge-free region. Presumably the potential could be written in the form given by Eq. (2-12-25) and our only job would be to find the coefficients A_n and B_n. We will now develop a beautiful and general method for determining these coefficients and incidentally gain a great deal of insight into the types of fields produced by various charge configurations.

We need to distinguish two cases, that where the charge is all at a smaller radius than the point at which we wish to determine φ and that where the charge is all at a larger radius than that point. Obviously the solution to any problem can be broken into a sum of these two types of solutions, and hence we lose no generality by treating only these situations.

We consider the first of these cases. Let r' and θ' be the coordinates referring to the charge distribution (see Fig. 2-21) and let r and θ be the coordinates at which we wish to evaluate the potential φ. (We assume that $r' \leqq r$ at all times.) In order that $\varphi \to 0$ as $r \to \infty$ we must require that $A_n = 0$ for all n. Hence

$$\varphi(r,\theta) = \sum_{n=0}^{\infty} \frac{B_n}{r^{n+1}} P_n(\cos\theta) \tag{2-14-1}$$

Our job is then to determine the coefficients B_n in terms of properties of the charge distribution. We recall our old expression for φ.

$$\varphi(r,\theta,\psi) = \int_{\substack{\text{charge} \\ \text{distribution}}} \frac{\rho(r',\theta')}{|\mathbf{r} - \mathbf{r}'|}\, dV' \tag{2-14-2}$$

The trick is, of course, to express $1/|\mathbf{r} - \mathbf{r}'|$ in a power series and then evaluate the set of individual terms. We remember that as long as $x < 1$, we can write

$$\frac{1}{\sqrt{1 + x}} = 1 - \frac{1}{2}x + \frac{1(3)}{2(4)}x^2 - \frac{1(3)(5)}{2(4)(6)}x^3 + \cdots \tag{2-14-3}$$

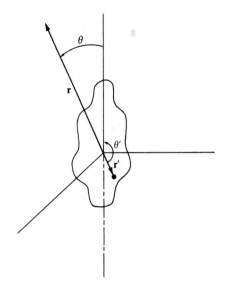

Fig. 2-21 We consider the potential at a point (r,θ) produced by an azimuthally symmetrical charge distribution $\rho(r',\theta')$, where $r' < r$.

Now

$$\frac{1}{|\mathbf{r} - \mathbf{r}'|} = \frac{1}{\sqrt{r^2 + r'^2 - 2\mathbf{r}\cdot\mathbf{r}'}}$$

$$= \frac{1}{r\sqrt{1 + \dfrac{r'^2}{r^2} - \dfrac{2\mathbf{r}\cdot\mathbf{r}'}{r^2}}} \tag{2-14-4}$$

Expanding by means of Eq. (2-14-3) and grouping terms with the same power of r'/r together, we have

$$\frac{1}{|\mathbf{r} - \mathbf{r}'|} = \frac{1}{r}\left[1 + \frac{\mathbf{r}\cdot\mathbf{r}'}{r^2} + \frac{3(\mathbf{r}\cdot\mathbf{r}')^2 - r'^2 r^2}{2r^4}\right.$$

$$\left. + \frac{5(\mathbf{r}\cdot\mathbf{r}')^3 - 3(\mathbf{r}\cdot\mathbf{r}')r'^2 r^2}{2r^6} + \cdots\right] \tag{2-14-5}$$

Our next step is to put Eq. (2-14-5) into a somewhat more pliable form. We first note that

$$\mathbf{r} = (r\sin\theta\cos\psi)\hat{\mathbf{i}} + (r\sin\theta\sin\psi)\hat{\mathbf{j}} + (r\cos\theta)\hat{\mathbf{k}} \tag{2-14-6}$$

$$\mathbf{r}' = (r'\sin\theta'\cos\psi')\hat{\mathbf{i}} + (r'\sin\theta'\sin\psi')\hat{\mathbf{j}} + (r'\cos\theta')\hat{\mathbf{k}} \tag{2-14-7}$$

Hence, in complete generality, $1/|\mathbf{r} - \mathbf{r}'|$ is a function of r, r', θ, θ', ψ, and ψ'. The fact that we are dealing with an azimuthally symmetrical charge distribution, however, permits us to evaluate the integral in Eq. (2-14-2) at $\psi = 0$ and apply the result at any value of ψ. Therefore, we need only consider Eq. (2-14-5) at $\psi = 0$. Inserting this into the above we find

$$\mathbf{r} \cdot \mathbf{r}' = rr'(\sin\theta \sin\theta' \cos\psi' + \cos\theta \cos\theta') \qquad (2\text{-}14\text{-}8)$$

$$(\mathbf{r} \cdot \mathbf{r}')^2 = r^2 r'^2(\sin^2\theta \sin^2\theta' \cos^2\psi' + \cos^2\theta \cos^2\theta'$$

$$+ 2\sin\theta \sin\theta' \cos\theta \cos\theta' \cos\psi') \qquad (2\text{-}14\text{-}9)$$

$$(\mathbf{r} \cdot \mathbf{r}')^3 = r^3 r'^3(\sin^3\theta \sin^3\theta' \cos^3\psi' + \cos^3\theta \cos^3\theta'$$

$$+ 3\sin^2\theta \sin^2\theta' \cos\theta \cos\theta' \cos^2\psi'$$

$$+ 3\sin\theta \sin\theta' \cos^2\theta \cos^2\theta' \cos\psi') \qquad (2\text{-}14\text{-}10)$$

and so forth.

Before proceeding, we again note that ρ has no ψ' dependence. Hence we can carry out the integration of Eq. (2-14-2) over ψ' first. This is equivalent to replacing the various powers of $\cos\psi'$ above by their average values before integration. The average values are

$$\langle\cos\psi'\rangle = 0 \qquad \langle\cos^2\psi'\rangle = \tfrac{1}{2} \qquad \langle\cos^3\psi'\rangle = 0 \qquad (2\text{-}14\text{-}11)$$

This leads to the following results when we average the various powers of $\mathbf{r} \cdot \mathbf{r}'$ over ψ':

$$\langle\mathbf{r} \cdot \mathbf{r}'\rangle_{av} = rr' \cos\theta \cos\theta'$$

$$\langle(\mathbf{r} \cdot \mathbf{r}')^2\rangle_{av} = r^2 r'^2(\cos^2\theta \cos^2\theta' + \tfrac{1}{2}\sin^2\theta \sin^2\theta') \qquad (2\text{-}14\text{-}12)$$

$$\langle(\mathbf{r} \cdot \mathbf{r}')^3\rangle_{av} = r^3 r'^3(\cos^3\theta \cos^3\theta' + \tfrac{3}{2}\sin^2\theta \sin^2\theta' \cos\theta \cos\theta')$$

Finally, to average Eq. (2-14-5) over ψ', we need to combine terms and observe that

$$\left\langle\frac{3(\mathbf{r} \cdot \mathbf{r}')^2 - r^2 r'^2}{2}\right\rangle_{av} = r^2 r'^2 P_2(\cos\theta)P_2(\cos\theta') \qquad (2\text{-}14\text{-}13)$$

$$\left\langle\frac{5(\mathbf{r} \cdot \mathbf{r}')^3 - 3(\mathbf{r} \cdot \mathbf{r}')r'^2 r^2}{2}\right\rangle_{av} = r^3 r'^3 P_3(\cos\theta)P_3(\cos\theta') \qquad (2\text{-}14\text{-}14)$$

Needless to say we can conjecture what the nth term in this series will be, and hence we write down immediately for Eq. (2-14-5) averaged over ψ':

$$\left\langle\frac{1}{|\mathbf{r} - \mathbf{r}'|}\right\rangle_{\substack{av \\ over\ \psi'}} = \frac{1}{r}\left[1 + \frac{r'}{r}P_1(\cos\theta)P_1(\cos\theta')\right.$$

$$\left. + \left(\frac{r'}{r}\right)^2 P_2(\cos\theta)P_2(\cos\theta')\right.$$

$$+ \left(\frac{r'}{r}\right)^3 P_3(\cos\theta)P_3(\cos\theta') + \cdots \Bigg]$$

$$= \frac{1}{r}\sum_{n=0}^{\infty}\left(\frac{r'}{r}\right)^n P_n(\cos\theta)P_n(\cos\theta') \qquad (2\text{-}14\text{-}15)$$

If we now finally insert this result into the integral for $\varphi(r,\theta)$ [Eq. (2-14-2)], we find

$$\varphi(r,\theta) = \sum_{n=0}^{\infty}\frac{P_n(\cos\theta)}{r^{n+1}}\int_{\substack{\text{charge}\\\text{distribution}}}\rho(r',\theta')(r')^n P_n(\cos\theta')\, dV' \qquad (2\text{-}14\text{-}6)$$

We conclude that the coefficients B_n of Eq. (2-14-1) are given by

$$B_n = \int_{\substack{\text{charge}\\\text{distribution}}}\rho(r',\theta')(r')^n P_n(\cos\theta')\, dV' \qquad (2\text{-}14\text{-}17)$$

The terms B_n are called the 2^n**-pole moments of the distribution.** For example,

$$B_0 = \int_{\substack{\text{charge}\\\text{distribution}}}\rho(r',\theta')\, dV'$$

$$= \text{total charge (or monopole moment)} \qquad (2\text{-}14\text{-}18)$$

$$B_1 = \int_{\substack{\text{charge}\\\text{distribution}}}\rho(r',\theta')z'\, dV' = \text{dipole moment} \qquad (2\text{-}14\text{-}19)$$

$$B_2 = \int_{\substack{\text{charge}\\\text{distribution}}}\rho(r',\theta')(r')^2 P_2(\cos\theta')\, dV'$$

$$= \text{quadrupole moment} \qquad (2\text{-}14\text{-}20)$$

Let us return to the beginning again and ask what $\varphi(r,\theta)$ would be like if the charge distribution were all at radius larger than r. (That is to say, $r' > r$ for all possible r'.) The expansion for $\varphi(r,\theta)$ that we must use now, in order that φ not diverge at $r = 0$, is

$$\varphi(r,\theta) = \sum A_n r^n P_n(\cos\theta) \qquad (2\text{-}14\text{-}21)$$

Equation (2-14-2) would still be appropriate for finding φ, but the expansion of $1/|\mathbf{r} - \mathbf{r}'|$ must now be in a power series in r/r'. Obviously everything we have done so far in expanding $1/|\mathbf{r} - \mathbf{r}'|$ would be valid if we just interchanged r and r'. Returning to Eq. (2-14-15) we now rewrite it as

$$\left\langle\frac{1}{|\mathbf{r}-\mathbf{r}'|}\right\rangle_{\substack{\text{av}\\\text{over }\psi'}} = \frac{1}{r'}\sum_{n=0}^{\infty}\left(\frac{r}{r'}\right)^n P_n(\cos\theta)P_n(\cos\theta') \qquad (2\text{-}14\text{-}22)$$

We conclude then that

$$A_n = \int_{\substack{\text{charge} \\ \text{distribution}}} \frac{\rho(r',\theta')}{(r')^{n+1}} P_n(\cos\theta')\, dV' \tag{2-14-23}$$

Amusingly enough, we can now find the potential even within the azimuthally symmetrical charge distribution itself, even though that is obviously *not* a charge-free region. We take the point (r,θ) at which we wish to find φ and break the charge distribution into two parts, that with $r' \leqq r$ and that with $r' \geqq r$. We add the two contributions together and lo and behold—we have φ. In general then, in a case of azimuthal symmetry,

$$\varphi(r,\theta) = \sum_{n=0}^{\infty} \left[r^n \int_{r'>r} \frac{\rho(r',\theta')P_n(\cos\theta')}{(r')^{n+1}}\, dV' \right.$$
$$\left. + r^{-n-1} \int_{r'<r} \rho(r',\theta')(r')^n P_n(\cos\theta')\, dV' \right] P_n(\cos\theta) \tag{2-14-24}$$

Let us try out this magnificent formula on a uniformly charged sphere of radius R (see page 33 and Fig. 2-2). The charge density ρ is a constant in this case and can be taken out of the integral. Equation (2-14-24) becomes [remembering that $\int_{-1}^{1} P_n(\cos\theta')\, d\cos\theta' = 0$ unless $n = 0$]

$$\varphi(r,\theta) = 4\pi\rho \int_r^R r'\, dr' + \frac{4\pi\rho}{r} \int_0^r (r')^2\, dr'$$

$$= 4\pi\rho \left(\frac{R^2}{2} - \frac{r^2}{6} \right) \tag{2-14-25}$$

A simple calculation using the fields we previously determined from Gauss' law will verify that Eq. (2-14-25) is right.

The really interesting applications of the development we have just carried out are in the field of atomic physics. As we shall see in the next section, careful calculation of the energy levels in atomic systems will permit us to determine important constants relating to the nuclear shape. In any case we are now prepared to deal in a systematic manner with any azimuthally symmetrical charge distribution.

2-15 THE INTERACTION ENERGY OF TWO NONOVERLAPPING AZIMUTHALLY SYMMETRIC CHARGE DISTRIBUTIONS; DETERMINATION OF NUCLEAR SHAPE

We will make use of what we have just learned to calculate the interaction energy of two nonoverlapping azimuthally symmetrical charge distributions. As we will see, this will have an immediate consequence in showing us how to determine the various multipole moments of the nucleus and hence will provide exceedingly valuable information about nuclear structure. We must first explain precisely what we mean by **interaction energy.** If we go back to Eq. (2-7-7) we recall that the total energy of any distribution of charge can be written as

$$U_{total} = \frac{1}{2} \int_{\substack{all \\ space}} \rho(\mathbf{r})\varphi(\mathbf{r}) \, dV \qquad (2\text{-}15\text{-}1)$$

Suppose that our charge distribution ρ can be broken into two distinct distributions ρ_1 and ρ_2. As an example of this we can take the atomic system where the nuclear charge distribution might be taken as ρ_1 and the electron charge distribution might be taken as ρ_2. In any case, we have

$$\rho(\mathbf{r}) = \rho_1(\mathbf{r}) + \rho_2(\mathbf{r}) \qquad (2\text{-}15\text{-}2)$$

Now each of these charge distributions is responsible for a portion of the potential at any given position. We can write

$$\varphi(\mathbf{r}) = \varphi_1(\mathbf{r}) + \varphi_2(\mathbf{r}) \qquad (2\text{-}15\text{-}3)$$

where

$$\varphi_1(\mathbf{r}) = \int_{\substack{all \\ space}} \frac{\rho_1(\mathbf{r}')}{|\mathbf{r} - \mathbf{r}'|} \, dV' \qquad (2\text{-}15\text{-}4)$$

and

$$\varphi_2(\mathbf{r}) = \int_{\substack{all \\ space}} \frac{\rho_2(\mathbf{r}')}{|\mathbf{r} - \mathbf{r}'|} \, dV' \qquad (2\text{-}15\text{-}5)$$

Thus we see that the total energy as evaluated through the integral Eq. (2-15-1) can be broken into four parts.

$$U_{total} = \frac{1}{2} \int_{\substack{all \\ space}} \rho_1(\mathbf{r})\varphi_1(\mathbf{r}) \, dV + \frac{1}{2} \int_{\substack{all \\ space}} \rho_2(\mathbf{r})\varphi_2(\mathbf{r}) \, dV$$

$$+ \frac{1}{2} \int_{\substack{all \\ space}} \rho_1(\mathbf{r})\varphi_2(\mathbf{r}) \, dV + \frac{1}{2} \int_{\substack{all \\ space}} \rho_2(\mathbf{r})\varphi_1(\mathbf{r}) \, dV \quad (2\text{-}15\text{-}6)$$

The first two terms in the total energy are just the self-energies of the two individual charge distributions. The third and fourth terms constitute the interaction energy. Now we can easily see that the third and fourth terms are equal:

$$\int_{\substack{all \\ space}} \rho_1(\mathbf{r})\varphi_2(\mathbf{r})\, dV = \iint_{\substack{all \\ space}} \frac{\rho_1(\mathbf{r})\rho_2(\mathbf{r}')\, dV\, dV'}{|\mathbf{r} - \mathbf{r}'|}$$

$$= \int_{\substack{all \\ space}} \rho_2(\mathbf{r}')\varphi_1(\mathbf{r}')\, dV' \qquad (2\text{-}15\text{-}7)$$

Hence we conclude that

$$U_{int} = \int_{\substack{all \\ space}} \rho_2(\mathbf{r})\varphi_1(\mathbf{r})\, dV \qquad (2\text{-}15\text{-}8)$$

Now, if the two distributions are azimuthally symmetrical about a common axis and arranged in such a way that ρ_2 is always outside ρ_1 (as in an atom), then our expression for the interaction energy can be written in a very simple way. We have for φ_1,

$$\varphi_1(r,\theta) = \sum_{n=0}^{\infty} \frac{B_n^{(1)}}{r^{n+1}} P_n(\cos\theta) \qquad (2\text{-}15\text{-}9)$$

where $B_n^{(1)}$ is the 2^n-pole moment of the inner distribution ρ_1. Substituting into Eq. (2-15-8), we obtain

$$U_{int} = \sum_{n=0}^{\infty} B_n^{(1)} A_n^{(2)} \qquad (2\text{-}15\text{-}10)$$

where, as usual,

$$A_n^{(2)} = \int_{\substack{all \\ space}} \frac{\rho_2(r,\theta) P_n(\cos\theta)}{r^{n+1}} \qquad (2\text{-}15\text{-}11)$$

In the event that our charge distribution is an atomic system the part corresponding to the electrons is generally well known. Hence the $A_n^{(2)}$ coefficients in the above expansion can be found. In addition to this there are basic limitations set on the possible values of n as the result of fundamental quantum-mechanical theorems. For example, the requirement that the laws of physics not change when we reverse the direction of flow of time in our equations implies that n cannot be odd. Furthermore, the maximum value of n is related to the total angular momentum of the system. Only total angular momentum values of $[\sqrt{l(l + 1)}h]/2\pi$ are allowed for a physical system, where l is an integer or half-integer and h is Planck's constant. The maximum value of n that we can have represented in a multipole expansion of the charge distribution may be shown on the basis of these fundamental theorems to be equal to $2l$. Hence by knowing the spin of the nucleus we can determine how many moments can possibly exist. Since the charge is known, the first unknown moment in Eq. (2-15-10) is the quadrupole moment. A measurement of this term will now tell us whether the nucleus is "pancake" shaped or "cigar" shaped. A pancake-shaped nucleus will have negative quadrupole moment and a cigar-shaped nucleus will have positive quadrupole moment. Of course, a sphere has no quadrupole moment.

2-16 THE ELECTROSTATIC STRESS TENSOR

In all our discussions of the methods for determining electric field and the energies associated with these fields we have paid very little attention to the forces exerted on charges and charge distributions. We will remedy this shortcoming now by showing in detail how the force on a distribution of charge can be calculated in either of two ways. We can carry out a volume integral over the charge distribution itself in which we multiply the charge within any infinitesimal volume dV by the field at that point. Alternatively, we will discover a method of converting our volume integral into a surface integral in which only the fields on a surface enclosing the distribution of interest need be known.

We begin, of course, with the force on the infinitesimal bit of charge in the volume dV. If ρ is the charge density in the volume and $\mathbf{E}$ is the electric field there, then

$$d\mathbf{F} = \rho \mathbf{E} \, dV \tag{2-16-1}$$

Integrating, we find for the total force on a given volume V of charge

$$\mathbf{F} = \int_V \rho \mathbf{E} \, dV \tag{2-16-2}$$

Let us apply this to a very simple problem. Consider a uniform sphere of charge of radius R and total charge Q (see Fig. 2-22). We would like to find the total force exerted by any one hemisphere of the sphere on the opposite hemisphere. For convenience we have set up our z axis as shown and evaluate

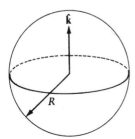

Fig. 2-22 A charge Q is spread uniformly over the volume of a sphere of radius R. We would like to find the force which the lower hemisphere exerts on the upper hemisphere.

the force on the upper hemisphere by the lower hemisphere. Obviously only the z component of this force will remain after integration:

$$F_z = \int_{\substack{\text{upper} \\ \text{hemisphere}}} \rho E_z \, dV \tag{2-16-3}$$

We recall from Eq. (2-2-6) that $E(r < R) = Qr/R^3$. Hence

$$E_z = \frac{Qr}{R^3} \cos \theta \tag{2-16-4}$$

where θ is the usual polar angle. The charge density ρ is just equal to $Q/\frac{4}{3}\pi R^3$. Substituting into Eq. (2-16-3), we obtain

$$F_z = \frac{3Q^2}{2R^6} \int_0^{\pi/2} \cos \theta \sin \theta \, d\theta \int_0^R r^3 \, dr$$

$$= \frac{3}{16} \frac{Q^2}{R^2} \tag{2-16-5}$$

So far we have added nothing new to our knowledge; we could have carried out this calculation 50 pages ago. We will now show that Eq. (2-16-2) can be converted into a surface integral where only a knowledge of the electric field on the surface is necessary. We recall that $\rho = (1/4\pi) \nabla \cdot \mathbf{E}$. Substituting into Eq. (2-16-2), we have

$$\mathbf{F} = \frac{1}{4\pi} \int_V \mathbf{E}(\nabla \cdot \mathbf{E}) \, dV \tag{2-16-6}$$

We next subdivide Eq. (2-16-6) into its components and make use of a simple vector identity and Gauss' theorem:

$$\mathbf{F} = \frac{1}{4\pi} \left[\int_V E_x(\nabla \cdot \mathbf{E})\hat{\mathbf{i}} \, dV + \int_V E_y(\nabla \cdot \mathbf{E})\hat{\mathbf{j}} \, dV + \int_V E_z(\nabla \cdot \mathbf{E})\hat{\mathbf{k}} \, dV \right]$$

$$= \frac{1}{4\pi} \left\{ \int_V [\nabla \cdot (E_x \mathbf{E}) - \mathbf{E} \cdot \nabla E_x]\hat{\mathbf{i}} \, dV + \int_V [\nabla \cdot (E_y \mathbf{E}) - \mathbf{E} \right.$$

$$\left. \cdot \nabla E_y]\hat{\mathbf{j}} \, dV + \int_V [\nabla \cdot (E_z \mathbf{E}) - \mathbf{E} \cdot \nabla E_z]\hat{\mathbf{k}} \, dV \right\}$$

$$= \frac{1}{4\pi} \int_S \mathbf{E}(\mathbf{E} \cdot \hat{\mathbf{n}}) \, dA - \frac{1}{4\pi} \int_V (\mathbf{E} \cdot \nabla)\mathbf{E} \, dV \tag{2-16-7}$$

The surface S, as usual, encloses the volume V. To evaluate the second integral on the right-hand side of Eq. (2-16-7), we use the vector identity

$$\nabla(\mathbf{E} \cdot \mathbf{E}) = 2(\mathbf{E} \cdot \nabla)\mathbf{E} + 2\mathbf{E} \times (\nabla \times \mathbf{E})$$

Since $\nabla \times \mathbf{E} = 0$, we can rewrite Eq. (2-16-6) into the form

$$\mathbf{F} = \frac{1}{4\pi} \int_S \mathbf{E}(\mathbf{E} \cdot \hat{\mathbf{n}}) \, dA - \frac{1}{8\pi} \int_V \nabla(E^2) \, dV \qquad (2\text{-}16\text{-}8)$$

We can now convert the second term into a surface integral by means of one of our standard identities. This leads to the result

$$\mathbf{F} = \frac{1}{4\pi} \int_S \left[\mathbf{E}(\mathbf{E} \cdot \hat{\mathbf{n}}) - \frac{E^2}{2} \hat{\mathbf{n}} \right] dA \qquad (2\text{-}16\text{-}9)$$

A good look at the integrand of Eq. (2-16-9) will convince us that it can be written as the product of a tensor $\mathbf{T}$ and the vector $\hat{\mathbf{n}}$. We write

$$\frac{1}{4\pi} \left[\mathbf{E}(\mathbf{E} \cdot \hat{\mathbf{n}}) - \frac{E^2}{2} \hat{\mathbf{n}} \right] = \mathbf{T}\hat{\mathbf{n}} \qquad (2\text{-}16\text{-}10)$$

where

$$\mathbf{T} = \frac{1}{4\pi} \begin{bmatrix} E_x{}^2 - \dfrac{E^2}{2} & E_x E_y & E_x E_z \\[2ex] E_x E_y & E_y{}^2 - \dfrac{E^2}{2} & E_y E_z \\[2ex] E_x E_z & E_y E_z & E_z{}^2 - \dfrac{E^2}{2} \end{bmatrix} \qquad (2\text{-}16\text{-}11)$$

We conclude then that

$$\mathbf{F} = \int_S \mathbf{T}\hat{\mathbf{n}} \, dA \qquad (2\text{-}16\text{-}12)$$

The tensor $\mathbf{T}$ is called the **electrostatic stress tensor**. As we see from Eq. (2-16-12), we can obtain the force on a given volume by integrating the product of the stress tensor and a unit normal over the surface bounding the volume. We will illustrate this technique by referring back to the problem we have just solved and again try to find the force on the upper hemisphere in Fig. 2-22.

The surface we choose is arbitrary as long as it encloses the entire upper hemisphere and no other charge. There are two natural choices for S. One is the complete xy plane and the other is the actual bounding surface of the hemisphere itself. We will only use one of these, the complete xy plane, and leave the other to the reader as an exercise.

On the xy plane, $E_z = 0$ and $\hat{\mathbf{n}} = -\hat{\mathbf{k}}$. Hence, on this plane,

$$\mathbf{T}\hat{\mathbf{n}} = \frac{1}{8\pi} E^2 \hat{\mathbf{k}}$$

Using $dA = 2\pi r\, dr$ in Eq. (2-16-9), we obtain

$$F = \frac{1}{4} \int_0^\infty E^2 r\, dr$$

$$= \frac{1}{4} \int_0^R \frac{Q^2 r^3}{R^6}\, dr + \frac{1}{4} \int_R^\infty \frac{Q^2}{r^3}\, dr$$

$$= \frac{3}{16} \frac{Q^2}{R^2} \tag{2-16-13}$$

Needless to say, this result is identical to the one we obtained earlier by more conventional means.

The utility of this stress-tensor technique will be most apparent in a situation where we know the electric fields surrounding a charged object but do not know the distribution of charge itself. By evaluating the stress tensor on a surrounding surface, we will indeed never have to determine this charge distribution if we want to know the force acting on it and nothing else.

PROBLEMS

2-1. A charge q is brought to a distance d from the center of a grounded ($\varphi = 0$), conducting sphere of radius R (see figure).

(a) Show that the potential at all points outside the sphere can be obtained by replacing it with an "equivalent" point charge at some place on the line between its center and q. (The equivalent charge need not have a magnitude of $-q$.)

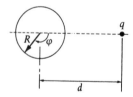

(b) What is the charge per unit area of the sphere as a function of the polar angle θ?

(c) What is the total charge of the sphere?

(d) How much force is exerted by the charge q on the sphere?

(e) Suppose the sphere in the above problem were uncharged, rather than at zero potential. What would the answers to parts (b) and (c) become?

2-2. A slab of dielectric of thickness t, length L, and width W is inserted between two plates of the same length and width with separation d. The plates are connected to a battery, with a potential difference V. Find the force on the dielectric when it has been inserted a distance y.

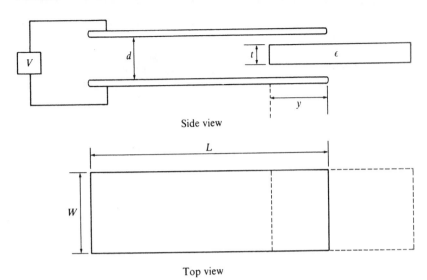

Side view

Top view

2-3. Find the energy stored in a uniform, spherical charge distribution of radius R and total charge Q.

2-4. A charge Q is deposited on a spherical conductor of radius R. What is the energy of the distribution?

2-5. Two long, concentric conducting cylinders have radii a and b, respectively, and are each of length l. The space between them is filled with material having

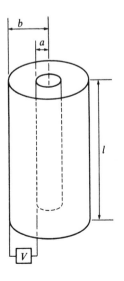

dielectric constant ε. If the potential difference between the cylinders is V, find the total energy stored in the fields between them.

2-6. Consider the plane interface between a region with dielectric constant $\varepsilon = 1.3$ and a region with dielectric constant 1.6. The electric field in the first region is at 45° to the surface. What is the direction of the electric field in the second region?

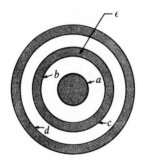

2-7. A spherical conductor of radius a is surrounded (as shown) by a concentric dielectric with dielectric constant ε, inner radius b, and outer radius c. These are in turn surrounded by a conducting shell of radius d. If a charge $+Q$ is placed on the inner conductor and $-Q$ on the outer conductor, find the electric field E and the potential φ at all points between the conductors.

2-8. A dielectric separates two conducting plates. The area of the plates is A and the separation between them is t. The dielectric constant varies linearly as

$$\varepsilon = \varepsilon_0 + kx \qquad \text{for } 0 \leqq x \leqq t$$

A charge $+Q$ is placed on one plate and $-Q$ on the other.
(a) Find $\mathbf{D}$ between the plates.
(b) Find $\mathbf{P}$ everywhere between the plates.
(c) Find $\mathbf{E}$ everywhere.
(d) What is the potential difference between the plates?

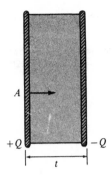

2-9. The space between two concentric conducting spherical shells is half-filled with material of dielectric constant ε, as shown. The radii of the shells are a and b, respectively. A charge Q is placed on the inner sphere and a charge $-Q$ is placed on the outer sphere. Find the fields $\mathbf{E}$, $\mathbf{D}$, and $\mathbf{P}$ at all points between the conductors.

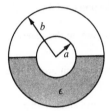

2-10. Find the six independent coefficients of capacitance for three concentric spherical conducting shells, having radii a, b, and c, respectively ($a < b < c$). Choose two sets of values for φ_a, φ_b, φ_c and show explicitly that Green's reciprocity theorem works.

2-11. A charge Q is brought from infinity to the neighborhood of an isolated uncharged conductor. The fields due to the induced charge distribution on the conductor do an amount of work W on Q as it is brought in. We now "freeze" the surface charge in place on the conductor and remove Q back to infinity. How much energy is stored in the remaining electric field distribution?

2-12. Consider an atomic system which consists of a proton at the center of a uniform negative spherical charge distribution of radius $R = 0.5 \times 10^{-8}$ cm. The total system is uncharged. A field $\mathbf{E}$ is now applied to the system causing it to become polarized.
  (a) Calculate the induced dipole moment.
  (b) Calculate the amount of work that the applied field has done in moving the proton from the center of the charge distribution to its new position.
  (c) Calculate the amount of field energy that would be present in a volume of

$\frac{4}{3}\pi R^3$ having constant field E within it. How does this compare with the answer to (b)?

2-13. A conducting spherical shell of radius R is cut into three segments, as shown, extending, respectively, from $\theta = 0$ to $\theta = 60°$, from $\theta = 60°$ to $\theta = 120°$,

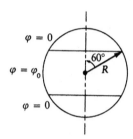

and from $\theta = 120°$ to $\theta = 180°$. They are insulated from one another. The uppermost and lowermost segments are grounded ($\varphi = 0$) and the central segment is held at potential φ_0.

(a) Find the first four terms in a multipole expansion of the potential for $r > R$.
(b) What is the total charge on the sphere?
(c) What is the quadrupole moment of the charge distribution on the sphere?
(d) What is the potential at the center of the sphere?

2-14. Making use of the stress tensor, demonstrate that the force per unit area on the surface of a conductor is $2\pi\sigma^2$ where σ is the surface charge per unit area.

2-15. Demonstrate that the capacitance of any conductor is always smaller than or equal to the capacitance of a conductor which can completely surround it.

2-16. A flat circular disk of radius R has a charge Q distributed uniformly over its area. Show that the potential φ at point (r,θ) where $r > R$ is given by

$$\varphi(r,\theta) = \frac{Q}{r}\left[1 - \frac{1}{4}\left(\frac{R}{r}\right)^2 P_2(\cos\theta) + \frac{1}{8}\left(\frac{R}{r}\right)^4 P_4(\cos\theta)\right.$$
$$\left. - \frac{5}{64}\left(\frac{R}{r}\right)^6 P_6(\cos\theta) + \cdots\right]$$

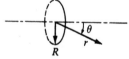

2-17. A charge Q is distributed uniformly along a line coincident with the z axis from $z = -a$ to $z = +a$. Show that the potential at a point (r,θ) where $r \geqq a$ is given by

$$\varphi(r,\theta) = \frac{Q}{r}\left[1 + \frac{1}{3}\left(\frac{a}{r}\right)^2 P_2(\cos\theta) + \frac{1}{5}\left(\frac{a}{r}\right)^4 P_4(\cos\theta) + \cdots\right]$$

3
Electromagnetism and Its Relation to Relativity

3-1 INTRODUCTION; THE MICHELSON-MORLEY EXPERIMENT

We begin our study of electromagnetism in a highly unorthodox way—
by deriving much of it from electrostatics, from the special theory of
relativity, and from the underlying hope that the laws of physics, when
looked at properly, are elegantly simple in their formulation. This is quite
antithetical to the usual notion that physics is an empirical science and
should be presented that way. The beauty of physics lies in the extent to
which seemingly complex and unrelated phenomena can be explained and
correlated through a high level of abstraction by a set of laws which are
usually amazing in their simplicity. In the history of this abstraction, no
triumph has been more spectacular than electromagnetic theory.

To develop electromagnetism in its most beautiful form we must begin with the special theory of relativity. (Historically, of course, the roles of electromagnetism and relativity were reversed, and relativity was derived to explain the fact that Maxwell's equations were not invariant under a galilean transformation.) We will first describe the Michelson-Morley experiment, which played a rather crucial role in the path to relativity.

At the turn of the century, one of the more intriguing problems for the experimental physicist was a measurement of the velocity of the "ether," that is, the medium which carried light. As we all know, the velocity of sound varies depending upon our velocity relative to the air. If v_s is the velocity of sound in still air and V is our velocity relative to the air, then we observe a velocity v_s' for the sound given by

$$v_s' = v_s - V \qquad\qquad (3\text{-}1\text{-}1)$$

Presumably, it was argued, light also has a fixed velocity relative to some medium which permeates all space, and if we find the velocity of that medium relative to the earth, we will be able to calculate the velocity of light in any direction as seen by our earthbound observer. To this end, Michelson and Morley designed an ingenious experiment which was to revolutionize the entire concept of space and time.

There were already some indications of trouble before Michelson and Morley came along. Historically all the basic laws of electromagnetism were already understood by that time and had been formulated by Maxwell in terms of his elegant set of equations. As we shall see, these equations embodied all electrostatics and magnetostatics as well as the possibility of producing electromagnetic radiation by accelerating charges. Now Maxwell's equations contain a constant c which plays *two* vital roles. On the one hand, it is the ratio between the electrostatic unit of charge and the electromagnetic unit of charge. In other words the magnetic field produced by a moving charge will depend on the ratio of its velocity to c as will the force felt by a charge moving in a magnetic field. On the other hand, c is also the velocity with which electromagnetic radiation will propagate. If indeed there were an ether, then the form of Maxwell's equations would change from reference frame to reference frame. In the reference frame of the ether itself the velocity of radiation could be considered a constant independent of direction. In any other frame of reference the velocity of light would depend on direction. Would that mean that the force between current loops would then depend on their orientation relative to the ether flow? Nothing in Maxwell's equations allows for this possibility at all, and they would obviously need extensive modification and correction. Notwithstanding these problems it was still considered of the utmost importance to make a direct measurement of the velocity of the ether relative to the earth. This Michelson and Morley set out to do.[1]

Inasmuch as the ether has no problem in passing through matter (it could not be removed from an evacuated container) and would hence not be dragged along by the earth, one would anticipate a variation in the velocity of light on the earth's surface as the earth traveled about the sun. The velocity of the earth relative to the sun is about 10^{-4} times the velocity of light in ether, and hence variations of this order were anticipated in the relative velocity of earth and ether over the course of a year.

An idealized sketch of the apparatus is shown in Fig. 3-1. It is assumed that the ether is moving to the right along the positive x axis with velocity V. A light beam, initially traveling in the positive x direction, strikes a half-silvered mirror A inclined at 45° to the x axis. Half the light continues through to a mirror B at a distance L_B and back again. The other half reflects into the positive y direction at mirror A, goes on to mirror D at a distance L_D, and then returns to A. The returning beams from both B and D recombine to produce beams going in the negative x and y directions.

[Now in practice the observer looking into A from either the negative x or negative y direction will see an interference pattern, as light which has passed through a given path on one arm meets light which has passed through a corresponding path on the other arm. Optical paths are made to vary somewhat from the "ideal" to produce a series of "fringes" as the observer looks into A. We will continue to assume, though, for the sake of

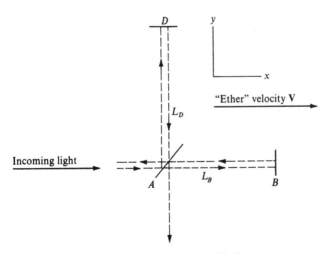

Fig. 3-1   An idealized sketch of the Michelson-Morley apparatus.

[1] The author takes no responsibility for the correctness of the historical exposition. Not having studied the history he has imagined what it must have been like and has arranged it to suit.

argument, that all beams are perfect, parallel beams traveling either $2L_B$ or $2L_D$ as they go through excursions $(A \rightarrow B \rightarrow A)$ or $(A \rightarrow D \rightarrow A)$.]

Let us calculate the time for the paths $A \rightarrow B \rightarrow A$ and $A \rightarrow D \rightarrow A$ in the classical manner, by imagining ourselves moving with the ether. We now see the light moving with velocity c, and our calculation requires only that we know how far the light must travel.

We see the light beam heading toward B, but we see B coming to meet it with velocity V. Hence

$$t_{A \rightarrow B} = \frac{L_B - Vt_{A \rightarrow B}}{c}$$

Similarly

$$t_{B \rightarrow A} = \frac{L_B + Vt_{B \rightarrow A}}{c}$$

Hence

$$t_{A \rightarrow B} = \frac{L_B}{c + V} \tag{3-1-2}$$

$$t_{B \rightarrow A} = \frac{L_B}{c - V} \tag{3-1-3}$$

and

$$t_{A \rightarrow B \rightarrow A} = \frac{2L_B}{c(1 - V^2/c^2)} \tag{3-1-4}$$

As we watch the light path $A \rightarrow D \rightarrow A$, it appears as in Fig. 3.2. The distance the light must travel is $2\sqrt{L_D^2 + V^2 t_{A \rightarrow D \rightarrow A}^2/4}$, and hence the time it takes must then be

$$t_{A \rightarrow D \rightarrow A} = \frac{2\sqrt{L_D^2 + V^2 t_{A \rightarrow D \rightarrow A}^2/4}}{c}$$

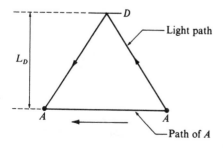

Fig. 3-2 The path of the light from A to D to A as seen from the "ether."

Thus

$$t_{A \to D \to A} = \frac{2L_D}{c\sqrt{1 - V^2/c^2}} \qquad (3\text{-}1\text{-}5)$$

If we were now to adjust L_B and L_D for maximum constructive interference, we would have

$$t_{A \to B \to A} = t_{A \to D \to A} \pm \frac{n\lambda}{c}$$

where λ is the wavelength of the light. Without loss of generality we can set $n = 0$ if L_B and L_D are about equal. We have then

$$L_D = \frac{L_B}{\sqrt{1 - V^2/c^2}} \qquad (3\text{-}1\text{-}6)$$

We can now turn the apparatus through 90° so that the ether is moving along L_D. The interference should change, of course, as we carry out this rotation. We have obviously arranged it, according to Eq. (3-1-6), so as to have L_D slightly longer than L_B. After rotation, however, this relationship will no longer do, in general, for maximum constructive interference.

We can investigate the sensitivity of our apparatus by asking how long the arms must be in order that maximum constructive interference changes to maximum destructive interference as we rotate. We assume that V is equal to the velocity of the earth, namely, $10^{-4}c$. In our rotated system the times are now given by

$$t'_{A \to D \to A} = \frac{2L_D}{c(1 - V^2/c^2)} \qquad (3\text{-}1\text{-}7)$$

and [making use of Eq. (3-1-6)]

$$t'_{A \to B \to A} = \frac{2L_B}{c\sqrt{1 - V^2/c^2}}$$

$$= \frac{2L_D}{c} \qquad (3\text{-}1\text{-}8)$$

Hence, letting the difference between these times be equal to half the wavelength divided by c, we have

$$\frac{2L_D}{c}\left(\frac{1}{1 - V^2/c^2} - 1\right) = \frac{\lambda}{2c}$$

This leads to the result

$$L_D = \frac{\lambda}{4}\frac{c^2}{V^2} \qquad (3\text{-}1\text{-}9)$$

If we now take $V = 10^{-4}c$ and $\lambda = 4 \times 10^{-5}$ cm, we find that L_D *must* be 10 meters long. This would be a formidable piece of hardware if it were built according to our model above. In practice the length is achieved by means of a long series of reflections, but the basic principle is unchanged. In any case, the observer watches the interference fringes carefully as the apparatus is rotated and looks for a shift in the pattern.

Can you imagine the excitement that Michelson and Morley felt when they observed *no change at all* in the interference pattern as they rotated the apparatus? The immediate explanation that probably occurred to them was that at this particular season the earth and the ether must be moving together. So they would have to wait 6 months to really be sure! It was probably a very difficult 6-month wait, but, needless to say, no change in the interference pattern was ever seen. Out of this experiment grew one of mankind's most beautiful ideas: Einstein's special theory of relativity.

3-2   THE LORENTZ TRANSFORMATION

Of course, there were many attempts at finding alternative solutions to the predicament posed by the Michelson-Morley experiment, but we will, in keeping with Einstein, take the very simplest. We shall assume that the velocity of light is a constant of nature and that observers in all frames of reference moving with constant velocity with respect to one another measure the same constant. This immediately explains the Michelson-Morley experiment and leads us directly to the Lorentz transformation for converting the coordinates (x,y,z,t) of an event in one frame Σ to the coordinates (x',y',z',t') of the same event in an equivalent frame Σ'. **Equivalent frames** will be taken to mean frames which move at constant velocity relative to each other.

To begin our derivation of the Lorentz transformation, let us assume that system Σ' is moving with velocity V along the x axis of system Σ (see Fig. 3-3a). Let us further assume that the x' axis lies superimposed along the x axis and that the y' and z' axes are parallel, respectively, to the y and z axes. Both clocks (t and t') will be set at zero when the origins of Σ and Σ' coincide.

We next show by trivial argument that $y' = y$ and $z' = z$. Let us take two absolutely identical twins and stand one along the y axis and the other along the y' axis and have the twins approach each other. Now, for the sake of argument, let us assume that twin John in the Σ system believes his brother Jim to be shorter. Then, since we assume neither system to be preferred, Jim must think that John is shorter. As they collide we have a most remarkable set of circumstances—John sees Jim's head plow into his own belly as he feels his own head plow into Jim's belly. Quite absurd!! There is only one sensible solution, each still sees the other as being of the

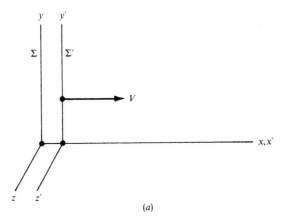

(a)

Fig. 3-3 (a) The coordinate system Σ' moves with velocity V along the x axis of the Σ coordinate system. John is riding along in the Σ system while Jim is in the Σ' system.

same height as himself. That is, dimensions transverse to the direction of motion are unaltered in a Lorentz transformation.

Now we come to the matter of clocks. We would like to examine how each of our twins sees the other's clock in relation to his own. Since the velocity of light c is a fundamental constant of nature, the "ideal" clock, so to speak, is a perfect stick of length L with a mirror at each end and a bit of light beam bouncing back and forth. Let each twin align his clock along the y (or y') axis. Each twin sees his own light beam take a time $\tau = 2L/c$ for the complete cycle. Meanwhile he sees the other twin's light take an amount of time t_{other}. Of course, t_{other} is larger than τ, since the light must travel along further (see Fig. 3-3b). The path of the other light is

$$\text{Path}_{\text{other}} = 2\sqrt{L^2 + \frac{(Vt_{\text{other}})^2}{4}}$$

Consequently

$$t_{\text{other}} = \frac{\text{path}_{\text{other}}}{c}$$

$$= \frac{2}{c}\sqrt{L^2 + \frac{V^2 t_{\text{other}}^2}{4}}$$

and, solving for t_{other},

$$t_{\text{other}} = \frac{2L}{c\sqrt{1 - V^2/c^2}} = \frac{\tau}{\sqrt{1 - V^2/c^2}} \qquad (3\text{-}2\text{-}1)$$

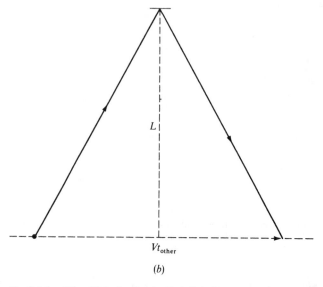

$$V t_{\text{other}}$$

(b)

Fig. 3-3 (cont'd)   (b) Path taken by Jim's light beam as seen by John.

Thus each sees the other's clock moving more slowly, by a factor of

$$\frac{1}{\sqrt{1 - V^2/c^2}}$$

Having developed a means of comparing clocks in two equivalent coordinate systems, we now attempt a comparison of lengths along the direction of relative motion. Let John, in the Σ system, lay his standard clock, of length L as he sees it, along the x axis. We presuppose now that Jim in the Σ' system measures the length of the rod as L', and we wish to relate L' to L. John (Σ) believes that the light takes a time $t = 2L/c$ for a complete cycle, and, as we have already determined, Jim thinks it takes longer, namely, $2L/c\sqrt{1 - V^2/c^2}$. Now let us visualize the entire process as seen by Jim (Σ') (see Fig. 3-3c). He sees the transit of light to the right as taking an amount of time t'_R. At the same time he sees the end of the rod moving toward the light beam with a velocity V, and hence the distance covered by the light beam in going from left to right (A to B') is just $L' - Vt'_R$. The light having traveled at a velocity c has thus taken a time $t'_R = (L' - Vt'_R)/c$, yielding

$$t'_R = \frac{L'}{c(1 + V/c)}$$

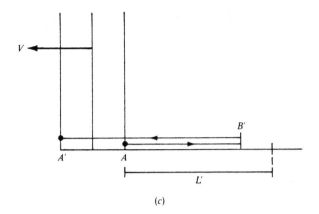

Fig. 3-3 (*cont'd*) (*c*) The Σ system as viewed by an observer in the Σ' system. L' is the apparent length of Σ's clock. The path indicated is followed by Σ's light beam in the course of one cycle.

Similarly we would find the time from B' to A' as

$$t'_L = \frac{L'}{c(1 - V/c)}$$

Using our information about clocks, we have

$$t'_L + t'_R = \frac{2L}{c} \frac{1}{\sqrt{1 - V^2/c^2}} = \frac{2L'}{c} \frac{1}{1 - V^2/c^2} \tag{3-2-2}$$

yielding

$$L' = \sqrt{1 - \frac{V^2}{c^2}}\, L \tag{3-2-3}$$

We thus come to the following conclusion. Each observer determines distances along the direction of relative motion to be foreshortened in the other coordinate system. Jim thinks that John is abnormally thin in one direction, and Jim, of course, thinks the same about John.

Having learned how to compare meter sticks and clocks, we are now ready to derive the full Lorentz transformation. Let us say that an event occurs at position x and at time t in the Σ system. The observer in the Σ' system reasons as follows:

1. When both clocks were at zero, then the point x would appear to him as

$$\sqrt{1 - \frac{V^2}{c^2}}\, x$$

2. During the time interval t' (before the event occurs as seen by Σ'), the Σ system moves in the negative x' direction by an amount Vt'.

3. Hence we conclude that

$$x' = \sqrt{1 - \frac{V^2}{c^2}}\, x - Vt'$$

Solving for x, we have

$$x = \frac{x' + Vt'}{\sqrt{1 - V^2/c^2}} \tag{3-2-4}$$

By symmetry we can also say

$$x' = \frac{x - Vt}{\sqrt{1 - V^2/c^2}} \tag{3-2-5}$$

Combining these equations, we have

$$t = \frac{t' + (V/c^2)x'}{\sqrt{1 - V^2/c^2}} \tag{3-2-6}$$

$$t' = \frac{t - (V/c^2)x}{\sqrt{1 - V^2/c^2}} \tag{3-2-7}$$

To make our notation simple, we define the symbols β and γ by

$$\beta = \frac{V}{c} \tag{3-2-8}$$

$$\gamma = \frac{1}{\sqrt{1 - \beta^2}} \tag{3-2-9}$$

Let us now set down the matrix

$$\mathbf{L} = \begin{bmatrix} \gamma & 0 & 0 & i\beta\gamma \\ 0 & 1 & 0 & 0 \\ 0 & 0 & 1 & 0 \\ -i\beta\gamma & 0 & 0 & \gamma \end{bmatrix} \quad \text{where } i = \sqrt{-1} \tag{3-2-10}$$

We will call the element in the μth row and νth column $L_{\mu\nu}$. If we now write

$$\begin{aligned} x_1 &= x & x_3 &= z \\ x_2 &= y & x_4 &= ict \end{aligned} \tag{3-2-11}$$

then the Lorentz transformation can simply be written as

$$x'_\mu = \sum_{v=1}^{4} L_{\mu v} x_v \qquad (3\text{-}2\text{-}12)$$

We note immediately the formal similarity with equations in three-dimensional space. For example, the analog of Eq. (1-2-3) is

$$\sum_{v=1}^{4} L_{\mu v} L_{\rho v} = \delta_{\mu \rho} \qquad (3\text{-}2\text{-}13)$$

In terms of our four-dimensional notation, the old rotation matrix **a** (see page 5) is now written as **R** :

$$\mathbf{R} = \begin{bmatrix} a_{11} & a_{12} & a_{13} & 0 \\ a_{21} & a_{22} & a_{23} & 0 \\ a_{31} & a_{32} & a_{33} & 0 \\ 0 & 0 & 0 & 1 \end{bmatrix} \qquad (3\text{-}2\text{-}14)$$

A pure rotation in three-dimensional space would then be expressed as

$$x'_\mu = \sum R_{\mu v} x_v \qquad (3\text{-}2\text{-}15)$$

Now **R** and **L** are both subsets of the set of all four-dimensional rotations (called the **Lorentz group**). If we wish to find the Lorentz transformation between two coordinate systems whose relative motion is along some direction other than the x axis, we can apply the succession of rotation, Lorentz transformation, and inverse rotation to achieve our ends.

The set of four quantities (x_1, x_2, x_3, x_4) is a four-vector in this four-dimensional space. As we anticipate, its length remains invariant under any Lorentz transformation:

$$x^2 + y^2 + z^2 - c^2 t^2 = x'^2 + y'^2 + z'^2 - c^2 t'^2 \qquad (3\text{-}2\text{-}16)$$

All sets of four quantities which transform as (x_1, x_2, x_3, x_4) under this group of transformations will be called **four-vectors**. Scalar products of four-vectors are made in the same way as before. If u and w are four-vectors, then the scalar product is

$$uw = \sum_{\mu=1}^{4} u_\mu w_\mu \qquad (3\text{-}2\text{-}17)$$

For convenience we will also adopt the convention that Greek subscripts (μ, v, etc.) refer always to the numbers 1, 2, 3, 4, while Latin subscripts (i, j, k, etc.) refer only to the numbers 1, 2, 3.

The definition of a tensor of the second rank in four dimensions is quite analogous to that in three dimensions. The 16 elements transform like the set of component products of two four-vectors. There is nothing analogous to the vector product in four dimensions, since, as we have seen, the vector product is a special tensor of the second rank in three dimensions. We shall have more to say about this shortly.

In any case, we have now developed enough formal machinery to enable us to transform the position and time of an event as seen by one observer into the position and time of the same event as seen by an equivalent observer. We can now take a look at the mechanics of a moving particle and see how the dynamical quantities which characterize its behavior can be transformed as we go from one to another of these coordinate systems.

We begin with the simplest dynamical quantity, its velocity, which we call $\mathbf{u}$ in the Σ system and $\mathbf{u}'$ in the Σ' system. As usual, Σ' moves with velocity V along the positive x axis. We have

$$u'_x = \frac{dx'}{dt'} = \frac{dx - V\,dt}{dt - (V/c^2)\,dx}$$

$$= \frac{u_x - V}{1 - Vu_x/c^2} \tag{3-2-18}$$

$$u'_y = \frac{1}{\gamma}\,\frac{u_y}{1 - u_x V/c^2} \tag{3-2-19}$$

$$u'_z = \frac{1}{\gamma}\,\frac{u_z}{1 - u_x V/c^2} \tag{3-2-20}$$

The most interesting of these velocity transformation formulae is Eq. (3-2-18) relating u_x and u'_x. In the old days, before we heard of relativity, we would have simply let $u'_x = u_x - V$. Thus if a particle was moving along the negative x axis with velocity $u_x = -\frac{3}{4}c$ and the Σ' system was moving along the positive x axis with velocity $\frac{3}{4}c$, we would naturally expect the Σ' observer to see the particle moving with velocity $1.5c$. It stands to reason! Now we say, instead, that $u'_x = -1.5c/(1 + \frac{9}{16}) = (-24/25)c$. Very odd indeed. In fact, try as we may, as long as $|u_x| < c$ and $|V| < c$ we can never have $|u'_x|$ greater than c.

We next consider the component of momentum of the particle that is transverse to the direction of relative motion of the two coordinate systems. This component must be invariant under Lorentz transformation from one of these systems to the other. Physically, this is easiest to understand by placing oneself first in one frame and then accelerating to the other frame. As we accelerate we see fictitious forces causing a change in momentum in the x (or x') direction but never in the y or z directions. (We could not even know how to define the y or z directions on the basis of the relative

velocity of the two coordinate systems above.) Making use of this requirement, and freeing ourself of the notion that mass is an invariant scalar quantity under a Lorentz transformation, we have

$$mu_y = m'u'_y \tag{3-2-21}$$

Inserting our previous result from Eq. (3-2-19), we have

$$\frac{m}{m'} = \frac{u'_y}{u_y} = \frac{1}{\gamma(1 - u_x V/c^2)} \tag{3-2-22}$$

The expression on the right can be simplified as follows. We first calculate
$1 - u'^2/c^2$ by inserting the expression

$$u'^2 = u'^2_x + u'^2_y + u'^2_z$$

$$= \frac{(u_x - V)^2 + u_y^2(1 - V^2/c^2) + u_z^2(1 - V^2/c^2)}{(1 - u_x V/c^2)^2}$$

Simplifying and combining terms yields

$$1 - \frac{u'^2}{c^2} = \frac{(1 - u^2/c^2)(1 - V^2/c^2)}{(1 - u_x V/c^2)^2}$$

or

$$\gamma\left(1 - \frac{u_x V}{c^2}\right) = \frac{\sqrt{1 - u^2/c^2}}{\sqrt{1 - u'^2/c^2}} \tag{3-2-23}$$

We will find this expression quite useful. Returning to Eq. (3-2-22), we can
now write

$$m\sqrt{1 - \frac{u^2}{c^2}} = m'\sqrt{1 - \frac{u'^2}{c^2}} \tag{3-2-24}$$

We see thus that the expression $m\sqrt{1 - u^2/c^2}$ is an invariant constant
under a Lorentz transformation. We will call it m_0, the value it has when
$u = 0$. We have then

$$m = \frac{m_0}{\sqrt{1 - u^2/c^2}} \tag{3-2-25}$$

The quantity m_0 is referred to as the **rest mass** of the object. Having found
the way to transform m as we go from one system to another, we are now
ready for the x component of momentum. As usual we write

$$p'_x = m'u'_x$$

$$= \frac{m_0}{\sqrt{1 - u^2/c^2}} \frac{u_x - V}{1 - u_x V/c^2} \tag{3-2-26}$$

But, using Eq. (3-2-23) again, we obtain

$$p'_x = \frac{m_0 \gamma}{\sqrt{1 - u^2/c^2}} (u_x - V) \tag{3-2-27}$$

Putting all of what we have learned together, we finally come up with the equations

$$p'_x = \gamma \left(p_x - \frac{V}{c} mc \right)$$

$$p'_y = p_y$$

$$p'_z = p_z \tag{3-2-28}$$

$$m'c = \gamma \left(mc - \frac{V}{c} p_x \right)$$

Lo and behold, the set of four entities (p_x, p_y, p_z, imc) transforms as a four-vector under a Lorentz transformation!

We now make a most remarkable discovery—that momentum conservation, if it is to be independent of coordinate system, implies energy conservation. Imagine a collision between particles ① and ② giving rise to particles ③ and ④. Momentum conservation tells us that

$$\mathbf{p}_1 + \mathbf{p}_2 = \mathbf{p}_3 + \mathbf{p}_4 \tag{3-2-29}$$

Let us form the four-vector $p_1 + p_2 - p_3 - p_4$:

$$p_1 + p_2 - p_3 - p_4 = [\mathbf{p}_1 + \mathbf{p}_2 - \mathbf{p}_3 - \mathbf{p}_4, i(m_1 + m_2 - m_3 - m_4)c]$$

$$= [0, i(m_1 + m_2 - m_3 - m_4)c] \tag{3-2-30}$$

If the first three components are to *remain* zero for *any* Lorentz transformation, then the fourth component must also be zero. Hence

$$m_1 + m_2 = m_3 + m_4 \tag{3-2-31}$$

To bring this equation into correspondence with classical physics we note that for small u

$$mc^2 = \frac{m_0 c^2}{\sqrt{1 - u^2/c^2}} \cong m_0 c^2 \left(1 + \frac{1}{2} \frac{u^2}{c^2} \right)$$

$$\cong m_0 c^2 + \tfrac{1}{2} m_0 u^2 \tag{3-2-32}$$

This is then equal, for small u, to a constant plus the kinetic energy of the mass. We then define the energy of our mass as mc^2 and observe that energy conservation is a direct consequence of momentum conservation.

We can derive a very useful relationship between energy and momentum by making use of Eq. (3-2-25). We let mc^2 be denoted by the symbol E and recall that

$$p^2 = m^2 u^2 = \frac{m_0{}^2 u^2}{1 - u^2/c^2}$$

Hence

$$E^2 = \frac{m_0{}^2 c^4}{1 - u^2/c^2} = p^2 c^2 + m_0{}^2 c^4 \qquad (3\text{-}2\text{-}33)$$

Another equally useful expression relates the velocity u to the ratio of pc and E. Making use of Eqs. (3-2-25) and (3-2-33), we can easily obtain

$$\beta = \frac{u}{c} = \frac{pc}{E} \qquad (3\text{-}2\text{-}34)$$

Before we leave the subject of relativistic kinematics, we will say a few words about a system of units which plays a predominant role in dealing with atomic, nuclear, and subnuclear phenomena. This system is based on the *electron volt* as a unit of energy. The volt is a unit of potential difference in the mks system and can be related to our unit of potential difference, the statvolt, by the equation

$$1 \text{ statvolt} = 10^{-8} c \text{ volts} \qquad (3\text{-}2\text{-}35)$$

where c is the velocity of light in *centimeters per second* ($10^{-8}c$ is about 300). The electron volt (eV) is then the amount of energy which corresponds to moving a charge equal to the charge of an electron (or proton) through a potential difference of 1 volt. Since the charge of the electron, e is equal to 4.8×10^{-10} esu, we have

$$1 \text{ eV} = \frac{4.8 \times 10^{-10}}{300} \text{ ergs} = 1.6 \times 10^{-12} \text{ ergs} \qquad (3\text{-}2\text{-}36)$$

Next we introduce the units of momentum and mass which go with the electron volt in common usage. Both are obtained from our energy units by allowing c to equal 1 in our kinematic equations. The unit of momentum is called the eV/c and the unit of mass is called the eV/c^2. The rest mass of a particle whose rest energy is $m_0 c^2$ eV is thus m_0 eV/c^2. (In Table 3-1 we list the rest masses of some of the more important elementary particles.) The momentum (in eV/c) of a particle whose mass and energy are known is obtained from Eq. (3-2-33):

$$p_{\text{in eV}/c} = \sqrt{E^2 - m_0{}^2}$$

where E is in eV and m_0 is in eV/c^2. The velocity of a particle is given by Eq. (3-2-34):

$$\beta = \frac{p}{E} \qquad (3\text{-}2\text{-}37)$$

where p is in eV/c and E is in eV. Although this system of units will seem odd to the student at first, he will soon become accustomed to the great advantage of not having to carry around c's everywhere.

Let us illustrate the methods discussed above by calculating the velocity of a proton which has acquired a kinetic energy of 500 MeV (a MeV is 10^6 eV). Referring to Table 3-1 we see that the mass of a proton is 938 MeV/c^2. Hence its rest energy is 938 MeV. Adding this to the kinetic energy, we obtain

$$E = 938 \text{ MeV} + 500 \text{ MeV} = 1438 \text{ MeV}$$

We next find the momentum p.

$$p = \sqrt{(1438)^2 - (938)^2} = 1090 \text{ MeV}/c$$

Finally we determine the velocity of the proton relative to the speed of light:

$$\beta = \frac{p}{E} = \frac{1090}{1438} = 0.758$$

TABLE 3-1   Tabulation of the masses of some elementary particles†

Particle	Symbol	Mass (in MeV/c^2)
Electron	$e^\pm$	0.511006 ± 0.000002
Muon	$\mu^\pm$	105.659 ± 0.002
Charged pion	$\pi^\pm$	139.578 ± 0.013
Neutral pion	π°	134.974 ± 0.013
Charged kaon	$\kappa^\pm$	493.82 ± 0.11
Neutral kaon	κ°	497.76 ± 0.16
Eta	η°	548.8 ± 0.6
Proton	p	938.256 ± 0.005
Neutron	n	939.550 ± 0.005
Lambda	Λ°	1115.57 ± 0.07
Sigma plus	Σ^+	1189.43 ± 0.17
Sigma zero	Σ°	1192.55 ± 0.11
Sigma minus	Σ^-	1197.42 ± 0.09
Xi zero	Ξ°	1314.7 ± 0.7
Xi minus	Ξ^-	1321.25 ± 0.18
Omega minus	Ω^-	1672.4 ± 0.6

†From compilation of Prof. A. Rosenfeld.

3-3 CHARGE DENSITY AND CURRENT DENSITY AS COMPONENTS OF A FOUR-VECTOR

We remember from elementary physics that current is somehow related to the movement of charge. In this section we will put this notion on a quantitative basis and then show that current density and charge density are closely related as components of a four-vector. This is not altogether surprising for we anticipate that stationary charge density as seen from one reference frame will appear to be moving in another reference frame.

It is simplest to begin with a collection of discrete charges which are sufficiently closely spaced so as to be effectively a continuum. We will assume that the charge q_i has a velocity $\mathbf{v}_i$, and we then define the current density $\mathbf{j}$ at a given point by the equation

$$\mathbf{j}_{esu} = \lim_{\Delta\tau \to 0} \frac{\sum_i q_i \mathbf{v}_i}{\Delta\tau} \tag{3-3-1}$$

where $\Delta\tau$ is a small volume element about the point of interest. (We have attached the subscript esu to indicate that the current density so defined is in electrostatic units.) Of course the charge density in this case is defined as before:

$$\rho = \lim_{\Delta\tau \to 0} \frac{\sum_i q_i}{\Delta\tau} \tag{3-3-2}$$

This definition of current density is quite in keeping with our intuition. If we want to determine the charge per unit time flowing past a bit of surface dA in the direction of the normal vector $\hat{\mathbf{n}}$, we just evaluate $\mathbf{j}_{esu} \cdot \hat{\mathbf{n}}\, dA$. The flux of $\mathbf{j}$ out of a given volume is just the charge leaving it per unit time.

$$\text{Rate at which charge is leaving } \Delta V = \int_{\substack{\text{surface} \\ \text{of } \Delta V}} \mathbf{j}_{esu} \cdot \hat{\mathbf{n}}\, dA \tag{3-3-3}$$

We now make a most important empirical observation. The total charge that a given object carries is an invariant with respect to a Lorentz transformation. Were this not the case we would have a very insane world indeed. As we heated a bit of matter, the average electron velocity would increase quite differently from the average nuclear velocity. Hence, if the negative charges of the electrons just canceled the positive nuclear charges at one temperature, they would not do so at all temperatures. The body would then change its overall apparent charge with changing temperature, and, remembering the basic strength of the electrostatic force (see page 34), disastrous consequences would follow. Fortunately, when the laws of physics were first set down, this problem was averted through the Lorentz invariance of total charge. (As we have seen, this invariance does *not* hold true for the total mass of an object. Its mass does increase as it speeds up. However, because the net gravitational force on matter is not the resultant of a delicate balance of opposites as it is in the case of electrostatics, the consequences are negligible.)

We now ask how j_{esu} and ρ transform under a Lorentz transformation. Since not all our charges are necessarily moving with the identical velocities, we must break our charge and current distribution into subdistributions. Each of these subdistributions will consist of all charges that do have more or less identical velocities. We can treat each of these subdistributions independently as we go from one frame of reference to another and add them together at the end.

For the typical case we have $j_{esu} = \rho v$ where v is the velocity of the subdistribution. We transform into the system Σ' moving with velocity v, in which case

$$j' = 0$$
$$\rho' = \rho_0 = \text{charge density in rest frame of subdistribution} \qquad (3\text{-}3\text{-}4)$$

When we transform back to the original system Σ, all distances along the transformation direction are compressed by the factor $\sqrt{1 - v^2/c^2}$, and, in order that the total charge remain constant, the charge density must increase by the factor $1/\sqrt{1 - v^2/c^2}$. Hence in the Σ system we have

$$\rho = \frac{\rho_0}{\sqrt{1 - v^2/c^2}} \qquad (3\text{-}3\text{-}5)$$

Since $j_{esu} = \rho v$ in this system, we also have

$$j_{esu} = \frac{\rho_0 v}{\sqrt{1 - v^2/c^2}} \qquad (3\text{-}3\text{-}6)$$

We notice that these equations are completely identical in velocity dependence with the previously derived equations for mass and momentum:

$$m = \frac{m_0}{\sqrt{1 - v^2/c^2}}$$

$$p = \frac{m_0 v}{\sqrt{1 - v^2/c^2}}$$

But the four quantities p_x, p_y, p_z, and imc transform as a four-vector under Lorentz transformation. The conclusion is clear. The four quantities $j_{x(esu)}$, $j_{y(esu)}$, $j_{z(esu)}$, $i\rho c$ must transform as a four-vector.

At this point it is convenient to make a change in our units. We introduce the so-called **gaussian units**, wherein ρ is expressed in electrostatic units as before but $\mathbf{j}$ is expressed in electromagnetic units. The definition of the **electromagnetic unit** is

$$\mathbf{j}_{emu} = \frac{\mathbf{j}_{esu}}{c} \tag{3-3-7}$$

We shall no longer carry the subscript emu or esu but tacitly assume that $\mathbf{j} = \mathbf{j}_{emu}$. The four quantities $j_x, j_y, j_z,$ and $i\rho$ constitute our new four-vector. Consider now how we would express conservation of charge within a given coordinate system. The rate of change of charge in a volume is equal to the charge entering the volume per unit time. Using Gauss' law, we can write this as

$$\nabla \cdot \mathbf{j} + \frac{1}{c} \frac{\partial \rho}{\partial t} = 0 \tag{3-3-8}$$

Note then that the left side of the equation is just the scalar product of the two four-vectors $[(\partial/\partial x, \partial/\partial y, \partial/\partial z, 1/ic(\partial/\partial t)]$ and $(j_x, j_y, j_z, i\rho)$ and is thus invariant under Lorentz transformation. If the conservation of charge holds in one coordinate system, it will hold in any equivalent system!

3-4 THERE MUST BE A "MAGNETIC FIELD"! (THE REQUIREMENT OF LORENTZ INVARIANCE IMPLIES A VECTOR POTENTIAL)

Until now our study of relativity has been largely limited to *kinematics*. That is to say, we have learned how to transform some of the variables that describe a dynamical system from one coordinate system to another, but we have not discussed at all the laws of physics which govern the behavior of these variables. We will now begin to do so with a very simple assumption. We will assume that the force on a charged particle in its own rest frame is given by the electric field as the particle sees it. Furthermore, we will assume that if the charge and current distributions that it sees in its own rest frame are not changing with time, then the electric field it sees will be determined, just as in the case of simple electrostatics, by Eq. (2-1-4).

The consequences of these innocent-sounding assumptions are absolutely fantastic. We will now show that the electric field *cannot* provide a complete description of the long-range forces resulting from charges and currents, but that there must in addition be other field quantities at each point in space which exert a velocity-dependent force on a *moving* charged particle. Thus we will deduce that there *must* be a magnetic type of force. Later we shall see how to express this force in terms of magnetic field.

We begin with an exceedingly simple problem. We consider an infinitely long cylindrical region of radius R which carries a constant current density $\mathbf{j}$ but no charge density ($\rho = 0$). We can achieve this by having an equal density of positive and negative charges moving in opposite directions.

A charge q is now projected with velocity $\mathbf{v}$ parallel to the current, at a distance r from its axis. We wish to find the force exerted on the charge (see Fig. 3-4).

We first note that nothing we have done in electrostatics would seem to indicate that there should be any force whatsoever on the charge q. Any force we discover now will be new. Of course, we anticipate from previous experience that there should be a "magnetic" force on q, but as we will now show, this state of affairs is completely deducible from electrostatics and the Lorentz transformation.

We begin by doing the obvious, jumping to a system in which the charge is at rest. In this system Σ' moving with velocity $\mathbf{v}$ with respect to the laboratory Σ, we see the current density $\mathbf{j}'$ and the charge density ρ' given by

$$j'_x = \gamma j_x = \frac{j_x}{\sqrt{1 - v^2/c^2}}$$

$$j'_y = j'_z = j_y = j_z = 0 \tag{3-4-1}$$

$$i\rho' = -i\beta\gamma j_x$$

That is to say, we see a *negative* charge density for $r \leqq R$ equal to

$$\rho' = \frac{-vj_x}{c\sqrt{1 - v^2/c^2}} \tag{3-4-2}$$

The electric field set up at the radius r is radial in direction and is just (from Gauss' law)

$$E_r' = + \frac{2\pi R^2 \rho'}{r} \tag{3-4-3}$$

If we let $I = \pi R^2 j_x =$ total current, we have

$$E_r' = - \frac{2Iv}{c\sqrt{1 - v^2/c^2}\, r} \tag{3-4-4}$$

Fig. 3-4   A charge q is projected with velocity $\mathbf{v}$ parallel to a current distribution $\mathbf{j}$. For $r \leqq R$, $\mathbf{j}$ = constant, $\rho = 0$. For $r > R$, $\mathbf{j} = 0$, $\rho = 0$.

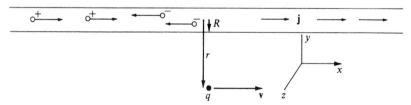

Now, the force on our charge q is just qE'_r in this system. But this force is also equal to $dp'_y/d\tau$ in the Σ' system where τ is the particle's proper time and p'_y is its transverse momentum. Hence

$$\frac{dp'_y}{d\tau} = -\frac{2Ivq}{c\sqrt{1 - v^2/c^2}\,r} \tag{3-4-5}$$

Going back to Σ, we note that

$$dp'_y = dp_y$$

and

$$dt = \frac{d\tau}{\sqrt{1 - v^2/c^2}} \tag{3-4-6}$$

Substituting back, we find the force on q in the laboratory.

$$F_y = \frac{dp_y}{dt} = -\frac{2Ivq}{cr} \tag{3-4-7}$$

As we see, this is just equal to what we would have predicted from magneto-statics, as we learned it in elementary physics. We would have said

$$F_y = -q\,\frac{v}{c}\,B \quad \text{and} \quad B = \frac{2I}{r}$$

So we have made a major discovery. A charged particle moving in the proximity of a current distribution has a force acting on it, even though there is no other charge density present! Unfortunately though, life is not quite so elementary, for if we try to project the charge q in the y direction (toward the axis of the current distribution), we run into trouble with our simple reasoning. If we jumped on board our particle, we would see no apparent change in the current density and we would see no charge density at all. This is because the current $\mathbf{j}$ is now transverse to our velocity of transformation and there is no longitudinal current or charge density in the laboratory. What do we do now? Do we conclude that the charge q will now have no force on it? Of course not! We have gone outside the area of validity of electrostatics because we now see a current distribution which is changing with time. It is approaching us with velocity $-\mathbf{v}$, and we have no right to expect that electrostatics will give us the full answer. So we must now go back and think out our whole problem from scratch.

To feel our way, let us recall the equation relating potential to charge distribution: $\nabla^2 \varphi = -4\pi\rho$. It is clear that this equation is not properly invariant under a Lorentz transformation. To make it so, the operator ∇^2 has to be replaced by something which transforms as a scalar in four-dimensional space. Obviously we want

$$\nabla^2 \rightarrow \frac{\partial^2}{\partial x^2} + \frac{\partial^2}{\partial y^2} + \frac{\partial^2}{\partial z^2} - \frac{1}{c^2}\frac{\partial^2}{\partial t^2} \equiv \square \tag{3-4-8}$$

The symbol on the right, called the **D'Alembertian**, reduces to the laplacian ∇^2 in the event that there is no time dependence to φ and has the right invariance under Lorentz transformation. The proper equation for φ is then

$$\square\varphi = -4\pi\rho \tag{3-4-9}$$

Since ρ is the fourth component of a four-vector, φ must also be so. Hence we must have three other components A_x, A_y, and A_z such that the set of objects $(A_x, A_y, A_z, i\varphi)$ is a four-vector and

$$\square A_x = -4\pi j_x \tag{3-4-10}$$

$$\square A_y = -4\pi j_y \tag{3-4-11}$$

$$\square A_z = -4\pi j_z \tag{3-4-12}$$

In the event that $\mathbf{j}$ and ρ are not time dependent, we can rewrite these equations in integral form just as in electrostatics.

$$\varphi = \int_{\substack{\text{all}\\\text{space}}} \frac{\rho(\mathbf{r}')}{|\mathbf{r} - \mathbf{r}'|}\, dV' \tag{3-4-13}$$

$$\mathbf{A} = \int_{\substack{\text{all}\\\text{space}}} \frac{\mathbf{j}(\mathbf{r}')}{|\mathbf{r} - \mathbf{r}'|}\, dV' \tag{3-4-14}$$

In summary then, the laws of electrostatics must be generalized by introducing a four-vector potential A_μ corresponding to the four-vector current j_μ such that

$$\sum_{\nu=1}^{4} \frac{\partial^2 A_\mu}{\partial x_\nu^2} = -4\pi j_\mu \tag{3-4-15}$$

This equation represents our first breakthrough in understanding how to generalize from electrostatics. We see that one component of potential is inadequate but that we must have four components in order that our old equations for electrostatics not be limited to one particular coordinate system. The Eqs. (3-4-15) have the property that if they hold for one system they will hold for *all* systems and reduce properly to the equation of electrostatics in the event that there is no time dependence in any given system.

3-5 THE ELECTRIC AND MAGNETIC FIELDS AS ELEMENTS OF A SECOND-RANK TENSOR

Our next step is to find the fields that exist at each point in space in terms of the four components of vector potential. Again we use what we have learned in electrostatics to point the way for us. We recall that in the absence of time dependence we have $\mathbf{E} = -\nabla\varphi$, which we can rewrite as

$$-iE_1 = \frac{\partial A_4}{\partial x_1}$$

$$-iE_2 = \frac{\partial A_4}{\partial x_2} \tag{3-5-1}$$

$$-iE_3 = \frac{\partial A_4}{\partial x_3}$$

It is clear then that the three components of electric field are really elements of a second-rank tensor in our four-dimensional space. Our job is now reduced to finding the complete tensor.

So far everything we have done has been entirely deductive, making use only of Coulomb's law, conservation of charge under Lorentz transformation, and the requirement of Lorentz invariance for our physical laws. We have now come to the end of this deductive path. At this point when the laws were being written, God had to make a decision. In general there are 16 components to a second-rank tensor in four dimensions. However, in analogy to three dimensions we can make a major simplification by choosing the completely antisymmetric tensor to represent our field quantities. Then we would have only 6 independent components instead of the possible 16. Under Lorentz transformation the tensor would remain antisymmetric and we would never have need for more than six independent components. Appreciating this and having a deep aversion to useless complication, God naturally chose[1] the antisymmetric tensor as His medium of expression. We define this tensor $F_{\mu\nu}$ as follows:

$$F_{\mu\nu} = \left(\frac{\partial A_\nu}{\partial x_\mu} - \frac{\partial A_\mu}{\partial x_\nu}\right) \tag{3-5-2}$$

---

[1] It has been pointed out to the author by Prof. D. Dorfan that God was actually quite constrained in His choice of the second-rank tensor. The demonstration that the tensor we are looking for (which we shall call **F**) must be totally antisymmetric goes as follows. We first observe that in the instantaneous rest frame of a particle of charge q the force on it must be equal to $q\mathbf{E}$ where $\mathbf{E}$ is the electric field it sees. The force is also equal to the rate of change of the particle's momentum **p** in this frame of reference. Hence we have, making use of Eq. (3-5-1),

$$\frac{dp_j}{d\tau} = \frac{-q}{i} F_{j4} \qquad \text{for } j = 1, 2, 3 \tag{3-5-1a}$$

Of course, anything we do for 1, 2, and 3 we also want to do for 4. Thus we must write

$$\frac{dp_\mu}{d\tau} = \frac{-q}{i} F_{\mu4} \qquad \text{for } \mu = 1, 2, 3, 4 \tag{3-5-1b}$$

Next we observe that the momentum four-vector in the particle's instantaneous rest frame is

Writing out the terms explicitly, we have

$$
\mathbf{F} = \begin{bmatrix}
0 & (\nabla \times \mathbf{A})_z & -(\nabla \times \mathbf{A})_y & i\left(\dfrac{\partial \varphi}{\partial x} + \dfrac{1}{c}\dfrac{\partial A_x}{\partial t}\right) \\[2ex]
-(\nabla \times \mathbf{A})_z & 0 & (\nabla \times \mathbf{A})_x & i\left(\dfrac{\partial \varphi}{\partial y} + \dfrac{1}{c}\dfrac{\partial A_y}{\partial t}\right) \\[2ex]
(\nabla \times \mathbf{A})_y & -(\nabla \times \mathbf{A})_x & 0 & i\left(\dfrac{\partial \varphi}{\partial z} + \dfrac{1}{c}\dfrac{\partial A_z}{\partial t}\right) \\[2ex]
-i\left(\dfrac{\partial \varphi}{\partial x} + \dfrac{1}{c}\dfrac{\partial A_x}{\partial t}\right) & -i\left(\dfrac{\partial \varphi}{\partial y} + \dfrac{1}{c}\dfrac{\partial A_y}{\partial t}\right) & -i\left(\dfrac{\partial \varphi}{\partial z} + \dfrac{1}{c}\dfrac{\partial A_z}{\partial t}\right) & 0
\end{bmatrix}
$$

$$(3\text{-}5\text{-}3)$$

where $\mathbf{A}$ is the three-vector (A_x, A_y, A_z). Notice now the remarkable fact that under a pure rotation the tensor $\mathbf{F}$ breaks into a polar vector and an axial vector. The components $(F_{23}, -F_{13}, F_{12})$ are really the three components of $\nabla \times \mathbf{A}$ and transform as an axial vector. The three components (F_{14}, F_{24}, F_{34}) are clearly equivalent to $i\left(\nabla\varphi + \dfrac{1}{c}\dfrac{\partial \mathbf{A}}{\partial t}\right)$, which is a polar vector. Hence in any given coordinate system it would appear as though there were two independent vectorial fields at each point in space. Only

---

$$p_\mu = (0,0,0,im_0 c) \tag{3-5-1c}$$

Thus Eq. (3-5-1b) can be rewritten in the form

$$\frac{dp_\mu}{d\tau} = \frac{q}{m_0 c} \sum_{\nu=1}^{4} p_\nu F_{\mu\nu} \tag{3-5-1d}$$

Both sides of Eq. (3-5-1d) now have the appearance of components of a vector. We next take the scalar product of this vector with the vector $2p$:

$$\sum_{\mu=1}^{4} 2p_\mu \frac{dp_\mu}{d\tau} = \frac{2q}{m_0 c} \sum_{\nu=1}^{4} \sum_{\mu=1}^{4} p_\nu F_{\mu\nu} p_\mu \tag{3-5-1e}$$

We observe that the left side of the equation can be rewritten as

$$\sum_{\mu=1}^{4} 2p_\mu \frac{dp_\mu}{d\tau} = \frac{d}{d\tau} \sum_{\mu=1}^{4} p_\mu^{\ 2} = \frac{d}{d\tau}(-m_0^{\ 2} c^2) = 0$$

We thus have

$$\sum_{\nu=1}^{4} \sum_{\mu=1}^{4} p_\nu F_{\mu\nu} p_\mu = 0 \tag{3-5-1f}$$

The left side of Eq. (3-5-1f) is a scalar. Hence the equation is true in any coordinate system which can be related to our instantaneous rest system by means of a Lorentz transformation. It is now left to the student to prove that if Eq. (3-5-1f) is to be true for any choice of system then $\mathbf{F}$ must be antisymmetric.

when we transform between them do the two fields become mixed together. In any case, let us return to Eq. (3-5-3) and make some important observations. The first of these is that $\nabla\varphi$ and $\dfrac{1}{c}\dfrac{\partial \mathbf{A}}{\partial t}$ are inseparable! Component by component they always occur together, and so a particle must experience their combination as though it were one type of force. Inasmuch as we identified $-\nabla\varphi$ as $\mathbf{E}$ in the case of electrostatics, we now write

$$\mathbf{E} = -\nabla\varphi - \frac{1}{c}\frac{\partial \mathbf{A}}{\partial t} \tag{3-5-4}$$

(This result embodies all of Faraday's famous law and is even more general. Its importance will become apparent in due course.)

We now make a definition. We define the magnetic field $\mathbf{B}$ as

$$\mathbf{B} \equiv \nabla \times \mathbf{A} \tag{3-5-5}$$

We have then

$$\mathbf{F} = \begin{bmatrix} 0 & +B_z & -B_y & -iE_x \\ -B_z & 0 & +B_x & -iE_y \\ +B_y & -B_x & 0 & -iE_z \\ iE_x & iE_y & iE_z & 0 \end{bmatrix} \tag{3-5-6}$$

Before we actually see how this magnetic field acts on a moving charge, we should examine in detail the behavior of the components of $\mathbf{F}$ as we go from one coordinate system to another through a Lorentz transformation. We will assume as usual that Σ' is moving along the positive x axis of Σ with velocity V. As usual we have

$$F'_{\mu\nu} = \sum_{\alpha,\beta=1}^{4} L_{\mu\alpha}L_{\nu\beta}F_{\alpha\beta} \tag{3-5-7}$$

Remembering that the Lorentz transformation matrix is given by Eq. (3-2-10), we have

$$\mathbf{F}' = \begin{bmatrix} 0 & \gamma(B_z - \beta E_y) & -\gamma(B_y + \beta E_z) & -iE_x \\ -\gamma(B_z - \beta E_y) & 0 & B_x & -i\gamma(E_y - \beta B_z) \\ \gamma(B_y + \beta E_z) & -B_x & 0 & -i\gamma(E_z + \beta B_y) \\ iE_x & i\gamma(E_y - \beta B_z) & i\gamma(E_z + \beta B_y) & 0 \end{bmatrix} \tag{3-5-8}$$

where $\beta = V/c$ and $\gamma = 1/\sqrt{1 - \beta^2}$.

We first observe that the component of either **E** or **B** along the direction of relative motion remains unchanged. This component is called the **longitudinal component** and in the present case is the component along the x direction.

$$B'_x = B_x$$
$$E'_x = E_x \tag{3-5-9}$$

As far as the transverse components of **E** are concerned, we note that

$$E'_y = \gamma E_y - \gamma \frac{V}{c} B_z$$

$$= \gamma E_y + \frac{\gamma}{c} (\mathbf{V} \times \mathbf{B})_y \tag{3-5-10}$$

$$E'_z = \gamma E_z + \frac{\gamma}{c} (\mathbf{V} \times \mathbf{B})_z$$

Hence we can write in general, for the component of **E** transverse to the relative motion of Σ and Σ',

$$E'_T = \gamma \left[\mathbf{E}_T + \left(\frac{\mathbf{V}}{c} \times \mathbf{B} \right)_T \right]$$

However, since $(\mathbf{V}/c) \times \mathbf{B}$ has no longitudinal component, we can simplify this to

$$\mathbf{E}'_T = \gamma \left[\mathbf{E}_T + \left(\frac{\mathbf{V}}{c} \times \mathbf{B} \right) \right] \tag{3-5-11}$$

Similarly we have

$$\mathbf{B}'_T = \gamma \left(\mathbf{B}_T - \frac{\mathbf{V}}{c} \times \mathbf{E} \right) \tag{3-5-12}$$

We can now easily find the force on a moving charge q moving with velocity **V** in the presence of electric and magnetic fields. We need merely transform to the rest system of the particle, set the force on the particle in this system equal to $q\mathbf{E}' = d\mathbf{p}'/d\tau$, and then go back to the original system, transforming $d\mathbf{p}'/d\tau$ as we change. (τ as we remember is time as measured in the particle's rest system.) We have for the transverse components

$$\frac{d\mathbf{p}'_T}{d\tau} = q\gamma \left(\mathbf{E}_T + \frac{\mathbf{V}}{c} \times \mathbf{B} \right) \tag{3-5-13}$$

But $\mathbf{p}'_T = \mathbf{p}_T$, since the transverse components of **p** are unchanged by a

Lorentz transformation. Also, from time dilation we have $t = \gamma\tau$. Hence

$$\frac{d\mathbf{p}_T}{dt} = q\mathbf{E}_T + q\frac{\mathbf{V}}{c} \times \mathbf{B} \tag{3-5-14}$$

Turning to the longitudinal component of $\mathbf{E}$, we have

$$\frac{dp'_L}{d\tau} = qE'_L = qE_L$$

or

$$\frac{\gamma dp'_L}{dt} = qE_L \tag{3-5-15}$$

But, making use of the Lorentz transformation and remembering that Σ' is the rest system of the particle, we have

$$p_L = \gamma p'_L + \beta\gamma m_0 c$$

Therefore $dp_L = \gamma \, dp'_L$ and hence

$$\frac{dp_L}{dt} = qE_L \tag{3-5-16}$$

Combining this with our previous result for the transverse components of the force, we have

$$\frac{d\mathbf{p}}{dt} = \mathbf{F} = q\mathbf{E} + q\frac{\mathbf{V}}{c} \times \mathbf{B} \tag{3-5-17}$$

This rather fundamental equation tells us precisely how electric and magnetic fields act on a moving charged particle. We can now, in principle, solve any dynamics problem involving the interaction of charged particles and currents if we can only find a procedure for determining these fields. We already know how to determine $\mathbf{E}$ from a static charge distribution. We also know how to evaluate $\mathbf{A}(\mathbf{r})$ if there is no time dependence to our currents, and as a result we know implicitly how to determine the magnetic fields which derive from static currents. In the chapters ahead we shall develop explicitly many of the techniques for determining the fields as a function of time and space. Before doing so, however, we must derive Maxwell's equations in all their beauty.

3-6 MAXWELL'S EQUATIONS

Our derivation of Maxwell's equations will again make use of the requirement of Lorentz invariance. We will begin with the old electrostatic equation

$\nabla \cdot \mathbf{E} = 4\pi\rho$ and ask what we must do in order that it hold in all systems. This will lead immediately to an equation relating the magnetic field $\mathbf{B}$ to the current density $\mathbf{j}$.

Let us first write $\nabla \cdot \mathbf{E} = 4\pi\rho$ in relativistic notation. Remembering that $F_{44} = 0$, we can rewrite $\nabla \cdot \mathbf{E}$ as

$$\nabla \cdot \mathbf{E} = \frac{\partial E_x}{\partial x} + \frac{\partial E_y}{\partial y} + \frac{\partial E_z}{\partial z}$$

$$= i\left(\frac{\partial F_{14}}{\partial x_1} + \frac{\partial F_{24}}{\partial x_2} + \frac{\partial F_{34}}{\partial x_3}\right)$$

$$= i\sum_{\mu=1}^{4} \frac{\partial F_{\mu 4}}{\partial x_\mu} = -i4\pi j_4 \tag{3-6-1}$$

But, we recall from Eqs. (1-6-6) to (1-6-11) that the product of a second-rank tensor and a vector transforms as a vector. Hence $\nabla \cdot \mathbf{E}$, as expected, is the fourth component of a vector, and we can now insist that Eq. (3-6-1) be generalized to the other components:

$$\sum_{\mu=1}^{4} \frac{\partial F_{\mu\nu}}{\partial x_\mu} = -4\pi j_\nu \tag{3-6-2}$$

Let us develop this explicitly for the first three components:

$$-\frac{\partial B_z}{\partial y} + \frac{\partial B_y}{\partial z} + \frac{1}{c}\frac{\partial E_x}{\partial t} = -4\pi j_x$$

$$+\frac{\partial B_z}{\partial x} - \frac{\partial B_x}{\partial z} + \frac{1}{c}\frac{\partial E_y}{\partial t} = -4\pi j_y \tag{3-6-3}$$

$$-\frac{\partial B_y}{\partial x} + \frac{\partial B_x}{\partial y} + \frac{1}{c}\frac{\partial E_z}{\partial t} = -4\pi j_z$$

Combining all three equations, we have

1. $\quad \nabla \times \mathbf{B} = \dfrac{1}{c}\dfrac{\partial \mathbf{E}}{\partial t} + 4\pi\mathbf{j} \tag{3-6-4}$

The fourth component, to repeat, gave us

2. $\quad \nabla \cdot \mathbf{E} = 4\pi\rho \tag{3-6-5}$

Since $\mathbf{B} = \nabla \times \mathbf{A}$ and the divergence of any curl is zero (see page 21), we have

3. $\quad \nabla \cdot \mathbf{B} = 0 \tag{3-6-6}$

And finally, from $\mathbf{E} = -\nabla\varphi - \dfrac{1}{c}\dfrac{\partial\mathbf{A}}{\partial t}$, we have

$$\nabla \times \mathbf{E} = -\frac{1}{c}\nabla \times \frac{\partial\mathbf{A}}{\partial t}$$

Inverting the order of differentiation, we get

4. $\quad\nabla \times \mathbf{E} = -\dfrac{1}{c}\dfrac{\partial\mathbf{B}}{\partial t}$ (3-6-7)

These then are the famous Maxwell equations. Note that the first and second of these are completely equivalent to but not nearly as aesthetically pleasing as Eq. (3-6-2). We would like to put the last two Maxwell equations in an equally beautiful form, but first we must introduce the so-called **completely antisymmetrical** tensor of the fourth rank, $\varepsilon_{\mu\nu\lambda\rho}$, defined as follows:

$$\varepsilon_{\mu\nu\lambda\rho} = 0 \qquad \text{if any two indices are equal}$$
$$\varepsilon_{1234} = 1$$
$$\varepsilon_{\mu\nu\lambda\rho} = 1 \qquad \text{if } \mu, \nu, \lambda, \text{ and } \rho \text{ are all different and can be obtained}$$
$$\text{from } 1,2,3, \text{ and } 4 \text{ by an even number of transpositions}$$
$$\varepsilon_{\mu\nu\lambda\rho} = -1 \qquad \text{if } \mu, \nu, \lambda, \text{ and } \rho \text{ are all different and can be obtained}$$
$$\text{from } 1,2,3, \text{ and } 4 \text{ by an odd number of transpositions}$$

We shall make use of two properties of $\varepsilon_{\mu\nu\lambda\rho}$, the demonstration of which we will leave to the reader.

1. The components of $\varepsilon_{\mu\nu\lambda\rho}$ transform into themselves under Lorentz transformation.

2. The sum $\sum\limits_{\lambda,\rho} \varepsilon_{\mu\nu\lambda\rho} B_{\lambda\rho}$ transforms as an element of a second-rank tensor under Lorentz transformation if $B_{\lambda\rho}$ is an element of a second-rank tensor. Making use of this, we define the tensor $\mathbf{G}$ by the equation

$$G_{\mu\nu} \equiv \frac{1}{2}i\sum \varepsilon_{\mu\nu\lambda\rho}F_{\lambda\rho} \tag{3-6-8}$$

where $F_{\lambda\rho}$ is an element of our field tensor.

The tensor $\mathbf{G}$ is antisymmetric and has elements

$$G_{12} = \tfrac{1}{2}i(\varepsilon_{1234}F_{34} + \varepsilon_{1243}F_{43})$$
$$= iF_{34}$$
$$G_{13} = \tfrac{1}{2}i(\varepsilon_{1324}F_{24} + \varepsilon_{1342}F_{42})$$
$$= -iF_{24}$$
$$G_{14} = \tfrac{1}{2}i(\varepsilon_{1423}F_{23} + \varepsilon_{1432}F_{32})$$
$$= iF_{23}$$

$$G_{23} = \tfrac{1}{2}i(\varepsilon_{2314}F_{14} + \varepsilon_{2341}F_{41})$$
$$\qquad = iF_{14}$$
$$G_{24} = \tfrac{1}{2}i(\varepsilon_{2413}F_{13} + \varepsilon_{2431}F_{31})$$
$$\qquad = -iF_{13}$$
$$G_{34} = \tfrac{1}{2}i(\varepsilon_{3412}F_{12} + \varepsilon_{3421}F_{21})$$
$$\qquad = iF_{12}$$

Summarizing, we have

$$\mathbf{G} = \begin{bmatrix} 0 & E_z & -E_y & iB_x \\ -E_z & 0 & E_x & iB_y \\ E_y & -E_x & 0 & iB_z \\ -iB_x & -iB_y & -iB_z & 0 \end{bmatrix} \tag{3-6-9}$$

Now, since $\mathbf{B} = \nabla \times \mathbf{A}$, we can write

$$\sum_\mu \frac{\partial G_{\mu 4}}{\partial x_\mu} = i\nabla \cdot \mathbf{B} = 0$$

For this to prevail in any coordinate system, we must have

$$\sum \frac{\partial G_{\mu v}}{\partial x_\mu} = 0 \qquad \text{for all } v \tag{3-6-10}$$

When we set $v = 1, 2, 3$, we just come up with $\nabla \times \mathbf{E} = -\dfrac{1}{c}\dfrac{\partial \mathbf{B}}{\partial t}$. Hence the four Maxwell equations can be rewritten as

$$\sum_\mu \frac{\partial F_{\mu v}}{\partial x_\mu} = -4\pi j_v$$
$$\sum_\mu \frac{\partial G_{\mu v}}{\partial x_\mu} = 0 \tag{3-6-11}$$

From the purely aesthetic point of view there is something wrong with these equations. Whereas the electric field can have a static source, namely, ρ, the magnetic field cannot. Wouldn't it be nice if somehow there were magnetic monopoles possible in nature so that the two equations would be "symmetric"? Suppose for a moment that such objects did exist. We could then define a so-called **magnetic charge density** $\rho^{(m)}$ and a corresponding **magnetic current density** $\mathbf{j}^{(m)}$. (The magnetic current density would actually be an axial three-vector field.) There would be a corresponding potential four-vector $A_\mu^{(m)}$ of which the first three components would form an axial three-vector. Finally, we would rewrite the second Eq. (3-6-11) as

$$\sum_{\mu=1}^{4} \frac{\partial G_{\mu v}}{\partial x_\mu} = 4\pi j_v^{(m)} \tag{3-6-12}$$

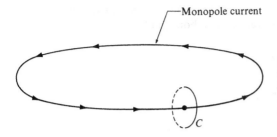

Monopole current

Fig. 3-5 A current of magnetic monopoles is circulating around a closed loop as shown. We evaluate $\oint \mathbf{E} \cdot d\mathbf{l}$ around the curve C which links the monopole current.

Now the possibility that magnetic monopoles exist has actually excited wide interest for another reason. It was pointed out by Prof. P. A. M. Dirac that their existence would explain the quantization of charge in a natural way.[1] This being the case a considerable amount of effort has been devoted to finding these monopoles. The efforts have been of two different sorts.

1. The fact that monopoles would be strongly acted upon by magnetic fields has been exploited by a number of experiments which have tried to pull them out of matter or out of the cosmic radiation by means of such fields.[2]
2. Before we describe the second type of experiment we must put Eq. (3-6-12) into more convenient form. If we let $v = 1, 2$, and 3, we obtain

$$\nabla \times \mathbf{E} = \frac{-1}{c} \frac{\partial \mathbf{B}}{\partial t} - 4\pi \mathbf{j}^{(m)} \qquad (3\text{-}6\text{-}13)$$

Let us next imagine that we have a steady continuous current of monopoles circulating around a closed loop as shown in Fig. 3-5. Nothing is changing with time, and so we are justified in removing the time-dependent term on the right side of Eq. (3-6-13). If the total magnetic "charge" in the loop is $Q^{(m)}$ and the time it takes for a given monopole to make it around is T, then the monopole current in the loop is just

$$I^{(m)} = \frac{Q^{(m)}}{cT}$$

[The factor c is required because our four-vector is $(j_x^{(m)}, j_y^{(m)}, j_z^{(m)}, i\rho^{(m)})$. That is, $\rho^{(m)}$ and $\mathbf{j}^{(m)}$ are measured in different units in complete analogy with ρ and $\mathbf{j}$ (see page 123).]

Let us next integrate $\mathbf{E} \cdot d\mathbf{l}$ around an imaginary closed curve C

[1] For a review of this subject in reasonably simple terms the reader is referred to an article by J. Schwinger in *Science*, **165**: 757 (Aug. 22, 1969).

[2] See, for example, E. M. Purcell, G. B. Collins, T. Fujii, J. Hornbostel, and F. Turkot, *Phys. Rev.*, **129**: 2326 (1963).

linking the loop. Using Stokes' theorem, we convert the line integral into a surface integral over the area bounded by C.

$$\oint_C \mathbf{E} \cdot d\mathbf{l} = \int_A \nabla \times \mathbf{E} \cdot \hat{\mathbf{n}} \, dA$$

$$= 4\pi \int_A \mathbf{j}^{(m)} \cdot \hat{\mathbf{n}} \, dA$$

$$= 4\pi I^{(m)}$$

$$= \frac{4\pi Q^{(m)}}{cT} \tag{3-6-14}$$

If a loop of wire were thus allowed to link the monopole current loop, then Eq. (3-6-14) would ensure that a current would flow in the loop. Indeed, it would be as though a battery with a potential difference of $4\pi Q^{(m)}/cT$ statvolts were placed in the wire loop.

This brings us then to the description of an actual experiment carried out by Prof. L. W. Alvarez[1] and collaborators at the University of California Lawrence Radiation Laboratory. They were interested in determining whether any monopoles might be trapped in pieces of moon rock brought back by the astronauts. After all, the moon's surface is being continuously bombarded by cosmic rays, and an occasional one might be a monopole. A small bottle of moon rock was placed on a track, as shown in Fig. 3-6, and made to go around and around with a time for 1 revolution of about 4 sec. Linking the track was a superconducting niobium coil with 1200 turns of wire. (A superconducting coil has no measurable resistance.) If there were a monopole trapped in the rock, then the electric current through the superconducting loop would change as a function of time. The rate of change of electric current will depend on the line integral of Eq. (3-6-14) as well as on the *inductance* of the coil, a property which we will discuss in due time. Suffice it to say that no monopole has yet been seen. We will return to this experiment in Sec. 5-1, after we have developed some more formal machinery, in order that we may be able to assess its true sensitivity. We will see at that time that it is capable of detecting the smallest quantum of magnetic charge consistent with Dirac's theory.

In any case, since no monopoles have as yet been seen, we will continue to write Maxwell's equations in the conventional form given by Eq. (3-6-11). Perhaps at some future time we will have need to change them.

We have now completed the foundation of our electromagnetic edifice. In the chapters ahead we will undertake an examination of the incredible richness of Maxwell's equations, culminating in the discovery that

---

[1] L. W. Alvarez et al., *Science*, **167** (3917): 701–703 (1970).

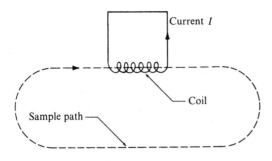

Current I

Coil

Sample path —

Fig. 3-6 A sketch of the sample path and superconducting coil used by Prof. L. W. Alvarez and collaborators to search for monopoles in moon rocks.

accelerating charges must radiate light. At no point should we lose sight though of the very simple assumptions that have gone into our derivation of this theory. Basically all we have used are Coulomb's law, the requirement of relativistic invariance, and a passion for simplicity and elegance. Remarkable indeed!

PROBLEMS

3-1. A neutral pion decays in about 10^{-16} sec into two photons (γ rays). Neutral pions can be produced in the laboratory by stopping negative pions in hydrogen and allowing them to be captured into atomic orbits about individual protons. When they reach the ground state of the atom, the negative pions are captured and the system undergoes the reaction

$$\pi^- + p \rightarrow \pi^\circ + n$$

(a) Ignoring the binding energy of the π^- in the atom before its capture, find the velocity of the π° emerging from the reaction.

(b) What is the kinetic energy of the emerging neutron?

(c) How far does the π° travel in the laboratory if it lives for a time of 10^{-16} sec as measured in its own rest system?

(d) What is the observed spectrum of gamma rays emerging from the hydrogen target?

3-2. An antiproton ($\bar{p}$) can be produced by firing energetic protons at other protons which are stationary in the laboratory. The reaction which takes place is

$$p + p \rightarrow p + \bar{p} + p + p$$

Find the threshold for this reaction. That is, find the energy of the least-energetic incoming proton that can initiate the reaction.

3-3. A Λ° particle lives for a time of 2.4×10^{-10} sec in its own frame of reference. How far does it travel in the laboratory if it is moving with velocity $0.5c$, $0.9c$, $0.99c$?

3-4. An electron traveling in a straight line along the x axis moves by a stationary observer as shown. The observer is located on the z axis with coordinates

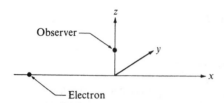

(0,0,1). Find and graph the z component of electric field seen by the observer for each of the following electron velocities: $0.6c$, $0.8c$, $0.9c$, $0.99c$.

3-5. Show that the components of $\varepsilon_{\mu\nu\lambda\rho}$ when transformed under a Lorentz transformation remain unaltered.

3-6. Show that if **B** is a second-rank tensor then $\sum\limits_{\lambda,\rho} \varepsilon_{\mu\nu\lambda\rho} B_{\lambda\rho}$ is an element of a second-rank tensor.

3-7. Show that $E^2 - B^2$ and $\mathbf{B} \cdot \mathbf{E}$ are each invariant under Lorentz transformation.

3-8. Let **E** and **B** be normal to each other and let $|\mathbf{E}| < |\mathbf{B}|$. Find the velocity **v** of a system within which $\mathbf{E} = 0$.

3-9. Consider a classical electron in a circular órbit of radius R about a proton. What is the magnetic field seen by an observer whose velocity coincides instantaneously with that of the electron?

3-10. Show that if we alter both **A** and φ by the transformation

$$\mathbf{A}' = \mathbf{A} + \nabla\psi$$

$$\varphi' = \varphi - \frac{1}{c} \frac{\partial\psi}{\partial t}$$

where ψ is any function of space and time, then the electric and magnetic fields are unaltered. This transformation is called a **gauge transformation.**

3-11. Consider a single magnetic monopole with unit magnetic "charge" traveling in a straight line down the x axis. An observer stationed at coordinate $(0,0,1)$ on the z axis will observe a pulse of electric field as the monopole goes by.
(a) Plot the magnitude of the y component of this field as a function of time for each of the following choices of monopole velocity: $\beta = 0.6$, $\beta = 0.8$, $\beta = 0.9$, $\beta = 0.99$.
(b) Plot the x and z components of the electric field for the same set of velocities.

4

Time-Independent
Current Distributions;
Magnetostatics

In this chapter we will develop a variety of techniques for determining the magnetic field **B** which arises as the result of a time-independent current distribution. We will then consider the dynamics of a moving charged particle or a current loop within this field with a number of interesting and practical results.

Basically there are three different ways in which we might approach the question of determining **B** as a function of position in space. We might attempt to find the vector potential **A** first, making use of either the differential equations (3-4-10) to (3-4-12) or the integral equation (3-4-14). We would then find **B** by determining $\nabla \times \mathbf{A}$. Alternatively we might begin with Eq. (3-4-14) in its general form, take its curl, and come up with a generally useful integral for **B** directly. Finally we might see if there are any simple integrals for **B** which might be useful in cases of high symmetry. As we shall see, each of these techniques is quite analogous to a corresponding technique that we have learned in electrostatics. Before we begin on this path, however, it is useful to say just a few words about the behavior of currents in typical conductors. We will develop the approximation known as Ohm's law.

4-1   AN ELEMENTARY DERIVATION OF OHM'S LAW

In most metals there is about one electron per atom that is relatively free to move about, being subject only to macroscopic electric fields and scattering on the individual lattice sites. The average velocity of an electron ($\bar{v}$) is of the order of 10^8 cm/sec and, inasmuch as the typical mean free path is of the order of 10^{-6} cm, there is very little change in the velocity due to any reasonable applied field in the time between collisions. Let λ be the mean free path. Then the time between collisions is $\Delta t = \lambda/v$. The average drift velocity gained in this time interval due to an applied field E is

$$v_d = \frac{1}{2}\left(\frac{eE}{m}\,\Delta t\right) = \frac{1}{2}\,\frac{eE}{m}\,\frac{\lambda}{\bar{v}} \tag{4-1-1}$$

After a collision it is assumed that the velocity of the electron is randomized in direction, and so the average induced current density is

$$j = \frac{Ne^2 E\lambda}{2m\bar{v}c} \quad \text{emu}$$

where N is the number of "free" electrons per cubic centimeter. We write this vectorially as

$$\mathbf{j} = \sigma\mathbf{E} \tag{4-1-2}$$

where

$$\sigma = \frac{Ne^2\lambda}{2m\bar{v}c} \tag{4-1-3}$$

is called the **conductivity**. (In actuality the drift speed is some 10 orders of magnitude lower than $\bar{v}$. Hence the simple approximation we made, that $\Delta t = \lambda/\bar{v}$, is quite reasonable.)

This simple proportionality between the applied field and the current density is called **Ohm's law**. It is most often applied to the case of a given piece of conductor of area A and length l. In that situation we see that the total current is $I = jA$ and the potential difference $\Delta\varphi$ across the conductor is given by El. Ohm's law then says that

$$E = \frac{\Delta\varphi}{l} = \frac{j}{\sigma} = \frac{I}{\sigma A}$$

Hence

$$\frac{\Delta\varphi}{I} = \frac{l}{\sigma A} \equiv R \qquad (4\text{-}1\text{-}4)$$

where R is called the **resistance**.

We can calculate the work done per unit time on the current $\mathbf{j}$ by the field $\mathbf{E}$. The work done per unit time on a charge q is just $\mathbf{F}\cdot\mathbf{v} = q\mathbf{E}\cdot\mathbf{v}$. Hence the work done on the current per unit volume per unit time is just

$$\frac{dW}{dt} = \sum q_i\mathbf{v}_i \cdot \mathbf{E} = \mathbf{j}_{\text{esu}}\cdot\mathbf{E} = c\mathbf{j}\cdot\mathbf{E} \qquad (4\text{-}1\text{-}5)$$

In the event that $\mathbf{j} = \sigma\mathbf{E}$, we can write

$$\frac{dW}{dt} = c\sigma E^2 \qquad (4\text{-}1\text{-}6)$$

We should comment at this point that the system of units we are using here for R and σ is such as to have the current always measured in abamperes and the potential in statvolts. The unit of resistance is thus the statvolt per abampere. Most texts when dealing with resistance will stay entirely with either esu or emu units. We have chosen to ignore that convention for the sake of overall consistency. Inasmuch as most resistors are labeled in ohms anyway, we need only know that

$$1 \text{ statvolt/abampere} = 29.98 \text{ ohms}$$

4-2 FINDING THE MAGNETIC FIELD THROUGH THE VECTOR POTENTIAL

Finding the vector potential $\mathbf{A}$ in magnetostatics is quite analogous with finding the potential φ in electrostatics. We can either carry out the integral

$$\mathbf{A}(\mathbf{r}) = \int_{\substack{\text{all}\\ \text{space}}} \frac{\mathbf{j}(\mathbf{r}')}{|\mathbf{r} - \mathbf{r}'|}\, dV'$$

or we can search for the unique solutions to $\nabla^2 A_i = -4\pi j_i$ subject to the appropriate boundary conditions. These methods will be most simply applied if we can make a direct analogy to an electrostatic problem that we solve easily and then adapt the solution appropriately. We will illustrate by application to two simple examples.

The first problem we solve is to find the magnetic field everywhere due to an infinite cylindrical uniform current distribution of radius R (see Fig. 4-1). We take the origin of our coordinate system on the axis of the cylinder and let the direction of current flow be along the z coordinate axis. The current density within the cylinder is $\mathbf{j}$.

The analogous problem is, of course, that of a uniform cylindrical charge distribution. We remember, making use of Gauss' law, that the electric field due to such a distribution is radial and has a magnitude E_r at a distance r from the axis given by

$$E_r = \begin{cases} 2\pi\rho r & \text{for } r < R \\[2mm] \dfrac{2\pi\rho R^2}{r} & \text{for } r > R \end{cases} \tag{4-2-1}$$

If we try to calculate the potential in the usual way by integrating out to infinity, we will come up with an infinite result because we are using an infinite cylinder. Since in the end it will be the fields and not the potentials that interest us, let us renormalize so as to have $\varphi = 0$ at $r = 0$. We have then

$$\varphi(r) = -\int_0^r E_r(r) \, dr$$

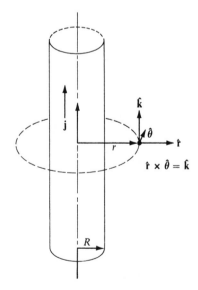

Fig. 4-1 We find the field around a cylinder of uniform current density.

This yields

$$\varphi(r) = \begin{cases} -\pi\rho r^2 & \text{for } r < R \\ -\pi\rho R^2\left(1 + 2\ln\dfrac{r}{R}\right) & \text{for } r > R \end{cases} \qquad (4\text{-}2\text{-}2)$$

Now back to the magnetostatic problem that really interests us. Since there is only a z component to the current, we have $A_x = A_y = 0$. We also have

$$A_z(\mathbf{r}) = \int_{\substack{\text{all} \\ \text{space}}} \frac{j_z(\mathbf{r}')}{|\mathbf{r} - \mathbf{r}'|} \, dV'$$

Our analogy tells us that

$$A_z(r) = \begin{cases} -\pi j r^2 & \text{for } r < R \\ -\pi j R^2\left(1 + 2\ln\dfrac{r}{R}\right) & \text{for } r > R \end{cases} \qquad (4\text{-}2\text{-}3)$$

Finally, we remember how to take the curl in cylindrical coordinates [see Eq. (2-9-16)]. We have in this case

$$\nabla \times \mathbf{A} = -\frac{\partial A_z}{\partial r} \, \hat{\boldsymbol{\theta}} \qquad (4\text{-}2\text{-}4)$$

Thus

$$\mathbf{B} = \begin{cases} 2\pi j r \, \hat{\boldsymbol{\theta}} & \text{for } r < R \\ \dfrac{2\pi j R^2}{r} \, \hat{\boldsymbol{\theta}} & \text{for } r > R \end{cases} \qquad (4\text{-}2\text{-}5)$$

As we shall soon see, this result could have been obtained much more rapidly through more sophisticated means (Ampere's law). Nevertheless it does serve as a useful first example.

As our second example we consider an infinite sheet of current having a thickness t and a constant current density. We set our coordinate system up with the x axis perpendicular to the sheet and the z axis along the current (see Fig. 4-2). The current density is then $\mathbf{j} = j\hat{\mathbf{k}}$ where $\hat{\mathbf{k}}$ is a unit vector in the z direction.

Again the vector potential is given by

$$\mathbf{A}(\mathbf{r}) = \int_{\text{sheet}} \frac{\mathbf{j}(\mathbf{r}') \, dV'}{|\mathbf{r} - \mathbf{r}'|} = j \int_{\text{sheet}} \hat{\mathbf{k}} \frac{dV'}{|\mathbf{r} - \mathbf{r}'|} \qquad (4\text{-}2\text{-}6)$$

The corresponding electrostatic problem would be a sheet with uniform

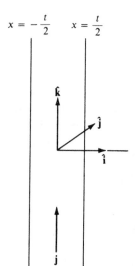

Fig. 4-2 We find the field due to an infinite sheet of uniform current density j and thickness t.

charge density ρ, and we would have had

$$\varphi(\mathbf{r}) = \rho \int_{\text{sheet}} \frac{dV'}{|\mathbf{r} - \mathbf{r}'|} \tag{4-2-7}$$

We solve the electrostatic problem by means of Gauss' law, obtaining

$$\mathbf{E} = \begin{cases} 2\pi\rho t\hat{\mathbf{i}} & \text{for } x \geqq \dfrac{t}{2} \\[2ex] 4\pi\rho x\hat{\mathbf{i}} & \text{for } -\dfrac{t}{2} \leqq x \leqq \dfrac{t}{2} \\[2ex] -2\pi\rho t\hat{\mathbf{i}} & \text{for } x < -\dfrac{t}{2} \end{cases} \tag{4-2-8}$$

If we set $\varphi = 0$ at $x = 0$, we have

$$\varphi = \begin{cases} \dfrac{\pi\rho t^2}{2} - 2\pi\rho tx & \text{for } x \geqq \dfrac{t}{2} \\[2ex] -2\pi\rho x^2 & \text{for } -\dfrac{t}{2} \leqq x \leqq \dfrac{t}{2} \\[2ex] \dfrac{\pi\rho t^2}{2} + 2\pi\rho tx & \text{for } x \leqq -\dfrac{t}{2} \end{cases} \tag{4-2-9}$$

Hence, by complete analogy, we can write down **A** for our original problem:

$$
\mathbf{A} = \begin{cases}
\left(\dfrac{\pi t^2}{2} - 2\pi t x\right) j\hat{\mathbf{k}} & \text{for } x \geq \dfrac{t}{2} \\[4mm]
-2\pi j x^2 \hat{\mathbf{k}} & \text{for } -\dfrac{t}{2} \leq x \leq \dfrac{t}{2} \\[4mm]
\left(\dfrac{\pi t^2}{2} + 2\pi t x\right) j\hat{\mathbf{k}} & \text{for } x \leq -\dfrac{t}{2}
\end{cases}
\tag{4-2-10}
$$

Finally, taking the curl of **A**, we have

$$
\mathbf{B} = \begin{cases}
2\pi t j \hat{\mathbf{j}} & \text{for } x \geq \dfrac{t}{2} \\[4mm]
4\pi j x \hat{\mathbf{j}} & \text{for } -\dfrac{t}{2} \leq x \leq \dfrac{t}{2} \\[4mm]
-2\pi t j \hat{\mathbf{j}} & \text{for } x \leq -\dfrac{t}{2}
\end{cases}
\tag{4-2-11}
$$

where $\hat{\mathbf{j}}$ is the unit vector in the y direction (not to be confused with current density).

As a rule we will find the techniques of this section to be of marginal utility because it will normally be no harder to calculate the field directly than it will be to go through the vector potential. Furthermore, there is no situation analogous to a set of conductors with fixed potentials to determine our boundary conditions. Nevertheless, when all else fails, this method may provide a path to the solution we seek.

4-3 THE BIOT-SAVART LAW

Making use of our expression for $\mathbf{A}(\mathbf{r})$, Eq. (3-4-14), we can proceed directly to find a general expression for $\mathbf{B}(\mathbf{r})$.

$$
\begin{aligned}
\mathbf{B}(\mathbf{r}) &= \nabla \times \mathbf{A}(\mathbf{r}) \\[3mm]
&= \nabla \times \int_{\substack{\text{all} \\ \text{space}}} \frac{\mathbf{j}(\mathbf{r}')\, dV'}{|\mathbf{r} - \mathbf{r}'|} \\[3mm]
&= \int_{\substack{\text{all} \\ \text{space}}} \nabla \times \frac{\mathbf{j}(\mathbf{r}')}{|\mathbf{r} - \mathbf{r}'|}\, dV' \\[3mm]
&= \int_{\substack{\text{all} \\ \text{space}}} \frac{\mathbf{j}(\mathbf{r}') \times (\mathbf{r} - \mathbf{r}')}{|\mathbf{r} - \mathbf{r}'|^3}\, dV
\end{aligned}
\tag{4-3-1}
$$

This result is called the **Biot-Savart law.** It is a simple prescription for finding the magnetic field **B** at any point **r** by integrating over the current distribution.

We often deal with situations where the actual current distribution within a wire itself is not important and can be averaged over. This situation occurs when the thickness of the wire is small compared with our distance from it. In these cases we can integrate over the cross-sectional area of the wire and replace the volume integral in Eq. (4-3-1) by a line integral along the wire. We have then

$$\mathbf{B}(\mathbf{r}) = \oint_{\substack{\text{around} \\ \text{wire}}} \frac{I\,d\mathbf{l}' \times (\mathbf{r} - \mathbf{r}')}{\left|\mathbf{r} - \mathbf{r}'\right|^3} \tag{4-3-2}$$

It is very tempting at this point to say that the little bit of wire $d\mathbf{l}'$ gives rise to a magnetic field $d\mathbf{B}(\mathbf{r})$ given by

$$d\mathbf{B}(\mathbf{r}) = \frac{I\,d\mathbf{l}' \times (\mathbf{r} - \mathbf{r}')}{\left|\mathbf{r} - \mathbf{r}'\right|^3} \tag{4-3-3}$$

If we then used Eq. (3-5-17) to find the force exerted by $d\mathbf{B}$ on a charge q at position **r** and moving with velocity **V**, we would find

$$d\mathbf{F} = \frac{Iq\mathbf{V}}{c} \times \frac{d\mathbf{l}' \times (\mathbf{r} - \mathbf{r}')}{\left|\mathbf{r} - \mathbf{r}'\right|^3} \tag{4-3-4}$$

The problem, as can be plainly seen, is that this force *violates Newton's third law.* It is not directed along a line joining the current element $d\mathbf{l}'$ with the charge q.

It is not possible for us to completely resolve this dilemma at this point in our course. We can however point out one serious problem. We are not entitled to write down Eq. (4-3-3) at all, since doing so presupposes that we can calculate the effect of a segment of wire as though the remainder of the wire did not exist. But if the remainder of the wire did not exist, then the current flowing in the segment would cause charge to pile up at its two ends and we would no longer be dealing with a magnetostatic problem. Now we can in principle break a current loop up in this way, and the charge pileups at the ends of the segments would just cancel when they were put back together to reform the loop. Hence, if we knew how to deal with the physically realistic case of a current segment with charging ends, we would be out of the woods. This knowledge will come in our next chapter.

In any case, let us return to Eq. (4-3-2), which is perfectly correct as long as our integration covers the entire current distribution. We will apply it to find the magnetic field along the axis of a circular current loop of radius R carrying current I. Referring to Fig. 4-3, we place the origin of our coordinate system at the center of the loop. We set the x axis normal to the

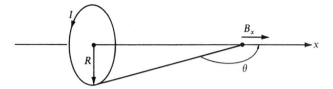

Fig. 4-3 We wish to find the magnetic field along the axis of a circular
current loop. By symmetry the field can have no y or z components.

loop so that when we look in the positive x direction the current appears
to be going clockwise. We see immediately that only the x component of
the field can exist after we have integrated over the loop, and it will be given
by

$$
\begin{aligned}
B_x &= \oint \frac{I\,dl}{R^2 + x^2} \sin\theta \\
&= \frac{2\pi I R^2}{(R^2 + x^2)^{\frac{3}{2}}} \\
&= \frac{2\pi I}{R} \sin^3\theta
\end{aligned}
\qquad (4\text{-}3\text{-}5)
$$

Making use of this result, we can find the magnetic field along the
axis of a solenoid having N turns per centimeter with each carrying a current
I. We choose a point along the axis at which we want to evaluate the field
and let θ_1 be the half-angle subtended by one end and θ_2 be the half-angle
subtended by the other end (see Fig. 4-4). The segment of the solenoid
lying between θ and $\theta + d\theta$ has a length

$$
dx = \frac{R}{\sin^2\theta}\,d\theta
$$

Hence the contribution to B is given by

$$
dB = \frac{2\pi NI\,dx}{R} \sin^3\theta = 2\pi NI \sin\theta\,d\theta
$$

Fig. 4-4 A long solenoid of radius R, having N turns per centimeter, each turn carrying
current I.

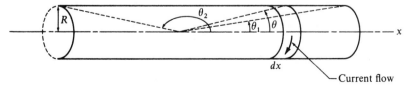

Integrating from θ_1 to θ_2, we have

$$B = 2\pi NI(\cos \theta_1 - \cos \theta_2) \tag{4-3-6}$$

We see then that the field at the center of a very long solenoid is approximately $4\pi NI$. The field at one end of the same solenoid is approximately $2\pi NI$.

Before we go on, we should get some feeling for magnitudes. The unit in which magnetic field is expressed is the gauss. Currents are measured in abamperes, and the only conversion factor we need to remember here is that 10 amperes = 1 abampere. Now, to take a reasonable example, we might choose N to be 10 turns per centimeter and I to be 1 abampere (10 amperes). In that case, the field in the middle of a long solenoid would be about 120 gauss. The largest fields that can be easily reached with normal electromagnets are in the range of 20,000 to 30,000 gauss. Superconducting and pulsed magnets can attain fields in the range of 100,000 to 200,000 gauss. Fields as high as 10^6 gauss have only rarely been achieved, in small volumes and with pulsed magnets. We shall shortly learn that the energy density of magnetic field is given by $(1/8\pi)B^2$, setting a rather clear limit on the field magnitudes that are attainable.

For comparison, we might have a look at electric fields which are also measured in gauss. (Remember that electric and magnetic fields are measured in the same units.) A useful conversion factor to remember is that 1 gauss = 300 volts/cm. The largest practical electric fields, in the neighborhood of 10^3 gauss, are thus considerably lower than the largest practical magnetic fields.

4-4 AMPERE'S LAW

When we are faced with a situation which has a high degree of symmetry, it is convenient to make use of Ampere's law for determining the magnitude of the magnetic field.

We begin with the magnetostatic equation

$$\nabla \times \mathbf{B} = 4\pi\mathbf{j} \tag{4-4-1}$$

Making use of Stokes' theorem, we can now evaluate the line integral of $\mathbf{B}$ around a closed curve:

$$\oint_C \mathbf{B} \cdot d\mathbf{l} = \int_{\substack{\text{surface} \\ \text{bounded} \\ \text{by } C}} \nabla \times \mathbf{B} \cdot \hat{\mathbf{n}} \, dA \tag{4-4-2}$$

The normal vector $\hat{\mathbf{n}}$ is chosen to point in the direction given by a right-hand rule as applied to C (see page 19). Inserting Eq. (4-4-1) into Eq. (4-4-2) yields

$$\oint_C \mathbf{B} \cdot d\mathbf{l} = 4\pi \int_{\substack{\text{surface} \\ \text{bounded} \\ \text{by } C}} \mathbf{j} \cdot \hat{\mathbf{n}} \, dA$$

$$= 4\pi(\text{current passing through surface bounded by } C) \quad (4\text{-}4\text{-}3)$$

We can now return to the problem we treated earlier, the uniform infinite, cylindrical current distribution (see Fig. 4-1). Once we decide that the field must be in the $\hat{\boldsymbol{\theta}}$ direction (by using the Biot-Savart law, for example), we can let C be a circle of radius r about the axis of the cylinder and apply Eq. (4-4-3) to determine the magnitude of $\mathbf{B}$. We have

$$\oint \mathbf{B} \cdot d\mathbf{l} = 2\pi r B$$

Hence, for $r \leq R$, we write

$$2\pi r B = (4\pi j)(\pi r^2)$$

and

$$B = 2\pi j r \quad (4\text{-}4\text{-}4)$$

For $r \geq R$, we have

$$2\pi r B = (4\pi j)(\pi R^2)$$

and

$$B = \frac{2\pi j R^2}{r} \quad (4\text{-}4\text{-}5)$$

These results are identical with what was found earlier [see Eq. (4-2-5)].

4-5 B AS THE GRADIENT OF A POTENTIAL FUNCTION

Under certain circumstances it is possible to express $\mathbf{B}$ as the gradient of a potential function φ_m. These circumstances require that the current distribution be in the form of closed loops and that the point at which the field is to be evaluated be in a current-free region. We shall derive an expression for φ_m explicitly for one simple current loop. In the event that there are many such loops to consider, the fields we obtain in this manner can be summed together.

Referring to Fig. 4-5a, we shall assume a current I in the loop C' and place an imaginary surface S' over the loop with normal vector $\hat{\mathbf{n}}'$ directed as shown. The vector $\mathbf{r}'$ refers then to a point on the loop and the vector $\mathbf{r}$ to the position at which we wish to evaluate $\mathbf{B}$. The vector potential

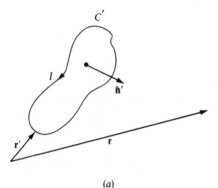

Fig. 4-5　(a) Finding the field at $\mathbf{r}$ due to a current loop.

(a)

at $\mathbf{r}$ is

$$\mathbf{A}(\mathbf{r}) = I \oint_{C'} \frac{d\mathbf{l}'}{|\mathbf{r} - \mathbf{r}'|} \tag{4-5-1}$$

At this point we make use of a simple mathematical identity. For any function f we can write

$$\oint_C f\, d\mathbf{l} = \int_{\substack{\text{surface} \\ \text{bounded} \\ \text{by } C}} (\hat{\mathbf{n}} \times \nabla f)\, dA \tag{4-5-2}$$

Taking $f = 1/|\mathbf{r} - \mathbf{r}'|$, we have

$$\mathbf{A}(\mathbf{r}) = I \int_{S'} \hat{\mathbf{n}}' \times \nabla' \frac{1}{|\mathbf{r} - \mathbf{r}'|}\, dA' \tag{4-5-3}$$

Taking the curl of $\mathbf{A}$ to find $\mathbf{B}$, we obtain

$$\mathbf{B}(\mathbf{r}) = I \nabla \times \int_{S'} \hat{\mathbf{n}}' \times \nabla' \frac{1}{|\mathbf{r} - \mathbf{r}'|}\, dA'$$

$$= I \int_{S'} \nabla \times \left(\hat{\mathbf{n}}' \times \nabla' \frac{1}{|\mathbf{r} - \mathbf{r}'|} \right) dA'$$

But $\nabla'(1/|\mathbf{r} - \mathbf{r}'|) = -\nabla(1/|\mathbf{r} - \mathbf{r}'|)$ and hence

$$\mathbf{B}(\mathbf{r}) = -I \int_{S'} \nabla \times \left(\hat{\mathbf{n}}' \times \nabla \frac{1}{|\mathbf{r} - \mathbf{r}'|} \right) dA' \tag{4-5-4}$$

We next make use of two rather simple vector identities to develop Eq. (4-5-4).

$$\nabla \times (A \times B) = A(\nabla \cdot B) - B(\nabla \cdot A) + (B \cdot \nabla)A - (A \cdot \nabla)B \quad (4\text{-}5\text{-}5)$$

$$\nabla(A \cdot B) = (A \cdot \nabla)B + (B \cdot \nabla)A + A \times (\nabla \times B) + B \times (\nabla \times A)$$
$$(4\text{-}5\text{-}6)$$

In these identities, the expression $(A \cdot \nabla)B$ means

$$(A \cdot \nabla)B = \left(A_x \frac{\partial}{\partial x} + A_y \frac{\partial}{\partial y} + A_z \frac{\partial}{\partial z} \right) B$$

$$= A_x \frac{\partial B}{\partial x} + A_y \frac{\partial B}{\partial y} + A_z \frac{\partial B}{\partial z} \qquad (4\text{-}5\text{-}7)$$

We first apply Eq. (4-5-5) to the integrand of Eq. (4-5-4). Remembering that $\hat{n}'$ does not depend on r, we obtain

$$\nabla \times \left(\hat{n}' \times \nabla \frac{1}{|r - r'|} \right) = \hat{n}' \left(\nabla^2 \frac{1}{|r - r'|} \right) - (\hat{n}' \cdot \nabla) \nabla \frac{1}{|r - r'|}$$
$$(4\text{-}5\text{-}8)$$

Since $r' \neq r$ (we are interested in the field in a current-free region), we can set $\nabla^2 (1/|r - r'|) = 0$, leaving us with

$$\nabla \times \left(\hat{n}' \times \nabla \frac{1}{|r - r'|} \right) = -(\hat{n}' \cdot \nabla) \nabla \frac{1}{|r - r'|}$$

Applying Eq. (4-5-6), we obtain

$$-(\hat{n}' \cdot \nabla) \nabla \frac{1}{|r - r'|} = -\nabla \left(\hat{n}' \cdot \nabla \frac{1}{|r - r'|} \right) \qquad (4\text{-}5\text{-}9)$$

Inserting this back into Eq. (4-5-4), we finally obtain an expression for $B(r)$:

$$B(r) = I\nabla \int_{S'} \hat{n}' \cdot \nabla \frac{1}{|r - r'|} \, dA'$$

$$= -I\nabla \int_{S'} \frac{\hat{n}' \cdot (r - r')}{|r - r'|^3} \, dA' \qquad (4\text{-}5\text{-}10)$$

Now we let $d\Omega'$ be the solid angle subtended by dA' with respect to the position r. We will choose $d\Omega'$ as positive if $\hat{n}' \cdot (r - r')$ is positive and negative if $\hat{n}' \cdot (r - r')$ is negative. In that case we can see that

$$d\Omega' = \frac{\hat{n}' \cdot (r - r')}{|r - r'|^3} \, dA' \qquad (4\text{-}5\text{-}11)$$

Hence we have proven our point.

$$\mathbf{B(r)} = -\nabla\varphi_m \qquad (4\text{-}5\text{-}12)$$

where

$$\varphi_m = I\Omega' \qquad (4\text{-}5\text{-}13)$$

and Ω' is the solid angle subtended by the surface S' with respect to $\mathbf{r}$.

Let us make use of this result to recalculate the magnetic field along the axis of a circular current loop. Referring back to Fig. 4-3, we note that the solid angle is just

$$\Omega' = 2\pi \int_\theta^\pi \sin\theta \, d\theta$$

$$= 2\pi(1 + \cos\theta) \qquad (4\text{-}5\text{-}14)$$

The gradient of Ω' is in the negative x direction, and hence $\mathbf{B}$ is in the positive x direction. We have then

$$B_x = -2\pi I \frac{\partial}{\partial x} \cos\theta$$

$$= 2\pi I \frac{\partial}{\partial x} \frac{x}{\sqrt{x^2 + R^2}}$$

$$= \frac{2\pi I R^2}{(x^2 + R^2)^{\frac{3}{2}}} \qquad (4\text{-}5\text{-}15)$$

Needless to say, this is the same result as we obtained earlier [see Eq. (4-3-5)].

The general results we have just obtained are particularly useful if we wish to find the magnetic field at a long distance from a small flat current loop of area A and current I. We have then (see Fig. 4-5b)

$$\varphi_m \cong IA\hat{n} \cdot \frac{\mathbf{r}}{r^3} \qquad (4\text{-}5\text{-}16)$$

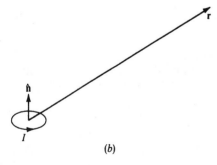

(b)

Fig. 4-5 (cont'd) (b) We find the magnetic field at large distances from a small flat current loop of area A carrying current I. The origin of our coordinate system is taken at the approximate center of the loop.

We define the magnetic moment of the loop as

$$\mathbf{m} = IA\hat{\mathbf{n}} \tag{4-5-17}$$

We then obtain the following simple approximation for φ_m:

$$\varphi_m = \mathbf{m} \cdot \frac{\mathbf{r}}{r^3} = -\mathbf{m} \cdot \nabla \frac{1}{r} \tag{4-5-18}$$

We can compare this with the expression we obtained for the electric potential of an electric dipole [see Eq. (2-6-7)]:

$$\varphi_e = -\mathbf{p} \cdot \nabla \frac{1}{r}$$

The similarity of these expressions leads us to conclude that the magnetic field distribution at large distances from a magnetic dipole is identical to the electric field distribution at large distances from an electric dipole. We need merely substitute $\mathbf{m}$ for $\mathbf{p}$ and $\mathbf{B}$ for $\mathbf{E}$ to obtain the corresponding results.

4-6 MAGNETIZATION (M) AND THE H FIELD

As we know, matter on the atomic level is made up of relatively stationary nuclei surrounded by electrons in various orbits. We shall discuss the electromagnetic properties of matter in much more detail in Chap. 10. In the meantime, though, we point out that there are three mechanisms whereby matter may acquire a macroscopic magnetic-moment distribution. (By macroscopic we mean averaged over a large number of atoms. This distribution is characterized by the magnetic moment per unit volume **M**, called the **magnetization.**) The three mechanisms are as follows:

1. The electrons' orbits or their intrinsic angular momenta may be arranged so as to give rise to a net magnetic moment within each atomic system. In general, thermal agitation will tend to randomize the directions of these moments, and only when an external field is applied so as to counteract this randomization do we have any macroscopic magnetization. Materials which behave in this way are called **paramagnetic.**
2. We shall soon learn that a changing magnetic field within a conductor induces currents which oppose this change. This phenomenon is called **diamagnetism.** The application of a magnetic field to a diamagnetic medium will induce currents within the atomic systems, and these in turn will lead to a macroscopic magnetic-moment density opposite in direction to the applied field.
3. Iron has two rather remarkable atomic properties that have far-reaching consequences with respect to its macroscopic magnetic behavior. First,

4 of the 26 electrons in an isolated atom of iron have their intrinsic angular momenta lined up. Second, within solid iron, there are very strong quantum-mechanical forces tending to make the intrinsic angular momenta of neighboring atoms line up. This results in domains of macroscopic size having net magnetizations corresponding to about two aligned electron moments per atom on the average. Applying a magnetic field causes those domains which are aligned in the same direction as the field to grow until the iron finally reaches a saturated state of magnetization. Typically, saturation occurs at fields in the neighborhood of 10,000 to 20,000 gauss. Needless to say, removal of the applied field does not lead to complete randomization of the domains. The residual magnetization can be quite large, as in the case of special permanent magnetic alloys, or it can be quite small, as in the case of soft iron. These phenomena are characteristic of ferromagnetism. Although iron is the best known of the ferromagnetic materials, it is not the only one; cobalt and nickel also exhibit the same properties.

In any case, all three mechanisms we have just described lead to a magnetization or magnetic moment per unit volume within materials. We now examine in detail how a magnetization distribution $\mathbf{M}(\mathbf{r}')$ leads to a magnetic field distribution $\mathbf{B}(\mathbf{r})$.

To begin with we must first find an expression for the vector potential at position $\mathbf{r}$ resulting from a dipole $\mathbf{m}$ at position $\mathbf{r}'$. To do so, we replace $\mathbf{m}$ by a tiny flat current loop and make use of Eq. (4-5-3). We change $\nabla'(1/|\mathbf{r} - \mathbf{r}'|)$ to $-\nabla(1/|\mathbf{r} - \mathbf{r}'|)$ and remove the integral sign. Remembering that $(I\,dA)\hat{\mathbf{n}} = \mathbf{m}$, we obtain

$$\mathbf{A}(\mathbf{r}) = -\mathbf{m} \times \nabla \frac{1}{|\mathbf{r} - \mathbf{r}'|} \tag{4-6-1}$$

Next we consider the vector potential $\mathbf{A}(\mathbf{r})$ arising from the magnetization $\mathbf{M}(\mathbf{r}')$. The little volume dV' has a magnetic moment $\mathbf{M}(\mathbf{r}')\,dV'$. Hence we write

$$\mathbf{A}(\mathbf{r}) = -\int_{\substack{\text{all}\\\text{space}}} \mathbf{M}(\mathbf{r}') \times \nabla \frac{1}{|\mathbf{r} - \mathbf{r}'|}\,dV'$$

$$= \int_{\substack{\text{all}\\\text{space}}} \mathbf{M}(\mathbf{r}') \times \nabla' \frac{1}{|\mathbf{r} - \mathbf{r}'|}\,dV'$$

$$= -\int_{\substack{\text{all}\\\text{space}}} \nabla' \times \frac{\mathbf{M}(\mathbf{r}')}{|\mathbf{r} - \mathbf{r}'|}\,dV' + \int_{\substack{\text{all}\\\text{space}}} \frac{\nabla' \times \mathbf{M}(\mathbf{r}')}{|\mathbf{r} - \mathbf{r}'|}\,dV' \tag{4-6-2}$$

We next make use of a simple theorem, the proof of which is left to the reader, to convert the first of these integrals to more convenient form. In general, for any vector function of space $\mathbf{W}(\mathbf{r})$, we can write

$$\int_{\text{vol } V} \nabla \times \mathbf{W}(\mathbf{r}) \, dV = \int_{\substack{\text{surface} \\ \text{of } V}} \hat{\mathbf{n}} \times \mathbf{W}(\mathbf{r}) \, dA \tag{4-6-3}$$

Applying this theorem and observing that $\mathbf{M} = 0$ at infinity, we note that

$$\int_{\substack{\text{all} \\ \text{space}}} \nabla' \times \frac{\mathbf{M}(\mathbf{r}')}{|\mathbf{r} - \mathbf{r}'|} \, dV' = \int_{\substack{\text{surface} \\ \text{at } \infty}} \hat{\mathbf{n}}' \times \frac{\mathbf{M}(\mathbf{r}')}{|\mathbf{r} - \mathbf{r}'|} \, dA' = 0 \tag{4-6-4}$$

Thus we have

$$\mathbf{A}(\mathbf{r}) = \int_{\substack{\text{all} \\ \text{space}}} \frac{\nabla' \times \mathbf{M}(\mathbf{r}')}{|\mathbf{r} - \mathbf{r}'|} \, dV' \tag{4-6-5}$$

Comparing this with our usual equation for $\mathbf{A}$ in terms of current density, we note that a magnetization distribution $\mathbf{M}$ can be replaced entirely by an equivalent current distribution $\mathbf{j}_M$ where

$$\mathbf{j}_M = \nabla \times \mathbf{M} \tag{4-6-6}$$

In general, all current distributions can be considered as consisting of two parts—that part which is unassociated with the magnetization and which we shall call $\mathbf{j}_F$ (free current) and that part which can be used to replace the magnetization distribution ($\mathbf{j}_M$). We write our basic equation for $\mathbf{B}$ as follows:

$$\nabla \times \mathbf{B} = 4\pi\mathbf{j} = 4\pi\mathbf{j}_F + 4\pi(\nabla \times \mathbf{M}) \tag{4-6-7}$$

Rewriting Eq. (4-6-7), we obtain

$$\nabla \times (\mathbf{B} - 4\pi\mathbf{M}) = 4\pi\mathbf{j}_F \tag{4-6-8}$$

We now define a new vector field $\mathbf{H}$ at each point in space as follows:

$$\mathbf{H} \equiv \mathbf{B} - 4\pi\mathbf{M} \tag{4-6-9}$$

We see then that $\mathbf{H}$ obeys the same differential equation with respect to the free currents as $\mathbf{B}$ obeys with respect to all the currents. That is,

$$\nabla \times \mathbf{H} = 4\pi\mathbf{j}_F \tag{4-6-10}$$

We can, of course, rewrite Ampere's law as

$$\oint_C \mathbf{H} \cdot d\mathbf{l} = 4\pi \int_{\substack{\text{surface} \\ \text{bounded} \\ \text{by } C}} \mathbf{j}_F \cdot \hat{\mathbf{n}} \, dA \tag{4-6-11}$$

Again, we must issue a caveat. The fact that $\mathbf{H}$ and $\mathbf{j}_F$ are related by the same differential equation as $\mathbf{B}$ and $\mathbf{j}$ does not mean that we calculate $\mathbf{H}$ from $\mathbf{j}_F$ in the same way as we calculated $\mathbf{B}$ from $\mathbf{j}$. There is, for example, no equivalent to the Biot-Savart law to make use of in calculating $\mathbf{H}$. Indeed, we can have $\mathbf{H}$ with no $\mathbf{j}_F$ at all—witness the case of a permanent bar magnet.

At this point we must interject a small bit of philosophy. It is customary to call $\mathbf{B}$ the magnetic induction and $\mathbf{H}$ the magnetic field strength. We reject this custom inasmuch as $\mathbf{B}$ is the truly fundamental field and $\mathbf{H}$ is a subsidiary artifact. We shall call $\mathbf{B}$ the magnetic field and leave the reader to deal with $\mathbf{H}$ as he pleases.

Returning to Eq. (4-6-9), we note that in many materials there is a proportionality between $\mathbf{H}$ and $\mathbf{M}$ expressed as $\mathbf{M} = \chi\mathbf{H}$ where χ is called the **magnetic susceptibility**. We can then write, letting $\mu = 1 + 4\pi\chi$,

$$\mathbf{B} = \mu\mathbf{H} \tag{4-6-12}$$

The constant μ is called the **magnetic permeability**.

As we have seen, we can calculate $\mathbf{B}$ everywhere by replacing $\mathbf{M}$ with its equivalent current distribution $\nabla \times \mathbf{M}$. However, we often find ourselves with a discontinuity in $\mathbf{M}$ at the interface between two materials, and it is useful to derive an expression for the equivalent surface current density at the discontinuity. We define what we mean by surface current density in a manner analogous to our definition of volume current density $\mathbf{j}$. If a set of charges q_i are moving on the surface within an area ΔA and with velocity $\mathbf{v}_i$, then the surface current density $\mathbf{k}$ is given by

$$\mathbf{k} = \lim_{\Delta A \to 0} \frac{\Sigma q_i \mathbf{v}_i}{\Delta A} \tag{4-6-13}$$

We now examine the interface between a region (I) with magnetization $\mathbf{M}_I$ and a region (II) with magnetization $\mathbf{M}_{II}$ (see Fig. 4-6). We place a gauss-

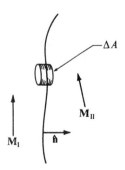

Fig. 4-6 We find the equivalent surface current at the boundary between two magnetized media.

ian surface of negligible thickness and area ΔA across the boundary and then evaluate $\mathbf{k}_M \, \Delta A$.

$$\mathbf{k}_M \, \Delta A = \int_{\text{vol}} \mathbf{j}_M \, dV \tag{4-6-14}$$

Remembering the theorem of Eq. (4-6-3), we can write

$$\mathbf{k}_M \, \Delta A = \int_{\text{vol}} \nabla \times \mathbf{M} \, dV$$

$$= \hat{\mathbf{n}} \times (\mathbf{M}_{\text{II}} - \mathbf{M}_{\text{I}}) \, \Delta A$$

Hence we conclude

$$\mathbf{k}_M = \hat{\mathbf{n}} \times (\mathbf{M}_{\text{II}} - \mathbf{M}_{\text{I}}) \tag{4-6-15}$$

where the normal $\hat{\mathbf{n}}$ goes from region I to region II.

We learned in our last section that it is possible to find a potential function φ_m such that $\mathbf{B} = -\nabla \varphi_m$, provided that we could divide the current distribution into loops and we stayed clear of the region within which the current was flowing. We will now see how a distribution of magnetization $\mathbf{M}$ leads to such a potential function. Moreover, we will see that if we use $\mathbf{H}$ rather than $\mathbf{B}$ then we will not be excluded from a region where the magnetization is not zero. We assume for the purpose of the present calculation that the free current is zero. The fields which the free currents produce can be added in separately.

The basic differential equations when $\mathbf{j}_F = 0$ are just

$$\nabla \times \mathbf{H} = 0$$
$$\nabla \cdot \mathbf{B} = 0 \tag{4-6-16}$$

The first of these equations permits us immediately to define a function φ_H at each point in space such that

$$\varphi_H(\infty) = 0 \quad \text{and} \quad \mathbf{H} = -\nabla \varphi_H$$

The function is obviously the line integral of $\mathbf{H}$ from the point in question to infinity.

$$\varphi_H(P) = \int_P^\infty \mathbf{H} \cdot d\mathbf{l} \tag{4-6-17}$$

The second equation can be rewritten as

$$\nabla \cdot \mathbf{B} = \nabla \cdot (\mathbf{H} + 4\pi\mathbf{M})$$

$$= -\nabla^2 \varphi_H + 4\pi\nabla \cdot \mathbf{M}$$

$$= 0$$

Hence we conclude that φ_H satisfies the Poisson equation

$$\nabla^2 \varphi_H = -4\pi\rho_m \tag{4-6-18}$$

where

$$\rho_m = -\nabla \cdot \mathbf{M} \tag{4-6-19}$$

We now recall the uniqueness theorems we learned in electrostatics. There is one and only one solution to Eq. (4-6-18) that goes to zero at infinity. Hence by analogy to electrostatics we can write

$$\varphi_H(\mathbf{r}) = \int_{\substack{\text{all} \\ \text{space}}} \frac{\rho_m(\mathbf{r}')}{|\mathbf{r} - \mathbf{r}'|} \, dV' \tag{4-6-20}$$

Thus we can replace a distribution of magnetization $\mathbf{M}$ by an equivalent "charge" distribution equal to $-\nabla \cdot \mathbf{M}$. We can then use this charge distribution to calculate $\mathbf{H}$ in precisely the same way as a normal charge distribution would be used to calculate $\mathbf{E}$.

In the event that we have a surface of discontinuity between two regions of different $\mathbf{M}$, we need to calculate the equivalent surface "charge" density σ_m. Referring back to Fig. 4-6, and making use of Gauss' theorem, we find

$$\sigma_m \, \Delta A = \int_{\text{vol}} \rho_m \, dV$$

$$= -\int_{\text{vol}} (\nabla \cdot \mathbf{M}) \, dV$$

$$= -\hat{\mathbf{n}} \cdot (\mathbf{M}_{\text{II}} - \mathbf{M}_{\text{I}}) \, \Delta A$$

Thus

$$\sigma_m = -\hat{\mathbf{n}} \cdot (\mathbf{M}_{\text{II}} - \mathbf{M}_{\text{I}}) \tag{4-6-21}$$

where $\hat{\mathbf{n}}$ points from region I to region II.

Before we apply these results, we will note the general boundary conditions for the behavior of $\mathbf{B}$ and $\mathbf{H}$ in crossing the interface between two regions of different magnetization. We assume no free current at the interface, and hence we have $\nabla \times \mathbf{H} = 0$ and $\nabla \cdot \mathbf{B} = 0$. By complete analogy with electrostatics (see page 50), we conclude that the normal component of $\mathbf{B}$ and the tangential component of $\mathbf{H}$ are continuous at the boundary.

We will illustrate these methods for finding the fields arising from a magnetization distribution by considering two rather simple examples.

As our first example we take a sphere of radius R and uniform magnetization **M**. We would like to find **B** and **H** for all points in space, both within and outside the sphere. For convenience we set our coordinate system up with its origin at the center of the sphere and its z axis pointing along **M**. We let the coordinates r' and θ' refer to a point within the sphere. The coordinates r and θ will refer then to the point at which we wish to determine the fields (see Fig. 4-7).

A cursory glance at our problem indicates that the simplest way to proceed would have us replace **M** by its equivalent charge distribution. We would then be able to make use of the techniques we learned in electrostatics for determining the potential in the event of cylindrical symmetry by means of a multipole expansion. The equivalent charge distribution is all on the surface and has magnitude given by

$$\sigma_m = \mathbf{M} \cdot \hat{\mathbf{n}} = M \cos \theta' \qquad (4\text{-}6\text{-}22)$$

Fig. 4-7 The magnetic field due to a uniform spherical distribution of magnetization **M**. Outside the sphere, **B** and **H** are equal and appear to come from a perfect dipole of magnitude $\frac{4}{3}\pi R^3 M$. Within the sphere $\mathbf{B} = 8\pi\mathbf{M}/3$ and $\mathbf{H} = -4\pi\mathbf{M}/3$ as shown.

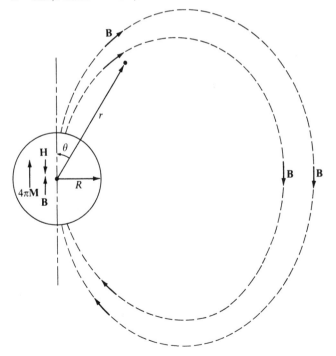

The "potential" φ_H at points outside the sphere is given by

$$\varphi_H(r \geq R) = \sum_{n=0}^{\infty} \frac{P_n(\cos\theta)}{r^{n+1}} \int_{\text{surface}} \sigma_m(\theta') R^n P_n(\cos\theta') \, dA' \qquad (4\text{-}6\text{-}23)$$

Remembering that $dA' = 2\pi R^2 \, d\cos\theta'$, letting $\mu' = \cos\theta'$, and making use of Eq. (4-6-22), we obtain

$$\varphi_H(r \geq R) = 2\pi M \sum_{n=0}^{\infty} \frac{P_n(\cos\theta) R^{n+2}}{r^{n+1}} \int_{-1}^{1} P_1(\mu') P_n(\mu') \, d\mu'$$

Applying the orthogonality condition for the Legendre polynomials, we obtain

$$\varphi_H(r \geq R) = \tfrac{4}{3}\pi R^3 M \frac{\cos\theta}{r^2} \qquad (4\text{-}6\text{-}24)$$

We conclude then that a sphere of uniform magnetization produces a pure dipole field outside itself. The fields outside the sphere are given by

$$\mathbf{H}(r \geq R) = \mathbf{B}(r \geq R) = -\nabla\varphi_H(r \geq R)$$

$$= -\left(\frac{\partial\varphi_H}{\partial r}\right)\hat{\mathbf{r}} - \frac{1}{r}\left(\frac{\partial\varphi_H}{\partial\theta}\right)\hat{\boldsymbol{\theta}}$$

$$= \frac{4\pi R^3 M}{3r^3}(2\cos\theta\,\hat{\mathbf{r}} + \sin\theta\,\hat{\boldsymbol{\theta}}) \qquad (4\text{-}6\text{-}25)$$

Within the sphere, making use of Eqs. (2-14-21) and (2-14-23), we find

$$\varphi_H(r \leq R) = \tfrac{4}{3}\pi M r \cos\theta = \tfrac{4}{3}\pi M z \qquad (4\text{-}6\text{-}26)$$

Hence the $\mathbf{H}$ field is a constant within the sphere:

$$\mathbf{H}(r \leq R) = -\nabla\varphi_H = -\tfrac{4}{3}\pi\mathbf{M} \qquad (4\text{-}6\text{-}27)$$

The $\mathbf{B}$ field inside is also constant and is given by

$$\mathbf{B}(r \leq R) = \mathbf{H}(r \leq R) + 4\pi\mathbf{M} = \frac{8\pi}{3}\mathbf{M} \qquad (4\text{-}6\text{-}28)$$

It is interesting to check the boundary conditions on $\mathbf{B}$ and $\mathbf{H}$. At the surface, the tangential $\mathbf{H}$ fields and the normal $\mathbf{B}$ field are continuous, viz.,

$$\mathbf{H}_T(\text{in}) = (\mathbf{H}\cdot\hat{\boldsymbol{\theta}})\hat{\boldsymbol{\theta}}$$

$$= (\tfrac{4}{3}\pi M \sin\theta)\hat{\boldsymbol{\theta}}$$

$$= \mathbf{H}_T(\text{out}) \qquad (4\text{-}6\text{-}29)$$

and

$$B_n(\text{in}) = \mathbf{B}(\text{in}) \cdot \hat{\mathbf{r}}$$

$$= \frac{8\pi}{3} M \cos \theta$$

$$= B_n(\text{out}) \tag{4-6-30}$$

We next consider a cylindrical bar magnet with uniform magnetization M throughout. We assume M to be parallel to the axis of the cylinder. As we have seen, the entire bar magnet can be replaced by a solenoid of current with surface current density given by $\mathbf{k}_m = \mathbf{M} \times \hat{\mathbf{n}}$ [see Eq. (4-6-15)]. The magnetic field along the axis of the bar magnet can thus be obtained directly from Eq. (4-3-6) if we replace NI by $|k_m|$. We have then

$$\mathbf{B}(\text{point on axis}) = 4\pi\mathbf{M}(\cos \theta_1 - \cos \theta_2) \tag{4-6-31}$$

where θ_1 and θ_2, as before, are the half-angles subtended by the two ends at the point of interest.

To obtain a qualitative feeling for the directions and magnitudes of B and H at other points both within or outside the magnet, we replace it with its equivalent charge distribution. Referring to Fig. 4-8, we see that the H field is opposite in direction from the B field within the magnet and is continuous everywhere except at the ends of the magnet. The B field is discontinuous on the circumference of the magnet but is continuous at the ends. The fact that B approaches $4\pi\mathbf{M}$ in the center as the magnet becomes longer is now apparent because of the weakening of the H field at that point. Needless to say, a detailed quantitative analysis of the fields, while not difficult, is a bit messy and will not be attempted here.

Fig. 4-8 A bar magnet with uniform magnetization gives rise to the H field shown in (a) and the B field shown in (b).

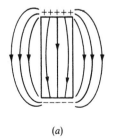 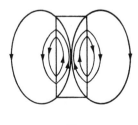

(a)   (b)

4-7 THE ENERGY OF A STATIC CURRENT DISTRIBUTION;
FORCE AND TORQUE ON A MAGNETIC DIPOLE

As we have seen, a static current distribution $\mathbf{j}$ leads directly to a vector potential, which in turn gives rise to a magnetic field $\mathbf{B}$. Naturally there is an energy that is associated with this current distribution or with the resultant fields, and, as we anticipate, this energy will be quite analogous to the electrostatic energy we found for a static charge distribution.

In evaluating the magnetostatic energy of a current distribution, we will ignore the electric field which is needed to keep the current flowing and whose energy we can already determine by means of Eq. (2-7-4) or Eq. (2-7-6). We consider then that a current density distribution $\mathbf{j}(\mathbf{r})$ leads to a vector-potential distribution $\mathbf{A}(\mathbf{r})$. Were we to multiply our current density at each point in space by the numerical factor α, then the vector potential at each point in space would be multiplied by the same factor α. Consider now that we begin with a given value of α and increase it by $d\alpha$. Assume also that we carry out this increase in a time dt. While we are changing from $\alpha\mathbf{A}$ to $(\alpha + d\alpha)\mathbf{A}$, we are creating an electric field everywhere in space equal to

$-\dfrac{1}{c}\left(\dfrac{d\alpha}{dt}\right)\mathbf{A}$ [see Eq. (3-5-4)]. The work being done on these charges by

the electric field per unit time is [see Eq. (4-1-5)]

$$\frac{dW}{dt} = \int_{\substack{\text{all}\\\text{space}}} (c\alpha\mathbf{j}) \cdot \left(-\frac{1}{c}\,\frac{d\alpha}{dt}\right)\mathbf{A}\,dV$$

$$= -\int_{\substack{\text{all}\\\text{space}}} \alpha\,\frac{d\alpha}{dt}\,(\mathbf{A}\cdot\mathbf{j})\,dV \tag{4-7-1}$$

Since these forces are acting to prevent us from building up the current, the internal energy is being increased as we do build it up. The total magnetostatic internal energy of our final current distribution is thus

$$U_B = \int_0^1 \alpha\,d\alpha \int_{\substack{\text{all}\\\text{space}}} \mathbf{j}\cdot\mathbf{A}\,dV$$

$$= \frac{1}{2}\int_{\substack{\text{all}\\\text{space}}} \mathbf{j}\cdot\mathbf{A}\,dV \tag{4-7-2}$$

The current density above includes everything, magnetization as well as free current. Accordingly, we can replace $\mathbf{j}$ by $(1/4\pi)(\nabla \times \mathbf{B})$. We obtain then

$$U_B = \frac{1}{8\pi} \int_{\substack{\text{all} \\ \text{space}}} (\nabla \times \mathbf{B}) \cdot \mathbf{A} \, dV$$

$$= \frac{1}{8\pi} \int_{\substack{\text{all} \\ \text{space}}} \nabla \cdot (\mathbf{B} \times \mathbf{A}) \, dV + \int_{\substack{\text{all} \\ \text{space}}} (\mathbf{B} \cdot \nabla \times \mathbf{A}) \, dV$$

$$= \frac{1}{8\pi} \int_{\substack{\text{surface} \\ \text{at } \infty}} (\mathbf{B} \times \mathbf{A}) \cdot \hat{\mathbf{n}} \, dA + \int_{\substack{\text{all} \\ \text{space}}} B^2 \, dV$$

Since **B** goes to zero at least as fast as $1/r^2$ and **A** goes to zero at least as fast as $1/r$, the first integral goes to zero at infinity. We are then left with

$$U_B = \frac{1}{8\pi} \int B^2 \, dV \qquad (4\text{-}7\text{-}3)$$

In the event that the magnetization **M** is proportional to **B**, we can calculate directly the work we need to do to set up a distribution of free current $\mathbf{j}_F$. The *total* vector potential **A** will now be proportional to $\mathbf{j}_F$, and the energy U_F would be given by

$$U_F = \frac{1}{2} \int_{\substack{\text{all} \\ \text{space}}} \mathbf{j}_F \cdot \mathbf{A} \, dV \qquad (4\text{-}7\text{-}4)$$

Applying Maxwell's equations then leads us to the result

$$U_F = \frac{1}{8\pi} \int_{\substack{\text{all} \\ \text{space}}} \mathbf{B} \cdot \mathbf{H} \, dV \qquad (4\text{-}7\text{-}5)$$

Returning to Eq. (4-7-2), we can express the total energy of a static current distribution in slightly different form. We remember that

$$\mathbf{A}(\mathbf{r}) = \int_{\substack{\text{all} \\ \text{space}}} \frac{\mathbf{j}(\mathbf{r}')}{|\mathbf{r} - \mathbf{r}'|} \, dV'$$

Hence

$$U_B = \frac{1}{2} \int\!\!\int_{\substack{\text{all} \\ \text{space}}} \frac{\mathbf{j}(\mathbf{r}') \cdot \mathbf{j}(\mathbf{r})}{|\mathbf{r} - \mathbf{r}'|} \, dV' \, dV \qquad (4\text{-}7\text{-}6)$$

It is interesting to consider the situation where our current distributions can be broken into N current loops. In that case Eq. (4-7-2) can be rewritten as

$$U_B = \frac{1}{2} \sum_{i=1}^{N} I_i \oint_{C_i} \mathbf{A} \cdot d\mathbf{l}_i \qquad (4\text{-}7\text{-}7)$$

where I_i is the current in the ith loop C_i. Using Stokes' theorem, we can change the line integrals into surface integrals.

$$U_B = \frac{1}{2} \sum_{i=1}^{N} I_i \int (\nabla \times \mathbf{A}) \cdot \hat{\mathbf{n}}_i \, dA_i$$

$$= \frac{1}{2} \sum_{i=1}^{N} I_i \int_{S_i} \mathbf{B} \cdot \hat{\mathbf{n}}_i \, dA_i \qquad (4\text{-}7\text{-}8)$$

We define the **magnetic flux** Φ_i through the ith loop by the equation

$$\Phi_i = \int_{S_i} \mathbf{B} \cdot \hat{\mathbf{n}}_i \, dA_i \qquad (4\text{-}7\text{-}9)$$

Because $\nabla \cdot \mathbf{B} = 0$, the choice of surface over which to carry out this integral is arbitrary, as long as the surface is bounded by C_i. The energy is thus given by

$$U_B = \frac{1}{2} \sum_{i=1}^{N} I_i \Phi_i \qquad (4\text{-}7\text{-}10)$$

The flux through the ith loop has a contribution to it arising from the current in each of the other loops as well as from the ith loop itself. The contribution to the flux through the ith loop by the current in the jth loop is clearly proportional to that current. Thus we can write

$$\Phi_i = \sum_{j=1}^{N} L_{ij} I_j \qquad (4\text{-}7\text{-}11)$$

The constants of proportionality L_{ij} are called the **coefficients of inductance**. In particular the coefficient L_{ii} is often called the **self-inductance** of the ith loop.

We notice a remarkable formal resemblance between the equations we are developing now and the ones we developed earlier for a set of charged conductors. We continue the analogy by proving that $L_{ij} = L_{ji}$. To find the coefficient L_{ij} we allow a current I to flow in the jth coil and evaluate the flux through the ith coil. We then divide the flux by I.

$$L_{ij} = \frac{1}{I} \int_{S_i} \mathbf{B} \cdot \hat{\mathbf{n}}_i \, dA_i$$

$$= \frac{1}{I} \oint_{C_i} \mathbf{A} \cdot d\mathbf{l}_i$$

But

$$A = \oint_{C_j} \frac{I \, d\mathbf{l}_j}{|\mathbf{r}_i - \mathbf{r}_j|}$$

Hence

$$L_{ij} = \oint_{C_i} \oint_{C_j} \frac{d\mathbf{l}_i \cdot d\mathbf{l}_j}{|\mathbf{r}_i - \mathbf{r}_j|} \tag{4-7-12}$$

Obviously $L_{ij} = L_{ji}$. Thus if a current I in the ith loop gives rise to a flux Φ through the jth loop, then the same current I in the jth loop will give rise to the flux Φ through the ith loop.

The energy of our system can now be written as

$$U_B = \frac{1}{2} \sum_{i=1}^{N} \sum_{j=1}^{N} L_{ij} I_i I_j \tag{4-7-13}$$

Back in our study of electrostatics we learned that a displacement of a set of conductors keeping the potential constant led to a change in energy which was equal and opposite to that obtained from the same displacement with the charges kept fixed. We now prove the "equivalent" theorem in magnetostatics. Let $(\delta U_B)_I$ be the change in energy of our system of current loops if we displace them, keeping the currents constant. Let $(\delta U_B)_\Phi$ be the change in energy if we displace the loops, keeping the flux constant. Then $(\delta U_B)_\Phi = -(\delta U_B)_I$.

The proof proceeds precisely as before.

$$(\delta U_B)_I = \frac{1}{2} \sum_{i,j} \delta L_{ij} I_i I_j$$

On the other hand,

$$(\delta U_B)_\Phi = \frac{1}{2} \sum_{i,j} \delta L_{ij} I_i I_j + \frac{1}{2} \sum_{i,j} L_{ij} \delta I_i I_j + \frac{1}{2} \sum_{i,j} L_{ij} I_i \delta I_j$$

$$= \frac{1}{2} \sum_{i,j} \delta L_{ij} I_i I_j + \sum_{i,j} L_{ij} I_i \delta I_j$$

But

$$\Phi_i = \sum_j L_{ij} I_j \quad \text{and} \quad \delta \Phi_i = \sum_j \delta L_{ij} I_j + \sum_j L_{ij} \delta I_j = 0$$

Hence

$$(\delta U_B)_\Phi = \frac{1}{2} \sum_{i,j} \delta L_{ij} I_i I_j - \sum_{i,j} \delta L_{ij} I_i I_j$$

$$= -\frac{1}{2} \sum_{i,j} \delta L_{ij} I_i I_j$$

$$= -(\delta U_B)_I \tag{4-7-14}$$

This result will be exceedingly useful in permitting us to evaluate the forces and torques on current loops (or magnetic dipoles) by the method of **virtual work**. In this method we imagine carrying out very small displacements of the system. For each displacement we can evaluate the amount by which the system's energy will be changed. If we take care to see that the only forces (or torques) which act during the displacement are those that we wish to evaluate, then we can determine them by equating the work that they do to the change in energy of the system.

Before we can apply this method to our current loops, we must first decide what, if anything, to keep constant in our displacement. To do this we will have to go a bit ahead of ourselves and make use of a result that we will derive in the next chapter, but which we are not altogether unfamiliar with. If a coil is displaced in such a way that the flux through it changes, then the integral of $\mathbf{E} \cdot d\mathbf{l}$ around the loop will not be zero within the reference frame of the loop while this change is taking place. (This, we remember, is how dynamos work.) Since there is a current in the loop, this electric field will do a certain amount of work on the system in addition to the work being done by the forces we are trying to evaluate. If, on the other hand, we carry out the displacement while keeping the flux through the loop constant, then the only forces doing work are the forces of interest and we need not concern ourselves with any other work on the system. Hence the quantity which is relevant in determining the forces and torques on our loops is $(\delta U_B)_\Phi$. (In the analogous electrostatic situation the relevant energy change for determining the forces on conductors was δU_Q, corresponding to a displacement in which the charges were kept fixed.)

We apply this technique to find the torque and the force on a magnetic dipole with dipole moment $\boldsymbol{\mu}$ in an external magnetic field. For convenience we will assume that the external field is provided by one current loop C_1. Needless to say, the result will be generalized to any applied field whatsoever. We approximate the dipole by a small current loop C_2 with area ΔA_2 and current I_2 such that $\boldsymbol{\mu} = (I_2 \, \Delta A_2)\hat{\mathbf{n}}_2$ (see Fig. 4-9).

Making use of Eq. (4-7-13), we have

$$U_B = \tfrac{1}{2}L_{11}I_1{}^2 + \tfrac{1}{2}L_{22}I_2{}^2 + L_{12}I_1I_2 \tag{4-7-15}$$

Now, in the displacement we are going to make (either translating or rotating the dipole), the self-inductance of C_1 or C_2 will not be changed. We want to evaluate $(\delta U_B)_\Phi$, but it is much easier to evaluate $(\delta U_B)_I$ and then take its negative. Hence we obtain

$$(\delta U_B)_I = (\delta L_{12})I_1I_2$$

$$= \delta\left(\frac{1}{I_1}\int_{S_2}\mathbf{B}_1 \cdot \hat{\mathbf{n}}_2 \, dA\right)I_1I_2 \tag{4-7-16}$$

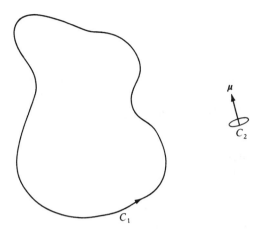

Fig. 4-9 The field produced by C_1 causes a torque and a force to act on the dipole represented by C_2.

where $\mathbf{B}_1$ is the field at C_2 due to C_1. Inasmuch as the loop C_2 is of infinitesimal size, we can remove the integral sign with the result that

$$(\delta U_B)_I = \delta(\mathbf{B}_1 \cdot I_2 \, \Delta A_2 \, \hat{\mathbf{n}}_2)$$

$$= I_2 \, \Delta A_2 \, \delta(\mathbf{B}_1 \cdot \hat{\mathbf{n}}_2)$$

$$= \mu \delta(\mathbf{B}_1 \cdot \hat{\mathbf{n}}_2) \tag{4-7-17}$$

Finally, we write down $(\delta U_B)_\Phi$.

$$(\delta U_B)_\Phi = -(\delta U_B)_I$$

$$= -\mu \delta(\mathbf{B}_1 \cdot \hat{\mathbf{n}}_2) \tag{4-7-18}$$

We are now ready to find the torque on the dipole. To do so we make a rotational displacement of the dipole in various directions (keeping the flux through it constant) and note the change in energy. The drop in energy for a given angular displacement $\delta\theta$ is greatest if we turn μ toward $\mathbf{B}_1$. Hence the torque acts to turn μ in that direction. The magnitude of the torque is given by

$$|\tau|\delta\theta = -\mu\delta(\mathbf{B}_1 \cdot \hat{\mathbf{n}}_2) = +\mu B_1 \sin\theta \, \delta\theta$$

We have then

$$\tau = \mu \times \mathbf{B}_1 \tag{4-7-19}$$

Next we can evaluate the force on the dipole. We move the dipole around, keeping the direction of $\hat{\mathbf{n}}_2$ fixed, and observe the change in energy. The force is in the direction in which the energy is decreasing most rapidly.

Thus

$$\mathbf{F} = \mu\nabla(\mathbf{B}_1 \cdot \hat{\mathbf{n}}_2)$$

$$= (\boldsymbol{\mu} \cdot \nabla)\mathbf{B}_1 \qquad (4\text{-}7\text{-}20)$$

Even though we have derived these expressions for an external field resulting from one external loop, they are completely general. We can add the external field due to any number of loops together; Eqs. (4-7-19) and (4-7-20) will still hold.

4-8 THE MOTION OF A CHARGED PARTICLE IN A CONSTANT MAGNETIC FIELD

Quite a while ago we discovered that a charged particle moving in a magnetic field experiences a force on it at right angles to its velocity [see Eq. (3-5-17)]. We will now take a closer look at this force and make use of it to calculate the orbit of the particle in a variety of interesting cases.

We begin with the very simplest, a constant field $\mathbf{B} = B_0\hat{\mathbf{k}}$, and assume that an observer sitting in the laboratory sees the charge as having mass m. (Needless to say, $m = m_0/\sqrt{1 - v_0^2/c^2}$ where v_0 is the velocity of the charge.) Inasmuch as the field does no work on the charge, its velocity and hence its mass remain constant.

Let $\mathbf{r} = x\hat{\mathbf{i}} + y\hat{\mathbf{j}} + z\hat{\mathbf{k}}$ be the position vector of the charged particle. Then, applying Eq. (3-5-17), we have

$$\frac{d^2\mathbf{r}}{dt^2} = \frac{qB_0}{mc}\frac{d\mathbf{r}}{dt} \times \hat{\mathbf{k}} = \frac{qB_0}{mc}\left(\frac{dy}{dt}\hat{\mathbf{i}} - \frac{dx}{dt}\hat{\mathbf{j}}\right) \qquad (4\text{-}8\text{-}1)$$

Separating the equation into component parts, we obtain

1. $\dfrac{d^2x}{dt^2} = \omega_0\dfrac{dy}{dt}$

2. $\dfrac{d^2y}{dt^2} = -\omega_0\dfrac{dx}{dt} \qquad (4\text{-}8\text{-}2)$

3. $\dfrac{d^2z}{dt^2} = 0$

where $\omega_0 = qB_0/mc$. The number ω_0 is often called the **cyclotron frequency**. The third equation yields

$$z = v_{0z}t + z_0 \qquad (4\text{-}8\text{-}3)$$

The second equation can be integrated once and then inserted into the first equation:

$$\frac{dy}{dt} = -\omega_0 x + \text{const} \qquad (4\text{-}8\text{-}4)$$

With no loss of generality, we can choose the origin of our coordinate system so as to set the constant equal to zero. Then

$$\frac{d^2x}{dt^2} = -\omega_0{}^2 x \tag{4-8-5}$$

and

$$x = R \cos (\omega_0 t + \delta) \tag{4-8-6}$$

Substituting back into Eq. (4-8-4) and integrating for y, we get

$$y = -R \sin (\omega_0 t + \delta) \tag{4-8-7}$$

The equations for x and y correspond to a circular path with a positive particle moving clockwise as seen from the positive z direction. The path in three dimensions is, of course, a helix. The radius of the helix can be determined from differentiating x with respect to time:

$$\frac{dx}{dt} = -R\omega_0 \cos (\omega_0 t + \delta)$$

Hence the maximum value of dx/dt is $R\omega_0$. If the particle speed is v_0, we have

$$R^2 \omega_0{}^2 + v_{0z}{}^2 = v_0{}^2$$

or

$$R = \frac{\sqrt{v_0{}^2 - v_{0z}{}^2}}{\omega_0} \tag{4-8-8}$$

If $v_{0z} = 0$, then

$$R = \frac{v_0}{\omega_0} = \frac{mcv_0}{qB_0} = \frac{p_0 c}{qB_0} \tag{4-8-9}$$

It is useful to put Eq. (4-8-9) into a numerical form corresponding to the situation where the charge q is that of an electron and its momentum is given in eV/c (see Sec. 3-2). We first observe that $p_0 c = \beta E$ where both p_0 and E are expressed in standard cgs units [see Eq. (3-2-34)]. This changes Eq. (4-8-9) into the form

$$R = \frac{\beta E_{(\text{ergs})}}{qB_0} \tag{4-8-10}$$

We next recall that the energy in electron volts of a particle having momentum expressed in eV/c and mass expressed in eV/c^2 is just

$$E_{(\text{eV})} = \sqrt{p_{(\text{eV}/c)}^2 + m_{0(\text{eV}/c^2)}^2} \tag{4-8-11}$$

Next we obtain its energy in ergs by multiplying with the conversion factor $e/299.8$ [where $299.8 = 10^{-8}$ (velocity of light in cm/sec)].

$$E_{(ergs)} = \frac{e}{299.8} \sqrt{p^2_{(eV/c)} + m^2_{0(eV/c^2)}} \qquad (4\text{-}8\text{-}12)$$

Finally, inserting Eq. (4-8-12) into Eq. (4-8-10) and remembering from Eq. (3-2-37) that $\beta\sqrt{p^2_{(eV/c)} + m^2_{0(eV/c^2)}} = p_{(eV/c)}$, we obtain

$$R = \frac{ep_{(eV/c)}}{299.8\,qB_0} \qquad (4\text{-}8\text{-}13)$$

If $q = e$, then we have the particularly simple result corresponding to an electron or proton (or any other particle with one quantum of charge).

$$R_e = \frac{p_{(eV/c)}}{299.8\,B_0} \qquad (4\text{-}8\text{-}14)$$

4-9 THE MOTION OF A CHARGED PARTICLE IN CROSSED ELECTRIC AND MAGNETIC FIELDS

The solutions to orbit problems involving both electric and magnetic fields are often expedited through the judicious use of the Lorentz transformation. We will illustrate this point by examining in some detail the behavior of a moving charged particle in the case where both the electric and magnetic fields are constant in space and time and at right angles to one another. We choose our coordinate system, as shown in Fig. 4-10, so as to have

$$\mathbf{B} = B_0\hat{\mathbf{k}} \qquad \mathbf{E} = E_0\hat{\mathbf{j}} \qquad (4\text{-}9\text{-}1)$$

Obviously, to solve the problem completely, we will have to specify both the initial position $\mathbf{r}_0$ and the initial velocity $\mathbf{v}_0$ of the particle. To simplify matters somewhat, we will let $\mathbf{v}_0$ lie along the x axis; generalizing to arbitrary $\mathbf{v}_0$ is straightforward and is left to the reader.

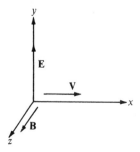

Fig. 4-10 To find the orbit of a particle in crossed electric and magnetic fields we transform to a system in which one of the fields is zero.

Now, it is not difficult to solve the problem in our given coordinate system. The differential equations must be treated carefully, inasmuch as m is no longer a constant of the motion as it is when no electric field is acting. We write

$$\frac{dm\mathbf{v}}{dt} = q\mathbf{E} + \frac{q\mathbf{v}}{c} \times \mathbf{B}$$

$$= q\left(E_0 - \frac{B_0}{c}\frac{dx}{dt}\right)\hat{\mathbf{j}} + \frac{qB_0}{c}\frac{dy}{dt}\hat{\mathbf{i}} \tag{4-9-2}$$

However, rather than solving these equations in a straightforward manner, we will use a bit of "trickery" which will serve to simplify the problem enormously.

Let us jump onto a coordinate system Σ' which is moving with velocity $\mathbf{V} = V\hat{\mathbf{i}}$ along the positive x axis (V is not in general the same as v_{0x} and is not to be confused with it). In Σ' we have

$$\mathbf{E}' = \gamma\left(E_0 - \frac{V}{c}B_0\right)\hat{\mathbf{j}}$$

$$\mathbf{B}' = \gamma\left(B_0 - \frac{V}{c}E_0\right)\hat{\mathbf{k}} \tag{4-9-3}$$

where

$$\gamma = \frac{1}{\sqrt{1 - V^2/c}}$$

Now it is clear that if $E_0 \neq B_0$ then there is some V for which either $\mathbf{E}'$ or $\mathbf{B}'$ is zero. To be specific, if $E_0 < B_0$ and $V = (E_0/B_0)c$, then $\mathbf{E}' = 0$. If $B_0 < E_0$ and $V = (B_0/E_0)c$, then $\mathbf{B}' = 0$. In either case the problem becomes much simpler in the new system with the indicated choice of V. We will solve the problem for the case where $E_0 < B_0$; the alternative case we leave to the reader.

In our new system we have only a magnetic field

$$\mathbf{B}' = \gamma\left(\frac{B_0^2 - E_0^2}{B_0}\right)\hat{\mathbf{k}} = B'\hat{\mathbf{k}} \tag{4-9-4}$$

The particle moves in a simple circle, at constant speed and hence with constant mass. The initial velocity is (making use of the velocity transformation equation)

$$v_{0x}'\hat{\mathbf{i}} = \frac{v_{0x} - V}{1 - v_{0x}V/c^2}\hat{\mathbf{i}} \tag{4-9-5}$$

The mass of the particle in this system is just

$$m' = \frac{m_0}{\sqrt{1 - (v'_{0x})^2/c^2}} \tag{4-9-6}$$

where m_0 is its rest mass. The frequency of rotation in this system is

$$\omega'_0 = \frac{qB'}{m'c} \tag{4-9-7}$$

The radius of the circle is, of course,

$$R' = \frac{v'_{0x}}{\omega'_0} \tag{4-9-8}$$

and the equation of the orbit is given by

$$x' = R' \sin \omega'_0 t'$$
$$y' = R' \cos \omega'_0 t' - R$$

(Remember that R' and ω'_0 are just *numbers* calculated according to the above equations. We assume that $x' = y' = 0$ at $t' = 0$.)

To go back to our original system we just substitute for x', y', t' according to the Lorentz transformation:

$$x' = \gamma(x - Vt)$$
$$y' = y$$
$$t' = \gamma\left(t - \frac{V}{c^2} x\right)$$

This will give us the precise, relativistically correct orbits in the Σ system.

In the circumstance that v_{0x} just happens to be equal to V, we have $v'_{0x} = 0$, and hence the particle is initially stationary in the Σ' frame. Since there is no electric field in the Σ' frame, the particle remains stationary. It thus continues to move with velocity $v_{0x}\hat{\mathbf{i}}$ in the Σ system, experiencing no net force. (Of course we expect this inasmuch as $q\mathbf{E} = -q\mathbf{v}/c \times \mathbf{B}$ at this point.)

4-10   LARMOR PRECESSION IN A MAGNETIC FIELD

We have already learned that a magnetic dipole with moment $\boldsymbol{\mu}$ in an external field $\mathbf{B}$ has a torque on it given by $\boldsymbol{\tau} = \boldsymbol{\mu} \times \mathbf{B}$ [see Eq. (4-7-19)]. Being basically the result of circulating charge, this magnetic moment is often accompanied by angular momentum $\mathbf{L}$ along the same axis as $\boldsymbol{\mu}$. In this case, as we shall see, the magnetic-moment vector will precess about the direction of applied field with what is known as the **Larmor frequency** ω_L.

We will illustrate this process by considering a classical electron in an orbit about a nucleus. Let us begin by turning off the applied field and determining the magnetic moment corresponding to the orbit, considering it as a current loop. If e is the charge of the electron (e is a negative number) then the current is just

$$I_{emu} = \frac{I_{esu}}{c} = \frac{e}{c} \times (\text{number of turns per second})$$

We let A be the area of the orbit. We then recall from Kepler's law that the rate at which area is swept out per unit time is a constant and is equal to $L/2m$ where L is the angular momentum of the electron and m is its mass. Hence.

$$I_{emu} = \frac{e}{c} \frac{1}{\text{period}}$$

$$= \frac{2}{c} \frac{dA/dt}{A}$$

$$= \frac{eL}{2mcA}$$

Remembering that the magnetic moment is equal to the current times the area, we conclude that

$$\mu = \frac{e\mathbf{L}}{2mc} \qquad\qquad (4\text{-}10\text{-}1)$$

We have then

$$\tau = \frac{d\mathbf{L}}{dt} = \frac{e}{2mc} \mathbf{L} \times \mathbf{B}$$

If we let the vector ω_L be defined by

$$\omega_L = -\frac{e\mathbf{B}}{2mc} \qquad\qquad (4\text{-}10\text{-}2)$$

we conclude that

$$\frac{d\mathbf{L}}{dt} = \omega_L \times \mathbf{L} \qquad\qquad (4\text{-}10\text{-}3)$$

This is the kinematical equation for the precession of the vector $\mathbf{L}$ about ω_L with angular frequency ω_L.

[In the case of a "spinning" elementary particle the relationship between **L** and μ is replaced by the more general one:

$$\mu = g \, \frac{e\mathbf{L}}{2mc} \tag{4-10-4}$$

The proportionality constant g is called the **gyromagnetic ratio**. For the electron, g is very close to 2.]

4-11 A METHOD OF MEASURING $g - 2$

The measurement of the gyromagnetic ratio of the electron or the muon has had considerable significance in formulating and testing the modern quantum theory of electrodynamics. Both these particles appear to have no important interactions with matter except through electromagnetism, and hence it is possible to calculate g quite accurately, making use of theories that have been developed in the last 25 years. In either case, g turns out to be equal to $2 + \delta$ where δ is a very small positive correction of the order of $\frac{1}{400}$. Needless to say, accurate measurements of $g - 2$ have been essential in determining the validity of the theory. So far there has been no significant discrepancy. As a means of illustrating the power of the relativistic techniques we have developed so far, we will describe in detail the first experiment[1] to determine the value of $g - 2$ for the muon.

We should begin by saying a few words about the muons themselves. Muons do not occur naturally in matter. They are the predominant component of the cosmic radiation at sea level, being the decay products of unstable pions and K mesons produced at high altitude, but they only live for about 2 μsec on the average in their own reference frame and then decay into electrons and neutrinos. Hence, to do experiments with them, one must begin by producing pions, preferably at an accelerator, and then allow these pions to decay. Now, in 1956 it was discovered[2] that pion decay does not conserve parity. (In simple terms, observing a pion decay in a mirror would not lead to the same set of physical laws as observing it directly.) In fact, negative muons are emitted with their angular momenta pointing along their directions of motion. Positive muons are emitted with their angular momenta pointing opposite to their directions of motion. Furthermore, if these positive muons were stopped in matter and allowed to decay, the direction of this angular momentum (or spin) at the moment of decay could be determined by the distribution in directions of the emitted decay electrons.

[1] G. Charpak et al., *Phys. Rev. Letters*, **6**:128 (1961); *Phys. Rev. Letters*, **1**:16 (1962).

[2] R. L. Garwin, L. M. Lederman, and M. Weinrich, *Phys. Rev.*, **105**:1415 (1957). J. L. Friedman and V. L. Telegdi, *Phys. Rev.*, **105**:1681 (1957).

This result leads quite naturally to a technique for determining $g - 2$. If the polarized muon is introduced into a magnetic field, the direction of its angular momentum will precess, as we have just shown. In addition, the muon itself will turn through its orbit. After spending a given amount of time in the field, the muon can be brought out and the angle between its spin and its direction of motion can be measured. As we shall prove, this angle is directly proportional to $g - 2$, and its measurement offers a very sensitive and direct test of the validity of our theory of quantum electrodynamics.

In Fig. 4-11 we illustrate the apparatus used in the $g - 2$ experiment. The muons were introduced into a large magnet, slowed down somewhat, and then trapped in orbits for awhile. After a number of turns they left the magnet, were stopped within a block of material, and their decay distribution was measured, yielding the directions of their final polarizations.

To follow what is going on, it would be most convenient to move along with a coordinate system attached to the muon. Inasmuch as the muon's own coordinate system is not an inertial system, we might imagine ourselves within a bit of difficulty here, but we can get out of it by a very simple

Fig. 4-11 A detailed drawing of the apparatus used for a measurement of $g - 2$ of the muon [G. Charpak et al., *Phys. Rev. Letters*, **6**: 128 (1961); *Phys. Rev. Letters*, **1**: 16 (1962)]. A muon enters the magnet and is passed through a beryllium block for energy degradation. It then passes through a large number of orbits until it emerges and is stopped in a target T. The numbers refer to scintillation counters that are used to detect either the muon or its decay electron.

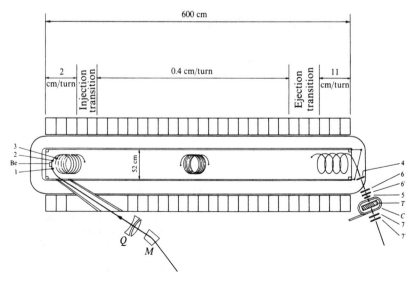

ruse. We will follow the muon's progress with a succession of inertial systems, each moving with the muon's speed, and tangential to its orbit at some point. We will then relate the observations within one coordinate system to those within the next system down the line until the particle finally leaves the field altogether.

We will make the simplifying assumption that the charged particle is moving with speed v at right angles to the field and that the motion all takes place within one plane. (Needless to say, this assumption is not precisely true in reality and a careful analysis of the orbits is necessary.) Subject to this assumption, an observer in each of our moving coordinate systems sees a magnetic field equal to $\gamma\mathbf{B}$ where $\mathbf{B}$ is the laboratory field at that point and $\gamma = 1/\sqrt{1 - v^2/c^2}$.

We begin by examining an infinitesimal portion of the particle's orbit as seen from both the laboratory system Σ and from the moving system Σ' which is tangential to the particle's orbit at this point (see Fig. 4-12). In either case the orbit appears to be a segment of arc, but within Σ' it appears much foreshortened and hence of smaller radius than in Σ. The sagitta of the arc is s as seen in Σ and s' as seen in Σ'. The chord of the arc is d as seen in Σ and d' as seen in Σ'. Obviously $s' = s$ and $d' = d/\gamma$. If $d\theta$ is the angle of arc as seen by Σ and $d\theta'$ is the angle of arc as seen by

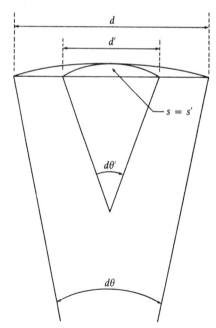

Fig. 4-12 A segment of the path of a particle in a magnetic field as seen by an observer in the laboratory (Σ) and by an observer with the velocity of the particle and tangential to its path (Σ').

Σ', then

$$\frac{d\theta'}{d\theta} = \frac{s'd}{sd'} = \gamma \tag{4-11-1}$$

Now what about the time to go from one end of the segment to the other? Since Σ' is the particle's rest frame at the moment, we can write

$$dt = \gamma\, dt' \tag{4-11-2}$$

Hence the rate of change of direction as seen by Σ' can be related to the rate of change of direction as seen by Σ.

$$\frac{d\theta'}{dt'} = \frac{\gamma^2\, d\theta}{dt} \tag{4-11-3}$$

But $d\theta/dt$ is just the cyclotron frequency of the particle, eB/mc, where B and m are as measured in the laboratory. This leads to

$$\frac{d\theta'}{dt'} = \frac{\gamma^2 eB}{\gamma m_0 c} = \frac{\gamma eB}{m_0 c} \tag{4-11-4}$$

Next we must find through what angle $d\varphi'$ the spin has precessed during the same time interval dt'. The magnetic field as seen by Σ' is γB, and hence the Larmor frequency is

$$\frac{d\varphi'}{dt'} = g\,\frac{e\gamma B}{2m_0 c} \tag{4-11-5}$$

Thus the angle between spin and orbit direction, as seen by Σ', has changed from one end of the arc to the other by an amount

$$d(\varphi' - \theta') = (g - 2)\,\frac{e\gamma B}{2m_0 c}\, dt'$$

$$= (g - 2)\,\frac{eB\, dt}{2m_0 c} \tag{4-11-6}$$

We now transform to the moving system which is tangential to the next segment of arc (Σ''). Since the relative velocity of Σ'' and Σ' is infinitesimal, the angle between spin and orbit at any given point on the orbit does not change with this transformation. Thus we have

$$(\varphi' - \theta')_{\substack{\text{end of} \\ \text{first segment}}} = (\varphi'' - \theta'')_{\substack{\text{beginning of} \\ \text{second segment}}} \tag{4-11-7}$$

Following the particle over the second arc, as seen from Σ'', we come up with the same Eq. (4-11-6) for the change in angle between orbit and spin direction. We continue this procedure until we leave the magnet. Adding together all the changes in angle between spin and direction of motion, we conclude that

ψ' = angle between direction of muon spin and direction of motion upon leaving the magnet, as seen by observer moving with muon

$$= (g - 2) \frac{e\overline{B}}{2m_0 c} t \tag{4-11-8}$$

where $\overline{B}$ is the average field experienced by the muon and t is the time spent in the magnet.

Finally, we must bring the muon to rest. The process of stopping the muon is equivalent to applying an electric field opposite to its direction of motion. An observer riding along with the muon sees the same electric field but no magnetic field at all. Consequently, he does not see any change in the muon's spin direction during the slowing-down process, and the final angle, in the laboratory, between the direction of the muon and the direction of its spin is given by Eq. (4-11-8).

The experiment we have just described was the first of a series carried out at the CERN laboratory in Geneva during the past 10 years. It is clear from an examination of Eq. (4-11-8) that a basic limitation on the accuracy of the measurement is the finite lifetime of the muon, about 2.2 μsec in its own reference frame. To make t, the laboratory time, as large as possible, one needs to increase the energy of the muon and thus make use of time dilation. The most recent experiment[1] uses muons having a γ of 12 and hence times of the order of 25 μsec. Typical values of ψ' are about 2000°. To avoid the difficulties of stopping such energetic muons, the experimenters actually looked at decays in flight. For a detailed description of the experiment, the reader should read the article cited in footnote 1. The experimental result has come quite close to that predicted by the theory.

$$\left(\frac{g - 2}{2}\right)_{\text{exp}} = 116,616 \pm 31 \times 10^{-8}$$

$$\left(\frac{g - 2}{2}\right)_{\text{theory}} = \frac{e^2}{2\pi\hbar c} + 0.7658 \frac{e^4}{\pi^2 \hbar^2 c^2} + \cdots$$

$$= 116,560 \times 10^{-8}$$

(In the above, e is the charge of the muon and $\hbar$ is Planck's constant divided by 2π. The combination of $e^2/\hbar c$, usually denoted by the symbol α, is dimensionless and goes by the name of **fine-structure constant**. Its value is about $\frac{1}{137}$.)

We should note that the value of $g - 2$ for the electron has also been determined by this method. A long series of experiments have been carried

[1] J. Bailey et al., *Phys. Rev. Letters*, **28B**: 287 (1968).

out by Prof. H. R. Crane and collaborators and are described in a recent issue of *Scientific American*.[1]

4-12 THE MAGNETIC STRESS TENSOR

We consider next the methods whereby we can determine the force exerted on a distribution of currents by magnetic fields. We will find, in complete analogy with electrostatics, that this force can be expressed either in terms of a volume integral requiring a knowledge of the current and the field at each point in the distribution or in terms of a surface integral in which only the field on the surface need be known. We will again find that the surface integral will have an integrand which is the product of a stress tensor **T** with the normal to the surface **n̂**.

We begin by going back to our expression for the force exerted on an elementary charge q by a magnetic field **B** [see Eq. (3-5-17)].

$$\mathbf{F} = \frac{q}{c}\, \mathbf{v} \times \mathbf{B} \tag{4-12-1}$$

For a distribution of current over a volume V, this becomes

$$\mathbf{F} = \int_V (\mathbf{j} \times \mathbf{B})\, dV \tag{4-12-2}$$

Remembering that $\mathbf{j} = (1/4\pi)\nabla \times \mathbf{B}$, we have

$$\mathbf{F} = \frac{1}{4\pi} \int_V (\nabla \times \mathbf{B}) \times \mathbf{B}\, dV \tag{4-12-3}$$

To convert this into a surface integral we take the components of **F** individually and transform them. For example,

$$
\begin{aligned}
F_x &= \frac{1}{4\pi} \int \hat{\mathbf{i}} \cdot (\nabla \times \mathbf{B}) \times \mathbf{B}\, dV \\[4pt]
&= \frac{1}{4\pi} \int [\hat{\mathbf{i}} \times (\nabla \times \mathbf{B})] \cdot \mathbf{B}\, dV \\[4pt]
&= \frac{1}{4\pi} \int_V \mathbf{B} \cdot \nabla B_x - \frac{1}{2} \frac{\partial B^2}{\partial x}\, dV \\[4pt]
&= \frac{1}{4\pi} \int_V \left[\frac{\partial B_x{}^2}{\partial x} + \frac{\partial B_x B_y}{\partial y} + \frac{\partial B_x B_z}{\partial z} - B_x(\nabla \cdot \mathbf{B}) - \frac{1}{2} \frac{\partial B^2}{\partial x} \right] dV
\end{aligned}
$$

$$\tag{4-12-4}$$

---

[1] H. R. Crane, *Scientific American*, January, 1968, p. 72.

We remember first that $\nabla \cdot \mathbf{B} = 0$. We next recall that the first three terms can be turned into a surface integral by following the same steps as we followed in proving Gauss' theorem. We finally obtain, for the component F_x,

$$F_x = \frac{1}{4\pi} \int_S (B_x^2 n_x + B_x B_y n_y + B_x B_z n_z) \, dA - \frac{1}{8\pi} \hat{\mathbf{i}} \cdot \int_V \nabla(B^2) \, dV$$

$$= \frac{1}{4\pi} \int_S \left[(B_x^2 n_x + B_x B_y n_y + B_x B_z n_z) - \frac{B^2 n_x}{2} \right] dA \qquad (4\text{-}12\text{-}5)$$

In summary then, for the three components of $\mathbf{F}$ we can write

$$\mathbf{F} = \int_S \mathbf{T} \hat{\mathbf{n}} \, dA \qquad (4\text{-}12\text{-}6)$$

where

$$\mathbf{T} = \frac{1}{4\pi} \begin{bmatrix} B_x^2 - \dfrac{B^2}{2} & B_x B_y & B_x B_z \\[2ex] B_x B_y & B_y^2 - \dfrac{B^2}{2} & B_y B_z \\[2ex] B_x B_z & B_y B_z & B_z^2 - \dfrac{B^2}{2} \end{bmatrix} \qquad (4\text{-}12\text{-}7)$$

A comparison with Eqs. (2-16-11) and (2-16-12) shows us a deep underlying similarity between electrostatics and magnetostatics. This is not altogether surprising though because, as we have learned, if we had begun with magnetic monopoles rather than charges, then the roles of $\mathbf{B}$ and $\mathbf{E}$ would have been reversed. When we use the stress tensor to find the force on a volume, we pay no heed to the charges and currents in the volume, only to the fields on its surface. Hence the electrostatic stress tensor must involve $\mathbf{E}$ in precisely the same way as the magnetostatic stress tensor involves $\mathbf{B}$.

We will illustrate the use of the magnetostatic stress tensor by means of an example. We have a long cylindrical thin-walled tube of radius R carrying current I as shown in Fig. 4-13. The force on the tube acts so as to tend to collapse it. To counteract this force we can pressurize the inside of the tube with pressure P. We ask then for the value of P needed to precisely balance out the magnetostatic force.

To solve this problem we pass a plane through the axis of the cylinder and call it the yz plane. We then calculate the magnetostatic force $\mathbf{F}$ exerted by the upper half ($x \geq 0$) of a unit length of the tube on the corresponding lower half ($x \leq 0$). We do this by integrating the stress tensor over the yz plane, remembering that $\hat{\mathbf{n}} = \hat{\mathbf{i}}$. We have then on the yz plane

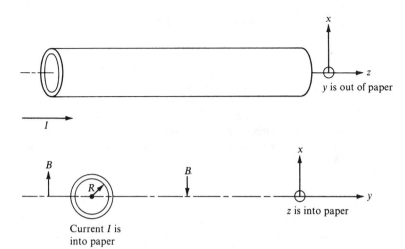

y is out of paper

z is into paper

Current I is
into paper

Fig. 4-13 We would like to find how much pressure to put on the inside of a tube of radius R to precisely counteract the magnetostatic force tending to collapse it.

$$
\mathbf{T} = \frac{1}{4\pi}
\begin{bmatrix}
\dfrac{B^2}{2} & 0 & 0 \\[2ex]
0 & \dfrac{-B^2}{2} & 0 \\[2ex]
0 & 0 & \dfrac{-B^2}{2}
\end{bmatrix}
\tag{4-12-8}
$$

$$
\mathbf{T}\hat{\mathbf{n}} = \frac{1}{8\pi} B^2 \hat{\mathbf{i}}
\tag{4-12-9}
$$

Integrating over a unit length, we have

$$
\mathbf{F} = \frac{1}{8\pi} \int_{-\infty}^{\infty} B^2 \hat{\mathbf{i}}\, dy
$$

Since $B = 0$ for $-R \leqq y \leqq R$, we find

$$
\mathbf{F} = \frac{1}{4\pi} \int_{R}^{\infty} B^2 \hat{\mathbf{i}}\, dy
$$

$$
= \frac{1}{4\pi} \int_{R}^{\infty} \left(\frac{2I}{y}\right)^2 \hat{\mathbf{i}}\, dy
$$

$$
= \frac{I^2}{\pi R} \hat{\mathbf{i}}
$$

To find P, we set the magnitude of $\mathbf{F}$ equal to $2PR$, yielding

$$P = \frac{I^2}{2\pi R^2} \qquad (4\text{-}12\text{-}10)$$

We shall come back to the stress tensor shortly when we deal with energy and momentum conservation in electromagnetism. At that time we will find that the electric and magnetic stress tensors coalesce to form a 3×3 portion of a four-dimensional electromagnetic energy-momentum tensor, the other components of which are the energy and momentum density contained in the fields.

PROBLEMS

4-1. A long straight conductor carries current I. It is in the form of a cylinder of radius R with an off-axis cylindrical hole of radius b, as shown. The distance between the axis of the cylinder and the axis of the hole is a. Find the magnetic field in the hole.

4-2. Making use of the stress tensor, prove that if the magnetic field is constant at all points on a surface surrounding a given object then there is no magnetic force acting on the object.

4-3. A circular toroid with rectangular cross section, as shown, is wound on a core

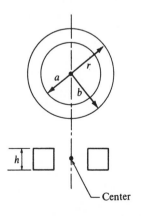

having permeability μ. N turns of wire are used, and a steady current I is put through the wire.

(a) Find the fields **H**, **B**, and **M** at radius r within the toroid.

(b) Find the vector potential **A** at the center of the toroid.

4-4. An electromagnetic crane is constructed of a U-shaped steel yoke with 1000 turns of wire carrying current I, as shown. The permeability of the steel is 1000. We would like to use it to lift a steel block of dimensions 30 cm × 30 cm × 120 cm and having the same permeability.

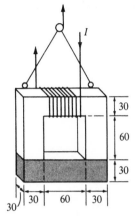

All dimensions in centimeters

Estimate the magnitude of I in order that we just be able to lift the block as shown (the density of steel can be taken as 8).

4-5. A long straight cylindrical solenoid of radius R is wound with N turns of wire per centimeter with each turn carrying current I. Find the pressure exerted on the inside of the solenoid at its midpoint.

4-6. It is possible to simulate a portion of the orbit of a charged particle in a magnetic field by means of a current-carrying wire under tension. Consider a short segment of such a wire under tension T in a magnetic field **B**. Demonstrate that the

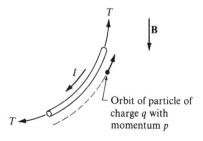

Orbit of particle of charge q with momentum p

path followed by the wire segment if it carries a current I corresponds to a portion of the orbit of a charge q having momentum p. (If q is positive, the direction of its motion along the orbit is opposite to that of the current.)

Derive an equation relating p and q to I and T.

4-7. An electromagnet is constructed with poles, as shown below. The field between the poles is in the direction indicated and is assumed to be constant at all points except at the boundaries where it drops to zero. A parallel beam of particles, each with mass m, velocity **v**, and charge q, enters the magnet at an angle of incidence φ and leaves the exit face with the same angle, having bent through 2φ. The fringing field of the magnet will cause the beam to be focused in the vertical direction. The magnet length is l.

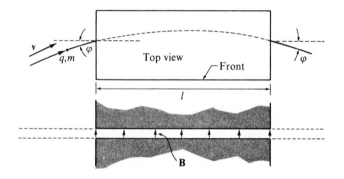

Find the approximate distance the beam travels, after leaving the magnet, before it comes to a vertical focus.

4-8. A block of conductor having height a and thickness b carries a current I, as shown. A magnetic field **B** is applied to the block, as shown. Assume that the current is carried by means of N electrons per cubic centimeter.

Find the potential difference that is developed between the top and the bottom of the block. How can the sign of this potential difference serve to determine the sign of the charge carriers? (The phenomenon described here is called the **Hall effect.**)

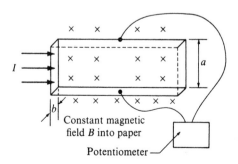

4-9. Two circular loops, each of radius R and carrying current I, are placed parallel to the yz plane and centered about the x axis at $x = -l$ and $x = +l$, respectively. Each produces a field in the $+x$ direction along its axis.

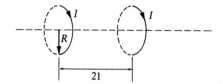

(a) Find an expression for the field along the x axis.

(b) Show that the first derivative of B_x with respect to x is zero at $x = 0$.

(c) For what ratio of l to R is the second derivative of B_x with respect to x also equal to zero at $x = 0$?

(d) Choose $R = 50$ cm and $I = 10$ abamperes. Fix l to conform with the conditions of part (c) above. Now plot B_x as a function of x from $x = -20$ cm to $x = +20$ cm on the x axis. (The circular loops considered above are called **Helmholtz coils.** They provide a cheap way of producing a fairly uniform but small magnetic field over a large volume.)

5
The Variation of the Electromagnetic Field with Time: Faraday's Law, Displacement Currents, the Retarded Potential

We enter now on what is undoubtedly the most exciting part of our voyage through the world of electromagnetism. While electrostatics and magneto-statics are interesting subjects, yielding results which are pretty and even surprising at times, they give no hint of the incredible beauty and richness of phenomenology that lies in store when we allow our currents and charge densities to change with time. We have already gotten a fleeting glimpse of the gold mine when we developed Maxwell's equations and observed that an electric field can result from a time-varying vector potential. In this chapter we will explore some of the simpler consequences of this observation, leaving for our next chapter the crowning prediction—*light*.

5-1 FARADAY'S LAW

We begin by reexamining the important discovery made by Faraday in the last century. He observed that it was possible to induce a current in a conducting loop by changing the flux through it. This effect could be accomplished in one of two ways, by moving the loop itself or by actually varying the magnetic field passing through the loop. In the first case, the cause of the current flow is the force exerted by the magnetic field on the moving charges in the wire. In the second case, the force is the result of a time-varying vector potential and its associated electric field. In either case, the integrated force per unit charge around the loop is proportional to the rate of change of magnetic flux Φ through the loop. This result is known as **Faraday's law.**

Before proceeding we will define what is meant by **electromotive force** $\mathscr{E}$. We must begin by choosing a direction around the loop C as shown in Fig. 5-1 and defining the normals to a surface covering the loop by means of the usual right-hand rule. As we said before, the loop is either stationary or moving. The force per unit charge on a charge which is fixed with respect to a given point on the loop is

$$\frac{\mathbf{F}}{q} = \mathbf{E} + \frac{\mathbf{v}}{c} \times \mathbf{B} \tag{5-1-1}$$

where $\mathbf{v}$ is the velocity of the given point on the loop, $\mathbf{B}$ is the local magnetic field, and $\mathbf{E}$ is the local electric field. If we integrate this around the loop, we obtain the electromotive force

$$\mathscr{E} = \oint_c \left(\mathbf{E} + \frac{\mathbf{v}}{c} \times \mathbf{B} \right) \cdot d\mathbf{l} \tag{5-1-2}$$

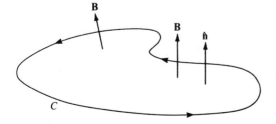

Fig. 5-1 We show that the emf around a curve C is proportional to the rate of change of flux through C.

Needless to say, the electromotive force will tend to make a current flow around the loop. In any case, the two portions of the integral above correspond to the two cases we discussed earlier. The first part, $\oint_c \mathbf{E} \cdot d\mathbf{l}$, is the result of the time-varying vector potential. The second part is patently the result of the loop's motion.

Consider then first the emf which results from varying the fields.

$$\mathscr{E}_{\substack{\text{loop} \\ \text{stationary}}} = \oint_c \mathbf{E} \cdot d\mathbf{l}$$

$$= -\frac{1}{c} \oint \frac{\partial \mathbf{A}}{\partial t} \cdot d\mathbf{l}$$

$$= -\frac{1}{c} \int_S \nabla \times \left(\frac{\partial \mathbf{A}}{\partial t} \right) \cdot \hat{\mathbf{n}} \, dA$$

where S is a surface covering the loop. Interchanging the order of differentiation and bringing the time derivative out of the integral, we obtain

$$\mathscr{E}_{\substack{\text{loop} \\ \text{stationary}}} = -\frac{1}{c} \frac{d}{dt} \int \nabla \times \mathbf{A} \cdot \hat{\mathbf{n}} \, dA$$

$$= -\frac{1}{c} \frac{d}{dt} \int \mathbf{B} \cdot \hat{\mathbf{n}} \, dA$$

$$= -\frac{1}{c} \left(\frac{d\Phi}{dt} \right)_{\substack{\text{due to changing} \\ \text{fields}}} \tag{5-1-3}$$

Next we consider the emf which results from a motion of the loop.

$$\mathscr{E}_{\substack{\text{due to} \\ \text{motion}}} = \oint_c \frac{\mathbf{v}}{c} \times \mathbf{B} \cdot d\mathbf{l}$$

But

$$\mathbf{v} \times \mathbf{B} \cdot d\mathbf{l} = -(\mathbf{v} \times d\mathbf{l}) \cdot \mathbf{B}$$

$$= -(\text{flux through area swept out by } d\mathbf{l} \text{ per unit time as a result of its motion})$$

Integrating around the loop, we again conclude

$$\mathscr{E}_{\substack{\text{due to} \\ \text{motion}}} = -\frac{1}{c} \left(\frac{d\Phi}{dt} \right)_{\substack{\text{due to motion} \\ \text{of loop}}}$$

We combine this result with Eq. (5-1-3) to conclude in general that

$$\mathscr{E}_{\text{total}} = -\frac{1}{c} \left(\frac{d\Phi}{dt} \right)_{\text{all sources}} \tag{5-1-4}$$

We can obtain another expression for $\mathscr{E}$ by recalling Eq. (4-7-11). We had shown that for a set of N loops

$$\Phi_i = \sum_{j=1}^{N} L_{ij} I_j$$

and hence

$$\mathscr{E}_i = \text{electromotive force around } i\text{th loop}$$

$$= -\frac{1}{c} \frac{d\Phi_i}{dt}$$

$$= -\frac{1}{c} \sum_{j=1}^{N} L_{ij} \frac{dI_j}{dt} \tag{5-1-5}$$

This result will be of great importance in the study of ac circuit theory.

We should offer one rather important comment at this point. Equation (5-1-3) relates the emf around a stationary loop to the rate of change of flux through the loop. Note though that there need be *no* magnetic field at all at the loop itself. Furthermore, the rate of change of flux is *simultaneous* with the emf. Both these points indicate to us that there is no direct causal relationship between changing magnetic field and electromotive force. Both are in fact caused by a time-varying vector potential which is the result of a time-varying current distribution. Needless to say a change in current density at some point must precede a related change in vector potential at some other point by enough time to allow for the transit of information at the velocity of light.

To develop some insight into Faraday's law we will consider several examples of its application. The first of these is a situation we are already quite familiar with. We have two long coaxial thin-walled conducting tubes with radii a and b, respectively (see Fig. 5-2). A current I flows down the inner tube and returns on the outer tube. We now allow the current to

Fig. 5-2 Two coaxial thin-walled cylinders carry a current I as shown. The inner cylinder has radius a and the outer cylinder has radius b. We change the current at a rate of dI/dt and ask for the electric field distribution as a function of radius.

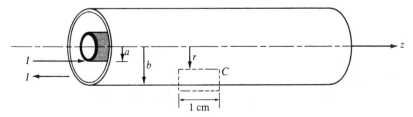

change at a rate of dI/dt and ask for the electric field distribution as a function of radius.

The simplest way of handling this problem is by finding the vector potential everywhere as a function of the current I. We consider the analogous electrostatic problem (see Sec. 4-2 for a discussion of this method of finding $\mathbf{A}$) where the inner conductor has a charge density per unit length of λ and the outer conductor has a charge density per unit length of $-\lambda$. The potential in this electrostatic problem is

$$\varphi = \begin{cases} -2\lambda \ln \dfrac{r}{b} & \text{for } a \leqq r \leqq b \\[2mm] 0 & \text{for } r \geqq b \\[2mm] -2\lambda \ln \dfrac{a}{b} & \text{for } r \leqq a \end{cases}$$

We can now write down the vector potential for the problem at hand by replacing λ by I and φ by A_z:

$$A_z(r) = \begin{cases} -2I \ln \dfrac{r}{b} & \text{for } a \leqq r \leqq b \\[2mm] -2I \ln \dfrac{a}{-b} & \text{for } r \leqq a \\[2mm] 0 & \text{for } r \geqq b \end{cases} \qquad (5\text{-}1\text{-}6)$$

Differentiating with respect to time, we find

$$E_z(r) = \begin{cases} \left(\dfrac{2}{c} \ln \dfrac{r}{b}\right) \dfrac{dI}{dt} & \text{for } a \leqq r \leqq b \\[2mm] \left(\dfrac{2}{c} \ln \dfrac{a}{b}\right) \dfrac{dI}{dt} & \text{for } r \leqq a \\[2mm] 0 & \text{for } r \geqq b \end{cases} \qquad (5\text{-}1\text{-}7)$$

Next let us find $E_z(r)$ by making use of Eq. (5-1-4) and evaluating Φ through the indicated curve C. We first use Ampere's law to determine $\mathbf{B}$ as a function of r.

$$2\pi r B = 4\pi I$$

$$B = \frac{2I}{r}$$

The flux through C is now determined by integrating B over the indicated area.

$$\Phi = \int_r^b \frac{2I}{r}\, dr = -2I \ln \frac{r}{b} \tag{5-1-8}$$

To determine $\mathscr{E}$ we note that the only contribution to $\mathscr{E}$ comes from the electric field $E_z(r)$ inside the outer conductor. Hence, for the indicated path C,

$$\mathscr{E} = E_z(r) \tag{5-1-9}$$

Combining Eqs. (5-1-8) and (5-1-9), we reproduce Eq. (5-1-7) for $E_z(r)$. Incidentally, the self-inductance per unit length of these coaxial cylinders can be obtained from Eqs. (5-1-9), (5-1-7), and (5-1-5). We take $r = a$ in these equations and find

$$\mathscr{E} = E_z(a) = \left(\frac{2}{c} \ln \frac{a}{b}\right) \frac{dI}{dt}$$

Also

$$\mathscr{E} = -\frac{1}{c} L \frac{dI}{dt}$$

Hence

$$L = 2 \ln \frac{b}{a} \tag{5-1-10}$$

As our second example let us consider a circular loop of conducting material placed in a plane perpendicular to an applied magnetic field B_0 (see Fig. 5-3a). The loop has radius b, self-inductance L, and a resistance around its circumference equal to R. We now reduce the applied field to zero with a linear time dependence:

$$B(t) = B_0 - kt \qquad \text{for } 0 \leq t \leq \frac{B_0}{k} \tag{5-1-11}$$

This change in magnetic field gives rise to a current in the loop. We would like to determine this current $I(t)$ as a function of time.

The total flux through the loop at any time is made up of two contributions. On the one hand, we have the applied flux, which is equal to $\pi b^2 B_0$ for $t \leq 0$, $\pi b^2 (B_0 - kt)$ for $0 \leq t \leq B_0/k$, and 0 for $t \geq B_0/k$.

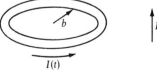

B_0

Fig. 5-3 (a) A loop of radius b having inductance L and resistance R is placed normal to a magnetic field as shown. The field is reduced to zero, leading to a current in the loop. [The indicated direction of $I(t)$ will be considered the positive direction.]

(a)

On the other hand, we have the flux which arises from the current in the loop itself and which is equal to $LI(t)$. The total flux is thus equal to

$$\Phi(t \leqq 0) = \pi b^2 B_0$$

$$\Phi\left(0 \leqq t \leqq \frac{B_0}{k}\right) = \pi b^2(B_0 - kt) + LI(t) \tag{5-1-12}$$

$$\Phi\left(t \geqq \frac{B_0}{k}\right) = LI(t)$$

Obviously, for $t \leqq 0$ there is no current flowing. During the time interval $0 \leqq t \leqq B_0/k$, Faraday's law tell us

$$\mathscr{E}\left(0 \leqq t \leqq \frac{B_0}{k}\right) = -\frac{1}{c}\frac{d\Phi}{dt}$$

$$= \frac{\pi b^2 k}{c} - \frac{L}{c}\frac{dI}{dt} \tag{5-1-13}$$

Now Ohm's law tells us (see Sec. 4-1) that

$$I(t) = \frac{\mathscr{E}}{R} \tag{5-1-14}$$

Substituting into Eq. (5-1-13), we find

$$\frac{L}{c}\frac{dI}{dt} + RI = \frac{\pi b^2 k}{c} \tag{5-1-15}$$

Solving for $I(t)$ and inserting the condition that $I(0) = 0$, we have

$$I(t) = \frac{\pi b^2 k}{Rc}(1 - e^{-Rct/L}) \qquad \text{for } 0 \leqq t \leqq \frac{B_0}{k} \tag{5-1-16}$$

For $t \geqq B_0/k$, $I(t)$ satisfies the differential equation

$$\frac{L}{c}\frac{dI}{dt} + RI = 0$$

Solving for $I(t)$ and requiring that $I(t)$ be continuous at $t = B_0/k$, we obtain

$$I(t) = \frac{\pi b^2 k}{Rc}(e^{RcB_0/Lk} - 1)e^{-Rct/L} \qquad \text{for } t \geqq \frac{B_0}{k} \tag{5-1-17}$$

In Fig. 5-3b we have plotted $I(t)$ as a function of time for three choices of k. As reasonable parameters we have taken $b = 10$ cm, $R = \frac{1}{300}$ statvolt/abampere, $L = 1000$ emu, and $B_0 = 1000$ gauss.

An interesting point which occurs to us as we examine the three curves in Fig. 5-3b is that the area appears to remain constant as we vary k.

To check this out we can integrate the expressions in Eqs. (5-1-16) and (5-1-17). We find that

$$\int_0^\infty I(t)\,dt = \begin{cases} \dfrac{\pi b^2 B_0}{Rc} & \text{emu} \\[3mm] \dfrac{\pi b^2 B_0}{R} & \text{esu} \end{cases} \qquad (5\text{-}1\text{-}18)$$

We can thus conclude that the total amount of charge passing any given point in the loop as we reduce the applied field to zero is dependent only on the initial flux through the loop and on its resistance. That this result is completely general for any loop whatsoever at any orientation with respect to the applied field is extremely simple to prove. We observe again that the flux through an arbitrary loop can be broken into the applied flux Φ_a and the flux due to the loop itself.

$$\Phi = \Phi_a + LI \qquad (5\text{-}1\text{-}19)$$

Fig. 5-3 (*cont'd*) (*b*) A plot of current versus time for various values of k ($R = \frac{1}{300}$ statvolt/ abampere, $b = 10$ cm, $L = 1000$ emu, $B = 1000$ gauss).

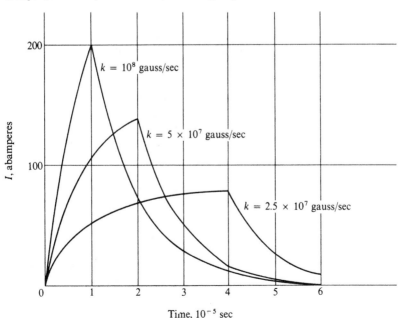

(*b*)

Hence in general

$$\mathscr{E} = IR = -\frac{1}{c}\frac{d\Phi}{dt}$$

$$= -\frac{1}{c}\frac{d\Phi_a}{dt} - \frac{L}{c}\frac{dI}{dt} \tag{5-1-20}$$

Integrating with respect to time, we have

$$\frac{L}{c}\int_0^\infty dI + R\int_0^\infty I\,dt = -\frac{1}{c}\int_0^\infty d\Phi_a \tag{5-1-21}$$

If we start with no current and end with no current, then $\int_0^\infty dI = 0$. Hence we can write

$$\int_0^\infty I\,dt = \frac{\Phi_a(\text{initial}) - \Phi_a(\text{final})}{Rc} \tag{5-1-22}$$

The observations we have just made serve as the basis behind the operation of the *flip-coil* method of magnetic field measurement. By using a ballistic galvanometer to integrate current and a coil whose resistance and physical dimensions are well known, we can map a magnetic field quite accurately.

Incidentally, we might look to see what happens if the resistance in our loop is equal to zero (this happens, of course, in the case of a superconducting loop). In that case, using Eq. (5-1-20), we find

$$\frac{d\Phi_a}{dt} + \frac{L}{c}\frac{dI}{dt} = 0 \tag{5-1-23}$$

Hence the total flux through the loop remains constant:

$$\Phi_a + LI = \text{const} \tag{5-1-24}$$

Before we complete this section, we should go back to the Alvarez magnetic monopole experiment that we described earlier (see Sec. 3-6) and evaluate its sensitivity. We had at that time developed an expression for the emf around a loop linking the path of the moon rock if the latter contained a monopole charge $Q^{(m)}$ [see Eq. (3-6-14)]. We can derive the same expression by noting that the total magnetic flux issuing from the monopole is just $4\pi Q^{(m)}$. Hence each time $Q^{(m)}$ passes through the loop, it changes the flux through it by an amount equal to $4\pi Q^{(m)}$. If the time for one pass through the loop is equal to T, then

$$\mathscr{E} = -\frac{1}{c}\frac{\Delta\Phi}{\Delta t} = -\frac{4\pi Q^{(m)}}{cT} \tag{5-1-25}$$

We now observe that if the loop contains n turns of superconducting wire with inductance L, and the sample passes through the loop N times, then the change in current ΔI can be obtained by integrating Eq. (5-1-23), leading to the result

$$\Delta I = \frac{4\pi NnQ^{(m)}}{L} \tag{5-1-26}$$

In the actual experiment the coil had 1200 turns of wire and a self-inductance of 78×10^6 emu. The detectors used were sensitive to a current change of 1.4×10^{-10} abampere. The magnitude of the basic Dirac monopole charge is about 68.5 times larger than the electron's charge in esu, and hence for $N = 400$ we would expect $\Delta I \cong 25 \times 10^{-10}$ abampere. The experiment was thus sufficiently sensitive to easily detect one Dirac monopole. Twenty-eight lunar samples were examined with a total mass of 8.4 kg. No monopoles were found.[1]

5-2 THE CONSERVATION OF ENERGY: THE POYNTING VECTOR

To derive the energy conservation laws from Maxwell's equations is remarkably simple. We take the scalar product of one equation with $\mathbf{E}$ and of another with $\mathbf{B}$, viz.,

$$\mathbf{E} \cdot (\nabla \times \mathbf{B}) = \frac{1}{c} \mathbf{E} \cdot \frac{\partial \mathbf{E}}{\partial t} + 4\pi \mathbf{j} \cdot \mathbf{E}$$

$$\mathbf{B} \cdot (\nabla \times \mathbf{E}) = -\frac{1}{c} \mathbf{B} \cdot \frac{\partial \mathbf{B}}{\partial t}$$

Subtracting the lower equation from the upper and using a simple vector identity, we have

$$-\nabla \cdot (\mathbf{E} \times \mathbf{B}) = \frac{1}{2c} \frac{\partial}{\partial t} (E^2 + B^2) + 4\pi \mathbf{j} \cdot \mathbf{E}$$

Multiplying by $c/4\pi$ and rewriting, we have

$$\nabla \cdot \left(\frac{c}{4\pi} \mathbf{E} \times \mathbf{B}\right) + \frac{\partial}{\partial t}\left(\frac{E^2 + B^2}{8\pi}\right) + c\mathbf{j} \cdot \mathbf{E} = 0 \tag{5-2-1}$$

We define the vector $\mathbf{S}$, called the **Poynting vector,** by

$$\mathbf{S} = \frac{c}{4\pi} \mathbf{E} \times \mathbf{B} \tag{5-2-2}$$

[1] For a more complete discussion see L. Alvarez, P. Eberhard, R. Ross, and R. Watt, *Science,* **167**: 701 (1970).

Integrating over any volume, we have

$$\int_{\substack{\text{surface} \\ \text{of vol}}} \mathbf{S} \cdot \hat{\mathbf{n}}\, dA + \frac{\partial}{\partial t} \int_{\text{vol}} \left(\frac{E^2 + B^2}{8\pi}\right) dV + \int_{\text{vol}} c\mathbf{j} \cdot \mathbf{E}\, dV = 0$$

$$(5\text{-}2\text{-}3)$$

Let us see if we can identify the various terms in this equation. The last term we have come across earlier in Eq. (4-1-5). We recall that $c\mathbf{j} \cdot \mathbf{E}$ is just the rate per unit volume per unit time at which mechanical work is being done on the charges by means of the electric field. Hence we have

$$\int_{\text{vol}} c\mathbf{j} \cdot \mathbf{E}\, dV = \text{rate of increase of mechanical energy of charges}$$

We also recall that $(E^2 + B^2)/8\pi$ is the energy density of the electric and magnetic fields. The second term in Eq. (5-2-3) is thus the rate at which field energy within the volume is increasing. Finally, the first term in Eq. (5-2-3) has the appearance of an outgoing flux and must be equal to the rate at which energy is leaving the volume per unit time. So everything just adds up right if we interpret $\mathbf{S}$ as being a vector which points along the direction in which energy is flowing and whose magnitude is equal to the flux of energy per unit time through a unit area normal to itself.

As we shall shortly see, Eq. (5-2-3) will be only *one* of the four energy-momentum conservation equations. We also anticipate that energy conservation and momentum conservation can be simply represented by just one single four-dimensional equation among proper four-dimensional quantities. Thus we shall succeed once again in unifying electromagnetic field equations by means of relativity. In the meantime though, we demonstrate a very simple application of the above by returning to our old standby, the long current-carrying wire (see Fig. 5-4). No fields are changing, and so

$$\frac{\partial}{\partial t} \int \frac{E^2 + B^2}{8\pi}\, dV = 0$$

Fig. 5-4 We apply the general energy conservation equation to a segment of current-carrying wire.

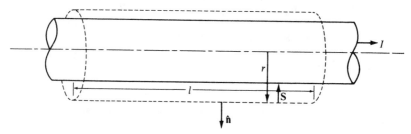

We draw an imaginary surface of radius r and length l about the wire, as shown. The wire segment enclosed is assumed to have a resistance R. To produce a current I, we need an electric field parallel to the wire of magnitude IR/l. The magnetic field at r circulates about the wire and has a magnitude equal to $2I/r$. Hence

$$|\mathbf{S}| = \frac{c}{4\pi} \frac{2I^2 R}{lr} \tag{5-2-4}$$

and

$$\int_{\text{surface}} \mathbf{S} \cdot \hat{\mathbf{n}}\, dA = -\frac{c}{4\pi} \frac{2I^2 R}{lr} 2\pi r l = -I^2 Rc$$

We see then that the amount of energy entering the volume per unit time is just equal to the ohmic dissipation within the wire.

5-3   MOMENTUM CONSERVATION IN ELECTROMAGNETISM

It would be quite straightforward to derive the momentum conservation relations by simple manipulations of Maxwell's equations. We choose, however, to use another approach, illustrating again the enormous power inherent in treating the electromagnetic field relativistically. We shall rewrite the energy conservation equation we have just obtained in terms of our four-dimensional representation. We will then change all subscript 4s to 1s, 2s, and 3s in turn, yielding the three momentum conservation equations. Simple inspection of these equations will show that they make sense and do indeed lead to the proper results when applied.

We begin by rewriting the energy conservation equation (5-2-1) in slightly different form.

$$\nabla \cdot \frac{-i\mathbf{S}}{c} + \frac{1}{ic} \frac{\partial}{\partial t}\left(\frac{E^2 + B^2}{8\pi}\right) = i\mathbf{j} \cdot \mathbf{E} = -\sum_{\mu=1}^{4} j_\mu F_{\mu 4} \tag{5-3-1}$$

The expression on the right is the fourth component of a four-vector. Hence we anticipate that $\left(\dfrac{-i\mathbf{S}}{c}, \dfrac{E^2 + B^2}{8\pi}\right)$ are four components of a second-rank tensor $T_{\mu\nu}$. We have

$$T_{14} = \frac{-iS_x}{c} \qquad T_{24} = \frac{-iS_y}{c} \qquad T_{34} = \frac{-iS_z}{c} \qquad T_{44} = \frac{E^2 + B^2}{8\pi}$$

$$\tag{5-3-2}$$

Let us see if we can write these components in terms of the field tensors **F** and **G**:

$$T_{14} = \frac{-iS_x}{c} = \frac{-i(\mathbf{E} \times \mathbf{B})_x}{4\pi}$$

$$= \frac{-i}{4\pi}(B_zE_y - B_yE_z) \tag{5-3-3}$$

Now there are two ways in which we can write T_{14}. Referring back to Eqs. (3-5-6) and (3-6-9), we find (since $F_{11} = F_{44} = 0$)

$$T_{14} = \frac{1}{4\pi}(F_{12}F_{24} + F_{13}F_{34})$$

$$= \frac{1}{4\pi}\sum_{\mu=1}^{4}F_{1\mu}F_{\mu4} \tag{5-3-4}$$

Alternatively we could write

$$T_{14} = \frac{1}{4\pi}(G_{12}G_{24} + G_{13}G_{34})$$

$$= \frac{1}{4\pi}\sum_{\mu=1}^{4}G_{1\mu}G_{\mu4} \tag{5-3-5}$$

Naturally, we expect that the prettiest combination of the above equations, namely, the one which is most symmetrical with respect to **E** and **B**, will match for all the components of Eq. (5-3-2):

$$T_{14} = \frac{1}{8\pi}\sum_{\mu=1}^{4}(F_{1\mu}F_{\mu4} + G_{1\mu}G_{\mu4}) \tag{5-3-6}$$

We can check out the other components against Eq. (5-3-2):

$$T_{24} = \frac{1}{8\pi}\sum_{\mu=1}^{4}(F_{2\mu}F_{\mu4} + G_{2\mu}G_{\mu4})$$

$$= \frac{-i}{8\pi}(2E_zB_x - 2E_xB_z)$$

$$= \frac{-i}{4\pi}(\mathbf{E} \times \mathbf{B})_y$$

$$T_{34} = \frac{1}{8\pi}\sum_{\mu=1}^{4}(F_{3\mu}F_{\mu4} + G_{3\mu}G_{\mu4})$$

$$= \frac{-i}{4\pi}(\mathbf{E} \times \mathbf{B})_z$$

$$T_{44} = \frac{1}{8\pi} \sum_{\mu=1}^{4} (F_{4\mu}F_{\mu 4} + G_{4\mu}G_{\mu 4})$$

$$= \frac{1}{8\pi} (E_x^2 + E_y^2 + E_z^2 + B_x^2 + B_y^2 + B_z^2)$$

$$= \frac{E^2 + B^2}{8\pi}$$

So it seems we have found the tensor we are looking for:

$$T_{\rho\sigma} = \frac{1}{8\pi} \sum_{\mu=1}^{4} (F_{\rho\mu}F_{\mu\sigma} + G_{\rho\mu}G_{\mu\sigma}) \tag{5-3-7}$$

Because both **F** and **G** are antisymmetric, **T** is symmetric:

$$T_{\rho\sigma} = T_{\sigma\rho} \tag{5-3-8}$$

The energy conservation equation can now be written as

$$\sum_{\rho=1}^{4} \frac{\partial T_{\rho 4}}{\partial x_\rho} = -\sum_{\rho=1}^{4} j_\rho F_{\rho 4} \tag{5-3-9}$$

If we replace 4 by 1, 2, and 3 we should obtain the momentum conservation equations

$$\sum_{\rho=1}^{4} \frac{\partial T_{\rho\sigma}}{\partial x_\rho} = -\sum_{\rho=1}^{4} j_\rho F_{\rho\sigma} \qquad \text{for } \sigma = 1, 2, 3 \tag{5-3-10}$$

We now rewrite this result in terms of **E**, **B**, and our charges and currents and see its physical significance. Let $\sigma = 1$, for example:

$$\left(\frac{\partial T_{11}}{\partial x} + \frac{\partial T_{21}}{\partial y} + \frac{\partial T_{31}}{\partial z}\right) - \frac{1}{c^2}\frac{\partial S_x}{\partial t} = (j_y B_z - j_z B_y) + \rho E_x$$

$$= (\mathbf{j} \times \mathbf{B})_x + \rho E_x \tag{5-3-11}$$

$$\int_V \left(\frac{\partial T_{11}}{\partial x} + \frac{\partial T_{21}}{\partial y} + \frac{\partial T_{31}}{\partial z}\right) dV = \int_V \frac{\partial}{\partial t}\left(\frac{S_x}{c^2}\right) dV$$

$$+ \int_V [(\mathbf{j} \times \mathbf{B})_x + \rho E_x] \, dV \tag{5-3-12}$$

Using Gauss' theorem, we can change the volume integral on the left into a surface integral. This yields

$$\int_S (T_{11}n_x + T_{21}n_y + T_{31}n_z) \, dA = \int_V \frac{\partial}{\partial t}\left(\frac{S_x}{c^2}\right) dV$$

$$+ \int_V [(\mathbf{j} \times \mathbf{B})_x + \rho E_x] \, dV \tag{5-3-13}$$

The integrand of the left side of Eq. (5-3-13) looks like the first component of the product of a three-dimensional tensor **T** and the normal vector n̂. In fact, this portion of the four-dimensional tensor **T** is nothing more than the sum of the electric and magnetic stress tensors we have studied earlier (see Secs. 2-16 and 4-12).

$$T_{11} = \frac{1}{4\pi}\left(E_x{}^2 + B_x{}^2 - \frac{E^2}{2} - \frac{B^2}{2}\right)$$

$$T_{22} = \frac{1}{4\pi}\left(E_y{}^2 + B_y{}^2 - \frac{E^2}{2} - \frac{B^2}{2}\right)$$

$$T_{33} = \frac{1}{4\pi}\left(E_z{}^2 + B_z{}^2 - \frac{E^2}{2} - \frac{B^2}{2}\right)$$

$$T_{12} = T_{21} = \frac{1}{4\pi}(E_x E_y + B_x B_y)$$

$$T_{13} = T_{31} = \frac{1}{4\pi}(E_x E_z + B_x B_z)$$

$$T_{23} = T_{32} = \frac{1}{4\pi}(E_y E_z + B_y B_z)$$

$$(5\text{-}3\text{-}14)$$

We generalize Eq. (5-3-13) then by writing down the complete momentum conservation equation

$$\int_S \mathbf{T}\hat{n}\, dA = \int_V \frac{\partial}{\partial t}\left(\frac{\mathbf{S}}{c^2}\right) dV + \int_V (\mathbf{j} \times \mathbf{B} + \rho\mathbf{E})\, dV \qquad (5\text{-}3\text{-}15)$$

The surface integral on the left can be thought of as the total momentum flowing into our volume through the surface per unit time. (Alternatively one might think of it as being the electromagnetic "force" exerted on our volume by the outside world.) The first integral on the right should be just equal to the rate of change of the field momentum within the volume. This would indicate that $\mathbf{S}/c^2$ is the momentum density of the electromagnetic field. Finally the second term on the right is nothing but the force on the charges and currents within the volume and is thus equal to the rate of change of mechanical momentum within the volume. We have thus derived the equation for momentum conservation in the presence of electromagnetic fields.

5-4 ELECTROMAGNETIC MASS

As a charged particle moves along through space with velocity **v**, the electromagnetic field it carries along has a momentum which depends upon its velocity. Inasmuch as we have learned that $\mathbf{S}/c^2$ is the momentum density of the field, we are now in a position to find the *total* field momentum for our moving particle. We shall see that it is proportional to both **v** and $1/\sqrt{1 - v^2/c^2}$, in precisely the same manner as the mechanical momentum. In the case of mechanical momentum, the constant of proportionality is called the **mechanical rest mass**. Similarly, we will call the constant of proportionality in evaluating the field momentum the **electromagnetic rest mass**. We shall then see that this is not exactly the same as the electromagnetic mass obtained by dividing the total field energy in the particle's rest system by c^2. This would be troublesome except for the obvious fact that we have left out another force entirely, the force which holds the charge together. In any case, there is no way of distinguishing mechanical mass from electromagnetic mass by applying forces to the system.

To simplify our considerations, we will make use of a specific example— a spherically symmetrical charge distribution of radius R moving with velocity v along the positive x axis (see Fig. 5-5). The direction of **S** is as shown in Fig. 5-5. Clearly, only the x component of $\mathbf{S}/c^2$ (the momentum density) is not averaged out as we carry out an integration over all space. So our job consists of evaluating S_x/c^2 everywhere and then integrating over all space to find p_x (electromagnetic).

The simplest way to proceed is to go over to the rest system of the particle (Σ'), calculate **T**′ in that system, transform **T**′ back to the laboratory system with velocity $-v$, examine $T_{14} = -iS_x/c$, and then integrate over space. We first write **T**′.

$$
\mathbf{T}' = \frac{1}{4\pi}
\begin{bmatrix}
E_x'^2 - \dfrac{E'^2}{2} & E_x'E_y' & E_x'E_y' & 0 \\[2ex]
E_x'E_y' & E_y'^2 - \dfrac{E'^2}{2} & E_y'E_z' & 0 \\[2ex]
E_x'E_z' & E_y'E_z' & E_z'^2 - \dfrac{E'^2}{2} & 0 \\[2ex]
0 & 0 & 0 & \dfrac{E'^2}{2}
\end{bmatrix}
\tag{5-4-1}
$$

where E' depends only on r'.

To transform to the laboratory, we multiply **T**′ on the left with **L** and on the right with $\tilde{\mathbf{L}}$ (the transpose of **L**) where

$$
\mathbf{L} =
\begin{bmatrix}
\gamma & 0 & 0 & -i\beta\gamma \\
0 & 1 & 0 & 0 \\
0 & 0 & 1 & 0 \\
i\beta\gamma & 0 & 0 & \gamma
\end{bmatrix}
$$

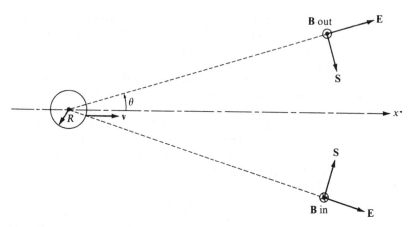

Fig. 5-5   We are interested in evaluating the electromagnetic momentum associated with a moving charge.

Carrying out the multiplication yields

$$T_{14} = \frac{i\beta\gamma^2}{4\pi}(E_x'^2 - E'^2) \tag{5-4-2}$$

and hence

$$S_x = \frac{\beta\gamma^2 c}{4\pi} E'^2 \sin^2 \theta' \qquad \text{where } \cos \theta' = \frac{E_x'}{E'} \tag{5-4-3}$$

To evaluate the total field momentum we integrate over all space.

$$p_x(\text{em}) = \frac{1}{c^2} \frac{\beta\gamma^2 c}{4\pi} \int_{\substack{\text{all} \\ \text{space}}} (E')^2 \sin^2 \theta' \, dV \tag{5-4-4}$$

But this is an integral much easier to evaluate in the Σ' system. We remember that because of the Lorentz contraction, $\gamma dV = dV'$. Hence

$$p_x(\text{em}) = \frac{\beta\gamma}{4\pi c} \int_{\substack{\text{all} \\ \text{space}}} (E')^2 \sin^2 \theta' \, dV'$$

$$= \frac{\beta\gamma}{4\pi c} 2\pi \int_0^\infty (E')^2 r'^2 \, dr' \int_0^\pi \sin^3 \theta' \, d\theta'$$

$$= \frac{2}{3} \frac{\beta\gamma}{c} \int_0^\infty (E')^2 r'^2 \, dr' \tag{5-4-5}$$

But the total electrostatic energy in the particle rest system is given by

$$U_0 = \frac{1}{2} \int_0^\infty (E')^2 r'^2 \, dr' \tag{5-4-6}$$

Hence we can write

$$p_x(\text{em}) = \frac{4}{3} \frac{\beta\gamma}{c} U_0 \tag{5-4-7}$$

We would have expected, in the absence of any other forces, to have obtained the relationship $p_x(\text{em}) = (\beta\gamma/c) U_0$.

Unfortunately, though, we have left out some factors which must play an important role in this problem. There has to be some additional force around to hold the charge together! Having omitted this force, which would presumably lead to a change in both the internal energy of the system and its momentum, we cannot expect to get a completely consistent theory. In any case, we have come up with the kind of velocity dependence we were looking for, and we are "almost" right in the relationship between rest energy and momentum.

Needless to say, there is no way in which one can experimentally separate the portions of the particle's self-energy which contribute to its mass. When we act on a charged particle with an external force of any sort, we act on the entire mass regardless of its origin.

5-5 THE DISPLACEMENT CURRENT

It is interesting to reexamine Ampere's law, now that we are permitting the charges and currents to be time dependent. We recall that Ampere's law was derived from one of Maxwell's equations, the one which related the curl of **B** to the current density. Reinserting the time-dependent terms which we dropped when we were studying magnetostatics, we have

$$\nabla \times \mathbf{B} = 4\pi\mathbf{j} + \frac{1}{c} \frac{\partial \mathbf{E}}{\partial t} \tag{5-5-1}$$

We observe that the effect of $\partial \mathbf{E}/\partial t$ is equivalent to that of an additional current density $\mathbf{j}_D$ given by

$$\mathbf{j}_D = \frac{1}{4\pi c} \frac{\partial \mathbf{E}}{\partial t} \tag{5-5-2}$$

This term is given the name of **displacement current density.**

Ampere's law can now be rewritten in the following form:

$$\oint_c \mathbf{B} \cdot d\mathbf{l} = 4\pi(I + I_D) \tag{5-5-3}$$

where

I_D = total displacement current through surface bounded by C

$$= \int_{\text{surface}} \mathbf{j}_D \cdot \hat{\mathbf{n}} \, dA \tag{5-5-4}$$

Needless to say, the total current given by $\mathbf{j} + \mathbf{j}_D$ must be divergence-less, or else the sum of $I + I_D$ would depend upon which surface we chose to cover our curve C. Let us check this out.

$$\nabla \cdot (\mathbf{j} + \mathbf{j}_D) = \nabla \cdot \mathbf{j} + \frac{1}{4\pi c} \nabla \cdot \frac{\partial \mathbf{E}}{\partial t}$$

$$= \nabla \cdot \mathbf{j} + \frac{1}{4\pi c} \frac{\partial}{\partial t} \nabla \cdot \mathbf{E}$$

$$= \nabla \cdot \mathbf{j} + \frac{1}{c} \frac{\partial \rho}{\partial t}$$

$$= 0 \text{ (conservation of charge)} \tag{5-5-5}$$

To observe the application of this new formulation of Ampere's law, we make use of the example of a parallel-plate capacitor which is charging at a constant rate. We allow the plates to have area A, as shown in Fig. 5-6.

We wish to evaluate the field along the indicated curve C. Now we have two choices for the surface over which we wish to do the current integration. We can choose a surface S_1 through the wire, in which case we will get

$$\oint_C \mathbf{B} \cdot d\mathbf{l} = 4\pi I \tag{5-5-6}$$

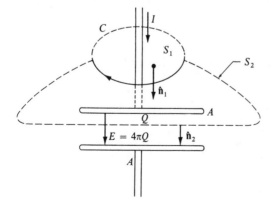

Fig. 5-6 We apply the modified Ampere's law to find the field around a wire charging a capacitor.

Alternatively, we can choose a surface S_2 through the capacitor. In this case we will get

$$\oint_c \mathbf{B} \cdot d\mathbf{l} = \frac{1}{c} \int \frac{\partial \mathbf{E}}{\partial t} \cdot \hat{\mathbf{n}} \, dA \tag{5-5-7}$$

But $\mathbf{E} \cdot \hat{\mathbf{n}} = 4\pi Q/A$ and hence

$$\frac{\partial \mathbf{E}}{\partial t} \cdot \hat{\mathbf{n}} = \frac{4\pi}{A} \frac{dQ}{dt} = \frac{4\pi}{A} cI \tag{5-5-8}$$

This leads us again to

$$\oint \mathbf{B} \cdot d\mathbf{l} = \frac{1}{c} \frac{4\pi}{A} cIA = 4\pi I \tag{5-5-9}$$

Thus, at least in this simple-minded case, the application of the modified Ampere's law leads to the anticipated result.

5-6 THE FOUR-VECTOR POTENTIAL AND HOW IT IS MODIFIED NOW THAT CURRENTS AND CHARGES ARE CHANGING WITH TIME

When we were deriving Maxwell's equations in Chap. 3, we found that the basic differential equations for $\mathbf{A}$ and φ had the form [see Eqs. (3-4-9) to (3-4-12)]

$$\left(\nabla^2 - \frac{1}{c^2} \frac{\partial^2}{\partial t^2} \right) \mathbf{A}(\mathbf{r},t) = -4\pi \mathbf{j}(\mathbf{r},t)$$

$$\left(\nabla^2 - \frac{1}{c^2} \frac{\partial^2}{\partial t^2} \right) \varphi(\mathbf{r},t) = -4\pi \rho(\mathbf{r},t) \tag{5-6-1}$$

We also wrote down the solutions in the event that there was no time dependence [see Eqs. (3-4-13) and (3-4-14)]:

$$\mathbf{A}(\mathbf{r}) = \int_{\substack{\text{all} \\ \text{space}}} \frac{\mathbf{j}(\mathbf{r}')}{|\mathbf{r} - \mathbf{r}'|} \, dV'$$

$$\varphi(\mathbf{r}) = \int_{\substack{\text{all} \\ \text{space}}} \frac{\rho(\mathbf{r}')}{|\mathbf{r} - \mathbf{r}'|} \, dV' \tag{5-6-2}$$

We will introduce time dependence by a rather intuitive guess and then go on to demonstrate that our guess does indeed satisfy Eq. (5-6-1).

Let us assume that a given current (or charge) at $\mathbf{r}'$ contributes to the potential at $\mathbf{r}$ in the same manner as when there was no time dependence but that the information travels from $\mathbf{r}'$ to $\mathbf{r}$ with a velocity c. Thus the bit of current at $\mathbf{r}'$ at time $t - |\mathbf{r} - \mathbf{r}'|/c$ gives rise to a bit of potential at position $\mathbf{r}$ and time t. Adding all contributions from all $\mathbf{r}'$ together, we have now

$$A(\mathbf{r},t) = \int_{\substack{\text{all}\\ \text{space}}} \frac{\mathbf{j}\left(\mathbf{r}', t - \dfrac{|\mathbf{r} - \mathbf{r}'|}{c}\right)}{|\mathbf{r} - \mathbf{r}'|} \, dV'$$

$$\varphi(\mathbf{r},t) = \int_{\substack{\text{all}\\ \text{space}}} \frac{\rho\left(\mathbf{r}', t - \dfrac{|\mathbf{r} - \mathbf{r}'|}{c}\right)}{|\mathbf{r} - \mathbf{r}'|} \, dV'$$

(5-6-3)

If these "solutions," which appear quite plausible on the surface, really do satisfy the appropriate differential equations, we will have achieved our goal. After this we can find all electric and magnetic fields as a function of space and time by means of the relations given in Eqs. (3-5-4) and (3-5-5).

$$\mathbf{E}(\mathbf{r},t) = -\nabla\varphi(\mathbf{r},t) - \frac{1}{c} \frac{\partial \mathbf{A}(\mathbf{r},t)}{\partial t}$$

$$\mathbf{B}(\mathbf{r},t) = \nabla \times \mathbf{A}(\mathbf{r},t)$$

Our first chore then is to demonstrate that $\varphi(\mathbf{r},t)$ and $\mathbf{A}(\mathbf{r},t)$ as we have written them are solutions of our differential equations. We tackle only $\varphi(\mathbf{r},t)$ explicitly. That $\mathbf{A}(\mathbf{r},t)$ satisfies our equation follows by complete analogy. Let us take $u = t - |\mathbf{r} - \mathbf{r}'|/c$. Now we will have need of the various derivatives of ρ:

$$\left.\begin{array}{l} \dfrac{\partial \rho}{\partial x} = \dfrac{\partial \rho}{\partial u} \dfrac{\partial u}{\partial x} \\[2mm] \dfrac{\partial \rho}{\partial y} = \dfrac{\partial \rho}{\partial u} \dfrac{\partial u}{\partial y} \\[2mm] \dfrac{\partial \rho}{\partial z} = \dfrac{\partial \rho}{\partial u} \dfrac{\partial u}{\partial z} \end{array}\right\} \quad \text{yielding } \nabla\rho = \frac{\partial \rho}{\partial u} \nabla u \qquad (5\text{-}6\text{-}4)$$

$$\nabla^2\rho = \nabla \frac{\partial \rho}{\partial u} \cdot \nabla u + \frac{\partial \rho}{\partial u} \nabla^2 u$$

$$= \frac{\partial^2 \rho}{\partial u^2} \nabla u \cdot \nabla u + \frac{\partial \rho}{\partial u} \nabla^2 u \qquad (5\text{-}6\text{-}5)$$

Also

$$\frac{\partial^2 \rho}{\partial t^2} = \frac{\partial^2 \rho}{\partial u^2} \qquad (5\text{-}6\text{-}6)$$

Now

$$\nabla u = -\frac{1}{c}\nabla|\mathbf{r}-\mathbf{r}'| = \frac{-(\mathbf{r}-\mathbf{r}')}{c|\mathbf{r}-\mathbf{r}'|} \qquad (5\text{-}6\text{-}7)$$

$$\nabla^2 u = -\frac{1}{c}\left[\nabla\frac{1}{|\mathbf{r}-\mathbf{r}'|}\cdot(\mathbf{r}-\mathbf{r}')+\frac{\nabla\cdot(\mathbf{r}-\mathbf{r}')}{|\mathbf{r}-\mathbf{r}'|}\right]$$

$$= -\frac{1}{c}\left(-\frac{1}{|\mathbf{r}-\mathbf{r}'|}+\frac{3}{|\mathbf{r}-\mathbf{r}'|}\right)$$

$$= \frac{-2}{c|\mathbf{r}-\mathbf{r}'|} \qquad (5\text{-}6\text{-}8)$$

$$\left(\nabla^2-\frac{1}{c^2}\frac{\partial^2}{\partial t^2}\right)\varphi(\mathbf{r},t) = \int\frac{\left(\nabla^2-\frac{1}{c^2}\frac{\partial^2}{\partial t^2}\right)\rho(\mathbf{r}',u)}{|\mathbf{r}-\mathbf{r}'|}\,dV'$$

$$+ 2\int\nabla\rho(\mathbf{r}',u)\cdot\nabla\frac{1}{|\mathbf{r}-\mathbf{r}'|}\,dV'$$

$$+ \int\left(\nabla^2\frac{1}{|\mathbf{r}-\mathbf{r}'|}\right)\rho(\mathbf{r}',u)\,dV' \qquad (5\text{-}6\text{-}9)$$

But we have already learned how to treat the third integral.

$$\nabla^2\frac{1}{|\mathbf{r}-\mathbf{r}'|} = 0 \qquad \text{except when } \mathbf{r}=\mathbf{r}'$$

We also know that

$$\int_{\substack{\text{all}\\ \text{space}}}\nabla^2\frac{1}{|\mathbf{r}-\mathbf{r}'|}\,dV' = -4\pi \qquad \text{(see page 32)}$$

Thus we can write

$$\int\nabla^2\frac{1}{|\mathbf{r}-\mathbf{r}'|}\rho(\mathbf{r}',u)\,dV = \rho(\mathbf{r},t)\int\nabla^2\frac{1}{|\mathbf{r}-\mathbf{r}'|}\,dV'$$

$$= -4\pi\rho(\mathbf{r},t) \qquad (5\text{-}6\text{-}10)$$

As far as the other integrals are concerned, we have

$$\frac{\left(\nabla^2-\frac{1}{c^2}\frac{\partial^2}{\partial t^2}\right)\rho(\mathbf{r}',u)}{|\mathbf{r}-\mathbf{r}'|} = -2\frac{\partial\rho}{\partial u}\frac{1}{|\mathbf{r}-\mathbf{r}'|^2} \qquad (5\text{-}6\text{-}11)$$

and

$$2\nabla\rho(\mathbf{r}'u) \cdot \nabla \frac{1}{|\mathbf{r} - \mathbf{r}'|} = 2 \frac{\partial\rho}{\partial u} \frac{1}{|\mathbf{r} - \mathbf{r}'|^2} \tag{5-6-12}$$

Hence the sum of the other integrals is zero, and we are left with

$$\left(\nabla^2 - \frac{1}{c^2} \frac{\partial^2}{\partial t^2}\right)\varphi(\mathbf{r},t) = -4\pi\rho(\mathbf{r},t)$$

Our proof is complete.

In our next chapter we will develop an intuitively appealing technique for evaluating this integral which is particularly useful in dealing with the field of a moving, small charge. We will then observe how the acceleration of a charge gives rise to a remarkable new phenomena, radiation.

PROBLEMS

5-1. Show that $\mathbf{A}$ and φ as given by Eq. (5-6-3) satisfy the so-called **Lorentz condition,** namely,

$$\nabla \cdot \mathbf{A} + \frac{1}{c} \frac{\partial\varphi}{\partial t} = 0$$

5-2. A circular loop of wire having radius b, resistance R, and self-inductance L is set with its plane perpendicular to a time-varying magnetic field $B(t) = B_0 \cos \omega t$.

(a) Develop and solve a differential equation describing the current through the loop as a function of time.

(b) How much energy is dissipated in the loop per unit time?

5-3. A resonant circuit is constructed by putting a capacitor and an inductor in series. The physical dimensions of both the capacitor and the inductor are completely known, and no dielectrics or magnetic materials are present in the

system. The circuit is now observed to resonate at a frequency ω. Show how the velocity of light c can be determined entirely in terms of the physical dimensions of the system and the resonant frequency ω.

5-4. A transmission line is made up of two long parallel perfect conductors of arbitrary cross section. Current flows down one conductor and returns on the other. The conductors are surrounded by vacuum. Show that the inductance per unit length L and the capacitance per unit length C are related by the equation $LC = 1$.

[Hint: Make use of the Lorentz transformation to move with some velocity v parallel to the conductors. Remember also that L is measured in emu and C is measured in esu. If L is measured in esu then $L_{esu} C_{esu} = 1/c^2$.]

5-5. Find the self-inductance per unit length of two coaxial thin-walled tubes, the inner one having radius a and the outer one having radius b.

5-6. Two parallel wires, each having radius a, are separated by a distance b. A current I goes down one and returns on the other. It is spread uniformly over the cross section of each wire. Find the inductance per unit length of the pair of wires.

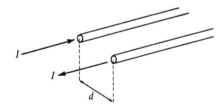

5-7. An electron travels in a circular orbit of radius R about a fixed proton. A magnetic field B_0 is now applied in the same direction as the electron's angular momentum. Find the change in the orbital magnetic moment of the electron as a result of the application of this magnetic field.

5-8. A charge q is moving in a circular orbit of radius R about the center of a cylindrically symmetrical magnet, as shown. Assume that the orbit of the charge lies in the median plane between the poles of the magnet and hence the only

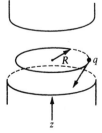

component of magnetic field that it sees is in the z direction. The field is now allowed to increase with time. Show that the particle will accelerate without any change in radius if the increase in the average field for $r \leqq R$ is twice the increase in the field at $r = R$. That is,

$$\frac{d\bar{B}(R)}{dt} = 2\frac{dB(R)}{dt}$$

where

$$\bar{B}(R) = \frac{2}{R^2} \int_0^R B(r)r\,dr$$

5-9.   A long, uniformly charged nonconducting cylinder of radius a carries a charge per unit length of λ. It is wound with N turns per centimeter of wire carrying a current I. This current gives rise to a magnetic field which we will consider as uniform throughout the cylinder.

(a) Find the Poynting vector as a function of distance from the axis of the cylinder. In which direction does it point?

(b) The momentum density of the electromagnetic field is given by S/c^2. Find the angular momentum per unit length of the electromagnetic field about the axis of the cylinder.

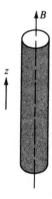

(c) The current I is now turned off. This gives rise to an induced electric field (by Faraday's law) which exerts a torque on the cylinder. Find this torque per unit length in terms of dI/dt.

(d) Integrate the result of part (c) to obtain the total change in the angular momentum per unit length of the cylinder. Compare this result with the answer to part (b).

5-10.   Consider a charge e and a magnetic monopole g situated a distance S apart.

(a) Calculate the total angular momentum of the electromagnetic field about an axis though g and e and show that it is totally independent of S.

(b) Set e equal to the magnitude of the electron charge and set the angular momentum of the field equal to the intrinsic angular momentum of the electron $(h/4\pi)$. How large is g?

6
Let There Be Light!

We now approach the high point in our study of electromagnetic theory. Although we have discovered many remarkable effects, none compares in importance with those in the area we are about to explore. We shall find that if we accelerate a charge it will emit radiation. Specifically, it will cause electric and magnetic fields to appear which decrease inversely as the distance from the charge to the *first power*. This is totally different in quality from what we have been used to, namely, fields which decrease inversely as the second power of the distance from their source. Furthermore, the new fields will be *transverse* fields. That is to say, their directions will be transverse to a line joining them to the accelerating charges. In addition these fields will appear to be moving with velocity c away from the accelerating charge. If we were not so familiar with radio waves and the electromagnetic nature of light, this would be an unbelievable discovery. We would head right out, as Hertz did, and try to produce these long-range electromagnetic disturbances, discovering radio communication in very short order. But, although we are somewhat jaded from long exposure to these phenomena, it will still be exciting to really understand their origins. This we now proceed to do.

6-1 A NEW WAY OF CALCULATING RETARDED POTENTIALS IN AN INTUITIVELY APPEALING MANNER

As we remember from our last chapter, one way of solving an electro-dynamics problem with time-varying currents is to calculate the retarded four-vector potential. For example, we would find $\varphi(\mathbf{r},t)$ by evaluating the integral

$$\varphi(\mathbf{r},t) = \int_{\substack{\text{all} \\ \text{space}}} \frac{\rho\left(\mathbf{r}', t - \dfrac{|\mathbf{r} - \mathbf{r}'|}{c}\right)}{|\mathbf{r} - \mathbf{r}'|} \, dV'$$

To evaluate this integral we must add together the contributions from charges at various distances from our position $\mathbf{r}$, always dividing by the distance $|\mathbf{r} - \mathbf{r}'|$ and always evaluating the charge at an earlier time $t' = t - |\mathbf{r} - \mathbf{r}'|/c$.

To evaluate this integral can be quite a labor. We will develop here an intuitively appealing method of performing the integration in physical terms and then apply it to find the potentials due to a moving point charge.

Let us say that we are interested in φ at position $\mathbf{r}$ and time t. Suppose at the instant t we started our clock going backward in time, with all the moving charges retracing their paths exactly. At exactly the same instant t, let us send a spherical "information-gathering" pulse out with velocity c, also moving backward in time (see Fig. 6-1). As the pulse reaches a given bit of charge at position $\mathbf{r}'$, it observes the charge as it was at time $t - |\mathbf{r} - \mathbf{r}'|/c$. But this is just what we want for our integral, so we count the charge, divide by $|\mathbf{r} - \mathbf{r}'|$, and go on.

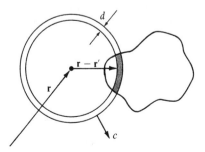

Fig. 6-1 A spherical shell expands with velocity c as our clock runs backward in time. As we encounter some charge, within the thickness d, we add it up and divide by $|\mathbf{r} - \mathbf{r}'|$, the radius of the shell.

To see how we proceed more clearly, it is perhaps better to go through our motions in a series of discrete steps and *add* rather than integrate. We let our pulse be the spherical shell shown in Fig. 6-1, with a shell thickness d. Every time our negative-running clock changes by d/c sec, we advance the shell by a distance d (thus the average shell velocity is c).

Now as we come upon some charge at a distance $|\mathbf{r} - \mathbf{r}'|$ from our starting point, it appears as it was a time $|\mathbf{r} - \mathbf{r}'|/c$ ago. While our shell is sitting there, we add together all the charge within its thickness d and divide by $|\mathbf{r} - \mathbf{r}'|$ (the radius of the shell). We put this number [it is just $(\rho \, dV)/|\mathbf{r} - \mathbf{r}'|$] into our adding machine and then sit back and relax until the shell moves on to its next position. Of course, as the shell moves ahead one position, the charges shift too, to the values they had at time $t - |\mathbf{r} - \mathbf{r}'|/c - d/c$. We repeat our calculation, add the result into our adding machine, and again wait for the shell to proceed.

Now isn't this sum just exactly what we want? We are just adding together all the $\rho \, dV$ at the appropriate earlier time and dividing by the distance $|\mathbf{r} - \mathbf{r}'|$. We use discrete steps, but that is no problem inasmuch as the thickness d can be made arbitrarily small, leading ultimately to an integral rather than a sum. It would seem then that we have found a new technique for evaluating both $\varphi(\mathbf{r}, t)$ and $\mathbf{A}(\mathbf{r}, t)$—a technique which, as we shall see, is especially useful in dealing with moving discrete charges.

6-2 THE POTENTIALS OF A SMALL MOVING CHARGE (LIENARD-WIECHERT POTENTIALS)

We wish to find the potentials at position $\mathbf{r}$ and time t due to a small moving charge at position $\mathbf{r}'$. By small we mean that we can neglect the variation of $1/|\mathbf{r} - \mathbf{r}'|$ as the imaginary spherical shell we have just described passes over our charge. Now we will carry forth our arguments, assuming the charge volume to be rectangular in shape and uniform in charge density. (Of course, any other charge distribution can be subdivided into such elements and their separate contributions can be added together.) If the original velocity of the charge at time t' was $\mathbf{v}'(t')$, we now give it a velocity $-\mathbf{v}'(t')$ for the purpose of doing our sum. As can be seen from Fig. 6-2, if $-\mathbf{v}'(t')$ is in the direction away from $\mathbf{r}$, then our expanding sphere will spend

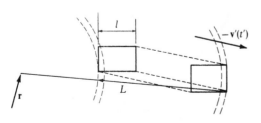

Fig. 6-2 The expanding sphere travels a distance L while still overlapping the charge.

more time overlapping the charge, and hence the contribution to the potentials will be larger than if the charge were stationary. Conversely, if the velocity $-\mathbf{v}'(t')$ will now be toward $\mathbf{r}$, the sphere will spend *less* time overlapping the charge, and the contribution to the potential will be smaller than if the charge were stationary. Referring again to Fig. 6-2, we see that our spherical shell travels for a total distance L while still overlapping the charge. Hence the potential we obtain is just

$$\varphi(\mathbf{r},t) = \frac{q}{|\mathbf{r} - \mathbf{r}'(t')|} \frac{L}{l} \tag{6-2-1}$$

where l is the length the charge appears to have. But

$$\frac{L - l}{\mathbf{v}' \cdot (\mathbf{r} - \mathbf{r}')/|\mathbf{r} - \mathbf{r}'|} = \frac{L}{c} \tag{6-2-2}$$

Hence

$$L\left\{1 - \frac{\mathbf{v}'(t') \cdot [\mathbf{r} - \mathbf{r}'(t')]}{c|\mathbf{r} - \mathbf{r}'|}\right\} = l$$

This leads us to the result

$$\varphi(\mathbf{r},t) = \frac{q}{|\mathbf{r} - \mathbf{r}'(t')|\left[1 - \dfrac{v'(t') \cdot \hat{\boldsymbol{\varepsilon}}'}{c}\right]} \tag{6-2-3}$$

where $\hat{\boldsymbol{\varepsilon}}'$ is a unit vector from $\mathbf{r}'(t')$ to $\mathbf{r}$.

$$\hat{\boldsymbol{\varepsilon}}' = \frac{\mathbf{r} - \mathbf{r}'(t')}{|\mathbf{r} - \mathbf{r}'(t')|} \tag{6-2-4}$$

Similarly we have for $\mathbf{A}(\mathbf{r},t)$,

$$\mathbf{A}(\mathbf{r},t) = \frac{q\mathbf{v}'(t')}{c|\mathbf{r} - \mathbf{r}'(t')|\left[1 - \dfrac{\mathbf{v}'(t') \cdot \hat{\boldsymbol{\varepsilon}}'}{c}\right]} \tag{6-2-5}$$

By means of these so-called **Lienard-Wiechert potentials,** we may now calculate the electric and magnetic fields due to our small moving charges. Although our usual concern will be with situations where $v \ll c$, we will for the moment carry along all orders in v/c so that in the future we can deal comfortably with charges that are moving at relativistic velocities. Our job then becomes one of differentiating these potentials. That is,

$$\mathbf{B}(\mathbf{r},t) = \nabla \times \frac{q\mathbf{v}'}{c|\mathbf{r} - \mathbf{r}'(t')|\left(1 - \dfrac{\mathbf{v} \cdot \hat{\boldsymbol{\varepsilon}}'}{c}\right)} \tag{6-2-6}$$

$$\mathbf{E}(\mathbf{r}, t) = -\nabla \frac{q}{|\mathbf{r} - \mathbf{r}'(t')| \left(1 - \frac{\mathbf{v}' \cdot \hat{\boldsymbol{\varepsilon}}'}{c} \right)}$$

$$- \frac{1}{c} \frac{\partial}{\partial t} \left[\frac{q\mathbf{v}'}{c|\mathbf{r} - \mathbf{r}'(t')| \left(1 - \frac{\mathbf{v}' \cdot \hat{\boldsymbol{\varepsilon}}'}{c} \right)} \right] \qquad (6\text{-}2\text{-}7)$$

6-3 DIFFERENTIATING THE LIENARD-WIECHERT POTENTIALS; THE RADIATION FIELD

We now undertake the rather laborious task of performing the differentiation called for in Eqs. (6-2-6) and (6-2-7). The difficulty lies in the complex implicit dependence of all the primed terms on $\mathbf{r}$ and t. We will prepare ourselves with a few preliminary derivatives.

$$t' = t - \frac{|\mathbf{r} - \mathbf{r}'|}{c} \qquad (6\text{-}3\text{-}1)$$

Now

$$\frac{\partial t'}{\partial t} = 1 - \frac{1}{c} \left(\frac{\partial}{\partial t'} |\mathbf{r} - \mathbf{r}'| \right) \frac{\partial t'}{\partial t}$$

and

$$\frac{\partial |\mathbf{r} - \mathbf{r}'|}{\partial t'} = -\mathbf{v}' \cdot \hat{\boldsymbol{\varepsilon}}'$$

Hence

$$\frac{\partial t'}{\partial t} = \frac{1}{1 - \frac{\mathbf{v}' \cdot \hat{\boldsymbol{\varepsilon}}'}{c}} \qquad (6\text{-}3\text{-}2)$$

Similarly

$$\nabla t' = -\frac{1}{c} \hat{\boldsymbol{\varepsilon}}' - \frac{1}{c} \left(\frac{\partial}{\partial t'} |\mathbf{r} - \mathbf{r}'| \right) \nabla t'$$

and hence

$$\nabla t' = \frac{-\hat{\boldsymbol{\varepsilon}}'}{c \left(1 - \frac{\mathbf{v}' \cdot \hat{\boldsymbol{\varepsilon}}'}{c} \right)} \qquad (6\text{-}3\text{-}3)$$

We will also find the following useful:

$$\frac{\partial \hat{\boldsymbol{\varepsilon}}'}{\partial t'} = \frac{\partial}{\partial t'}\left(\frac{\mathbf{r} - \mathbf{r}'}{|\mathbf{r} - \mathbf{r}'|}\right) = \frac{(\mathbf{v}' \cdot \hat{\boldsymbol{\varepsilon}}')(\mathbf{r} - \mathbf{r}')}{|\mathbf{r} - \mathbf{r}'|^2} - \frac{\mathbf{v}'}{|\mathbf{r} - \mathbf{r}'|}$$

or

$$\frac{\partial \hat{\boldsymbol{\varepsilon}}'}{\partial t'} = \frac{\hat{\boldsymbol{\varepsilon}}' \times (\hat{\boldsymbol{\varepsilon}}' \times \mathbf{v}')}{|\mathbf{r} - \mathbf{r}'|} \tag{6-3-4}$$

We can now evaluate

$$\frac{\partial}{\partial t'}\left[|\mathbf{r} - \mathbf{r}'|\left(1 - \frac{\mathbf{v}' \cdot \hat{\boldsymbol{\varepsilon}}'}{c}\right)\right] = -\mathbf{v}' \cdot \hat{\boldsymbol{\varepsilon}}'\left(1 - \frac{\mathbf{v}' \cdot \hat{\boldsymbol{\varepsilon}}'}{c}\right)$$
$$- |\mathbf{r} - \mathbf{r}'|\frac{\mathbf{a}'}{c} \cdot \hat{\boldsymbol{\varepsilon}}' - \frac{\mathbf{v}'}{c} \cdot \left[\hat{\boldsymbol{\varepsilon}} \times (\hat{\boldsymbol{\varepsilon}}' \times \mathbf{v}')\right]$$

or

$$\frac{\partial}{\partial t'}\left[|\mathbf{r} - \mathbf{r}'|\left(1 - \frac{\mathbf{v}' \cdot \hat{\boldsymbol{\varepsilon}}'}{c}\right)\right] = \frac{-|\mathbf{r} - \mathbf{r}'|\mathbf{a}' \cdot \hat{\boldsymbol{\varepsilon}}'}{c} - \mathbf{v}' \cdot \hat{\boldsymbol{\varepsilon}}' + \frac{v'^2}{c}$$

which yields

$$\frac{\partial}{\partial t'}\left[|\mathbf{r} - \mathbf{r}'|\left(1 - \frac{\mathbf{v}' \cdot \hat{\boldsymbol{\varepsilon}}'}{c}\right)\right] = \frac{-|\mathbf{r} - \mathbf{r}'|}{c}(\mathbf{a}' \cdot \hat{\boldsymbol{\varepsilon}}') - \mathbf{v}' \cdot \left(\hat{\boldsymbol{\varepsilon}}' - \frac{\mathbf{v}'}{c}\right)$$
$$\tag{6-3-5}$$

where

$$\mathbf{a}' = \frac{d\mathbf{v}'}{dt'} \tag{6-3-6}$$

We are now ready to take the gradient of φ:

$$-\nabla\varphi = -(\nabla\varphi)_{t' \text{ const}} - \frac{\partial\varphi}{\partial t'}\nabla t'$$

$$-(\nabla\varphi)_{t' \text{ const}} = \frac{q(\hat{\boldsymbol{\varepsilon}}' - \mathbf{v}'/c)}{|\mathbf{r} - \mathbf{r}'|^2\left(1 - \frac{\mathbf{v}' \cdot \hat{\boldsymbol{\varepsilon}}'}{c}\right)^2}$$

$$-\frac{\partial\varphi}{\partial t'} = \frac{q}{|\mathbf{r} - \mathbf{r}'|^2\left(1 - \frac{\mathbf{v}' \cdot \hat{\boldsymbol{\varepsilon}}'}{c}\right)^2}\left\{\frac{\partial}{\partial t'}\left[|\mathbf{r} - \mathbf{r}'|\left(1 - \frac{\mathbf{v}' \cdot \hat{\boldsymbol{\varepsilon}}'}{c}\right)\right]\right\}$$

$$= \frac{q}{|\mathbf{r} - \mathbf{r}'|^2\left(1 - \frac{\mathbf{v}' \cdot \hat{\boldsymbol{\varepsilon}}'}{c}\right)^2}\left[\frac{-|\mathbf{r} - \mathbf{r}'|}{c}(\mathbf{a}' \cdot \hat{\boldsymbol{\varepsilon}}) - \mathbf{v}' \cdot \left(\hat{\boldsymbol{\varepsilon}} - \frac{\mathbf{v}'}{c}\right)\right]$$

Combining our terms, we have

$$-\nabla\varphi = \frac{q\left[\hat{\varepsilon}'\left(1 - \frac{v'^2}{c^2}\right) - \frac{\mathbf{v}'}{c}\left(1 - \frac{\mathbf{v}}{c}\cdot\hat{\varepsilon}'\right)\right]}{|\mathbf{r} - \mathbf{r}'|^2\left(1 - \frac{\mathbf{v}'\cdot\hat{\varepsilon}'}{c}\right)^3}$$

$$+ \frac{q\hat{\varepsilon}'(\mathbf{a}'\cdot\hat{\varepsilon}')}{c^2|\mathbf{r} - \mathbf{r}'|\left(1 - \frac{\mathbf{v}'\cdot\hat{\varepsilon}'}{c}\right)^3} \quad (6\text{-}3\text{-}7)$$

Next we calculate

$$-\frac{1}{c}\frac{\partial\mathbf{A}}{\partial t} = -\frac{1}{c}\frac{\partial\mathbf{A}}{\partial t'}\frac{\partial t'}{\partial t}$$

$$-\frac{1}{c}\frac{\partial\mathbf{A}}{\partial t'} = -\frac{\partial}{\partial t'}\frac{q\mathbf{v}'}{c^2|\mathbf{r} - \mathbf{r}'|\left(1 - \frac{\mathbf{v}'\cdot\hat{\varepsilon}'}{c}\right)}$$

$$= \frac{-q\mathbf{a}'}{c^2|\mathbf{r} - \mathbf{r}'|\left(1 - \frac{\mathbf{v}'\cdot\hat{\varepsilon}'}{c}\right)} + \frac{q\mathbf{v}'}{c^2|\mathbf{r} - \mathbf{r}'|^2\left(1 - \frac{\mathbf{v}'\cdot\hat{\varepsilon}'}{c}\right)^2}$$

$$\left[-\frac{|\mathbf{r} - \mathbf{r}'|}{c}(\mathbf{a}'\cdot\hat{\varepsilon}') - \mathbf{v}'\cdot\left(\hat{\varepsilon}' - \frac{\mathbf{v}'}{c}\right)\right]$$

Hence we have

$$-\frac{1}{c}\frac{\partial\mathbf{A}}{\partial t} = \frac{-q\frac{\mathbf{v}'}{c}\left[\frac{\mathbf{v}'}{c}\cdot\left(\hat{\varepsilon}' - \frac{\mathbf{v}'}{c}\right)\right]}{|\mathbf{r} - \mathbf{r}'|^2\left(1 - \frac{\mathbf{v}'\cdot\hat{\varepsilon}'}{c}\right)^3} - \frac{q\mathbf{a}'}{c^2|\mathbf{r} - \mathbf{r}'|\left(1 - \frac{\mathbf{v}'\cdot\hat{\varepsilon}'}{c}\right)^2}$$

$$- \frac{q\frac{\mathbf{v}'}{c}(\mathbf{a}'\cdot\hat{\varepsilon}')}{c^2|\mathbf{r} - \mathbf{r}'|\left(1 - \frac{\mathbf{v}'\cdot\hat{\varepsilon}'}{c}\right)^3}$$

or, rewriting slightly,

$$-\frac{1}{c}\frac{\partial\mathbf{A}}{\partial t} = \frac{-q\frac{\mathbf{v}'}{c}\left[\frac{\mathbf{v}'}{c}\cdot\left(\hat{\varepsilon}' - \frac{\mathbf{v}'}{c}\right)\right]}{|\mathbf{r} - \mathbf{r}'|^2\left(1 - \frac{\mathbf{v}'\cdot\hat{\varepsilon}'}{c}\right)^3} + \frac{-q\mathbf{a}' + q\hat{\varepsilon}'\times\left(\mathbf{a}'\times\frac{\mathbf{v}'}{c}\right)}{c^2|\mathbf{r} - \mathbf{r}'|\left(1 - \frac{\mathbf{v}'\cdot\hat{\varepsilon}'}{c}\right)^3}$$

$$(6\text{-}3\text{-}8)$$

Combining $-\nabla\varphi$ and $-\dfrac{1}{c}\dfrac{\partial \mathbf{A}}{\partial t}$, we have

$$\mathbf{E}(\mathbf{r},t) = \frac{-q[\mathbf{a}' - (\mathbf{a}' \cdot \hat{\boldsymbol{\varepsilon}}')\hat{\boldsymbol{\varepsilon}}'] + q\hat{\boldsymbol{\varepsilon}}' \times \left(\mathbf{a}' \times \dfrac{\mathbf{v}'}{c}\right)}{c^2|\mathbf{r}-\mathbf{r}'|\left(1 - \dfrac{\mathbf{v}'\cdot\hat{\boldsymbol{\varepsilon}}'}{c}\right)^3}$$

$$+ \frac{q\left(\hat{\boldsymbol{\varepsilon}}' - \dfrac{\mathbf{v}'}{c}\right)\left(1 - \dfrac{v'^2}{c^2}\right)}{|\mathbf{r}-\mathbf{r}'|^2\left(1 - \dfrac{\mathbf{v}'\cdot\hat{\boldsymbol{\varepsilon}}'}{c}\right)^3} \quad (6\text{-}3\text{-}9)$$

Next, we calculate the magnetic field $\mathbf{B} = \nabla \times \mathbf{A}$:

$$\nabla \times \mathbf{A} = (\nabla \times \mathbf{A})_{t'\,\text{const}} - \left(\frac{\partial \mathbf{A}}{\partial t'}\right) \times \nabla t' \quad (6\text{-}3\text{-}10)$$

Now

$$(\nabla \times \mathbf{A})_{t'\,\text{const}} = \frac{-q\left(\hat{\boldsymbol{\varepsilon}}' - \dfrac{\mathbf{v}'}{c}\right) \times \dfrac{\mathbf{v}'}{c}}{|\mathbf{r}-\mathbf{r}'|^2\left(1 - \dfrac{\mathbf{v}'\cdot\hat{\boldsymbol{\varepsilon}}'}{c}\right)^2}$$

$$-\frac{\partial \mathbf{A}}{\partial t'} = \frac{-q\mathbf{a}' + q\hat{\boldsymbol{\varepsilon}}' \times \left(\mathbf{a}' \times \dfrac{\mathbf{v}'}{c}\right)}{c|\mathbf{r}-\mathbf{r}'|\left(1 - \dfrac{\mathbf{v}'\cdot\hat{\boldsymbol{\varepsilon}}'}{c}\right)^2} - \frac{q\mathbf{v}'\left[\dfrac{\mathbf{v}'}{c} \cdot \left(\hat{\boldsymbol{\varepsilon}}' - \dfrac{\mathbf{v}'}{c}\right)\right]}{|\mathbf{r}-\mathbf{r}'|^2\left(1 - \dfrac{\mathbf{v}'\cdot\hat{\boldsymbol{\varepsilon}}'}{c}\right)^2}$$

and

$$-\frac{\partial \mathbf{A}}{\partial t'} \times \nabla t' = \frac{q\mathbf{a}' \times \hat{\boldsymbol{\varepsilon}}' + q\hat{\boldsymbol{\varepsilon}}' \times \left[\hat{\boldsymbol{\varepsilon}}' \times \left(\mathbf{a}' \times \dfrac{\mathbf{v}'}{c}\right)\right]}{c^2|\mathbf{r}-\mathbf{r}'|\left(1 - \dfrac{\mathbf{v}'\cdot\hat{\boldsymbol{\varepsilon}}'}{c}\right)^3}$$

$$+ \frac{q\left(\dfrac{\mathbf{v}'}{c} \times \hat{\boldsymbol{\varepsilon}}'\right)\left[\dfrac{\mathbf{v}'}{c} \cdot \left(\hat{\boldsymbol{\varepsilon}}' - \dfrac{\mathbf{v}'}{c}\right)\right]}{|\mathbf{r}-\mathbf{r}'|^2\left(1 - \dfrac{\mathbf{v}'\cdot\hat{\boldsymbol{\varepsilon}}'}{c}\right)^3}$$

We finally have then

$$
\mathbf{B} = \frac{q\left(\dfrac{\mathbf{v'}}{c} \times \hat{\boldsymbol{\varepsilon}}'\right)\left(1 - \dfrac{v'^2}{c^2}\right)}{|\mathbf{r} - \mathbf{r'}|^2 \left(1 - \dfrac{\mathbf{v'} \cdot \hat{\boldsymbol{\varepsilon}}'}{c}\right)^3} + \hat{\boldsymbol{\varepsilon}}' \times \frac{-q\mathbf{a'} + q\left[\hat{\boldsymbol{\varepsilon}}' \times \left(\mathbf{a'} \times \dfrac{\mathbf{v'}}{c}\right)\right]}{c^2|\mathbf{r} - \mathbf{r'}|\left(1 - \dfrac{\mathbf{v'} \cdot \hat{\boldsymbol{\varepsilon}}'}{c}\right)^3}
$$

$$(6\text{-}3\text{-}11)$$

Examining our expressions for $\mathbf{E}$ and $\mathbf{B}$, we note that they each divide naturally into two parts—that which is independent of acceleration and drops off as $1/|\mathbf{r} - \mathbf{r'}|^2$ and that which is proportional to the acceleration of the charge and drops only as $1/|\mathbf{r} - \mathbf{r'}|$. The terms independent of $\mathbf{a}$ can, as expected, be obtained directly by carrying out a Lorentz transformation upon the electric field of a static charge q. The terms proportional to $\mathbf{a}$ are the ones which interest us now; they are the so-called **radiation fields**. We thus will ignore all terms that are proportional to $1/|\mathbf{r} - \mathbf{r'}|^2$ in our present discussion.

We first note that both $\mathbf{E}$ and $\mathbf{B}$ are perpendicular to $\hat{\boldsymbol{\varepsilon}}'$ and to each other. This is characteristic of the so-called **transverse** nature of electromagnetic radiation. The magnitude of $\mathbf{E}$ is the same as the magnitude of $\mathbf{B}$—a highly desirable result since otherwise we could transform either $\mathbf{E}$ or $\mathbf{B}$ away by an appropriate Lorentz transformation.

We also note the remarkable fact that this electromagnetic field seems to be propagating with time at a velocity c. That is, a given acceleration at time t' leads to fields at position $\mathbf{r}$ and at time t such that $|\mathbf{r} - \mathbf{r'}|/c = t - t'$. As t increases, the relevant field appears further and further away from where the acceleration was that gave rise to it.

6-4 ENERGY RADIATION: NONRELATIVISTIC TREATMENT

To see what happens insofar as the flow of energy is concerned, we can evaluate the Poynting vector $\mathbf{S}$.

$$
\mathbf{S} = \frac{c}{4\pi} \mathbf{E} \times \mathbf{B} = \frac{c}{4\pi} \mathbf{E} \times (\hat{\boldsymbol{\varepsilon}}' \times \mathbf{E}) = \frac{c}{4\pi} E^2 \hat{\boldsymbol{\varepsilon}}' \qquad (6\text{-}4\text{-}1)
$$

We will make the nonrelativistic approximation that the velocity of our charge is much less than c. This assumption serves to simplify our work considerably and is a perfectly valid assumption in all the applications we will be considering.

To evaluate $\mathbf{S}$, we must first have another look at $\mathbf{E}$. We have, for $v/c \ll 1$,

$$
\mathbf{E} = \frac{-q\mathbf{a}_p'}{c^2|\mathbf{r} - \mathbf{r'}|} \qquad (6\text{-}4\text{-}2)
$$

where

$$\mathbf{a}_p' = \mathbf{a}' - (\mathbf{a}' \cdot \hat{\mathbf{\epsilon}}')\hat{\mathbf{\epsilon}}' \tag{6-4-3}$$

As we can see, $\mathbf{a}_p'$ is the projected acceleration perpendicular to the line of sight $\hat{\mathbf{\epsilon}}'$. [Intuitively one should think of this as follows. The electric field is directly proportional (with proportionality constant $-q/c^2$) to the acceleration that we would "see" as we watch the particle. The acceleration we see must obviously appear as it was at an earlier time t'. Furthermore, its apparent magnitude goes down as $1/|\mathbf{r} - \mathbf{r}'|$ as we go further and further away from the location of the acceleration. Finally, what we would see is just the projected acceleration, that part which is at right angles to our line of sight. This "intuitive" approach to understanding the radiation field due to an accelerating charge will be particularly useful when we begin to examine coherent interference effects due to the organized motion of large assemblies of such charges.]

We wish to calculate the total energy radiated per unit time by a charge with acceleration $\mathbf{a}'$ at time t'. Its effect is felt at a distance R away at a time R/c later. We draw a sphere of that radius about our charge, as shown in Fig. 6-3. The magnitude of the electric field at (R, θ) is just

$$|\mathbf{E}(R,\theta)| = \frac{qa' \sin \theta}{c^2 R} \tag{6-4-4}$$

and hence

$$|\mathbf{S}| = \frac{q^2 a'^2 \sin^2 \theta}{4\pi c^3 R^2} \tag{6-4-5}$$

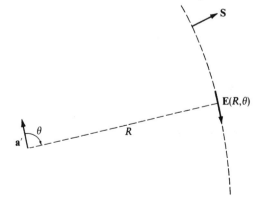

Fig. 6-3 We evaluate the energy lost by an accelerating particle by integrating the flux of the Poynting vector around a sphere of radius R.

The total amount of energy which was radiated per unit at time t' is then

$$\frac{du}{dt} = \frac{q^2 a'^2}{4\pi c^3} \int_0^\pi (2\pi \sin^3 \theta)\, d\theta$$

$$= \frac{2}{3} \frac{q^2 a'^2}{c^3} \tag{6-4-6}$$

This expression will be encountered quite often in studying the electromagnetic manifestations of particle acceleration and is known as the **Larmor formula.**

6-5 POLARIZATION

One of the immediate consequences of the transverse nature of electromagnetic radiation is the fact that only *two* polarization states are available for each radiation direction. That is, we need to choose only two orthogonal unit vectors, at right angles to $\hat{\varepsilon}'$, and we shall then be able to express our electric radiation field entirely in terms of components along these two vectors.

To illustrate then the origin of the various types of optical polarization we have learned about in elementary physics, it is useful to examine in detail the radiation pattern due to a charge moving in a circle, with $v \ll c$. In Fig. 6-4 we sketch a coordinate system relative to the center of the circular path taken by the charge.

The charge moves counterclockwise as viewed from the $+z$ direction. As we observe the charge from the various directions, we see the acceleration

Fig. 6-4 A charged particle moving in a circle gives rise to circularly polarized radiation along the axis of its motion and linearly polarized radiation at right angles to this axis. In other directions the "light" is elliptically polarized.

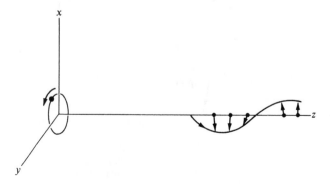

vector carrying out its periodic motion with the rotational frequency of the particle. From along the $+z$ axis, the acceleration vector always appears to have the same length and rotates about in a circle. Hence we have circularly polarized light going out along the positive and negative z directions. At any given point along the $+z$ axis, the electric field vector appears to rotate in a counterclockwise manner as we look toward the charge. We will call this **right-hand circularly polarized radiation,** since if the right thumb points in the direction of propagation (along the $+z$ axis), the fingers follow the direction in which **E** changes with time. We note, of course, that if we explore the pattern of **E** along the $+z$ axis at a *fixed* time, it looks like a *left-handed* screw.

Along the negative z axis the electric field has the spatial distribution corresponding to a *right-handed* screw and rotates at any given point in a clockwise manner as we look toward the charge. Naturally, this is called **left-hand circularly polarized radiation.**

In the xy plane the radiation is plane polarized with the electric vector always lying in this plane.

6-6 THE SCATTERING OF RADIATION BY A FREE ELECTRON

If our radiation field is now incident upon a charge, it will obviously "shake" it. That is to say, the electric field will cause the charge to accelerate and this in turn will cause the charge itself to radiate. In this way we can say that the charge has effectively "scattered" some of the incident radiation. It is remarkably simple for us to calculate the fraction of the incident energy per square centimeter at the charge that is reradiated in this manner. This fraction is called the **Thomson cross section** when evaluated for an electron in the limit that its velocity is considered small compared with c. (The term **cross section** is a natural one to use when asking for the fraction of the energy flux per square centimeter that is affected by a particle. We imagine that we replace the particle with a disk of a given cross-sectional area. If the radiation passes through it, it is affected; otherwise it is not. The symbol for cross section is invariably σ.)

$$\sigma_{\text{Thomson}} = \frac{\text{energy radiated by electron per unit time}}{\text{magnitude of Poynting vector of incident radiation}}$$

$$(6\text{-}6\text{-}1)$$

Now, the acceleration of the electron is just given by

$$a = \frac{F}{m_e} = \frac{eE}{m_e} \qquad (6\text{-}6\text{-}2)$$

and hence the energy radiated by the electron per unit time is just

$$\frac{du}{dt} = \frac{2e^2 a^2}{3c^3} = \frac{2}{3}\frac{e^4 E^2}{m^2 c^3} \tag{6-6-3}$$

The magnitude of the Poynting vector for the incident radiation is given by

$$|\mathbf{S}| = \frac{c}{4\pi} E^2 \tag{6-6-4}$$

Thus we have

$$\sigma_{\text{Thomson}} = \frac{8\pi}{3}\frac{e^4}{m_e^2 c^4} = \frac{8\pi}{3} r_0^2 \tag{6-6-5}$$

where we define $r_0 \equiv$ classical radius of electron $= e^2/m_e c^2$. Incidentally, we note that the classical radius of the electron is of the same order as the radius the electron would have if its mass were largely electromagnetic in origin (see page 200).

6-7 MATHEMATICAL SUPPLEMENT: COMPLETENESS AND ORTHOGONALITY

As we become more and more sophisticated in our study of physics, we will find increasingly more numerous the occasions when we make use of *complete* sets of *orthogonal* functions. (In fact, when we come to the study of quantum mechanics, we will discover that the sets of solutions to given boundary-value problems are all this type.) We have already come across one such set of functions, the Legendre polynomials $P_n(\cos\theta)$ that we studied in electrostatics. Other common examples are the harmonic functions $\sin kx$ and $\cos kx$ and the Bessel functions. Let us then explain in general terms what we mean by completeness and orthogonality.

A set of functions $g_i(x)$ will be considered **complete** over a given range of x if any reasonable[1] function $f(x)$ defined over this range can be expanded as a linear sum:

$$f(x) = \sum_{i=0}^{\infty} A_i g_i(x) \tag{6-7-1}$$

The functions will be considered mutually **orthogonal** over the range $x_{\min} \leqq$

---

[1] For a precise definition of what we call reasonableness the reader is referred to any standard mathematical text covering functional analysis and eigenfunction expansions. All functions which we are likely to come across in physics are *reasonable*.

$x \leqq x_{max}$ if

$$\int_{x_{min}}^{x_{max}} g_i^*(x)g_j(x) \, dx = 0 \qquad \text{for } i \neq j \tag{6-7-2}$$

where g_i^* is the complex conjugate of g_i.

6-8 MATHEMATICAL SUPPLEMENT: FOURIER SERIES AND FOURIER INTEGRAL

Perhaps the most useful complete set are the harmonic functions $\sin kx$ and $\cos kx$. We begin by considering only the interval $0 \leqq x \leqq a$ over which we wish to approximate the function $f(x)$. Since we do not care about $f(x)$ outside this interval, we are free to take it periodic with a periodicity given by

$$f(x) = f(x \pm a)$$

Thus we need only make use of those values of k which make $\sin kx$ and $\cos kx$ periodic with this same period. Thus

$$ka = 2\pi n \qquad \text{where } n \text{ is an integer}$$

We have then, for our expansion,

$$f(x) = \sum_{n=0}^{\infty} A_n \sin \frac{2\pi n}{a} x + \sum_{n=0}^{\infty} B_n \cos \frac{2\pi n}{a} x \tag{6-8-1}$$

That the individual functions are orthogonal can be seen from inspection:

$$\int_0^a \sin \frac{2\pi n}{a} x \sin \frac{2\pi m}{a} x \, dx = 0 \qquad \text{unless } n = m$$

$$\int_0^a \cos \frac{2\pi n}{a} x \cos \frac{2\pi m}{a} x \, dx = 0 \qquad \text{unless } n = m$$

$$\int_0^a \sin \frac{2\pi n}{a} x \cos \frac{2\pi m}{\pi} x \, dx = 0 \qquad \text{for all } n,m$$

That they form a *complete* set can only be asserted at this time and not proven. For a detailed proof the reader is again referred to the mathematical literature. Our problem then becomes simply one of finding the coefficients of $\sin \frac{2\pi n}{a} x$ and $\cos \frac{2\pi n}{a} x$. We multiply both sides of our expansion for $f(x)$ by $\sin \frac{2\pi m}{a} x$ and integrate from 0 to a.

$$\int_0^a f(x) \sin \frac{2\pi m}{a} x \, dx = \sum_{n=0}^{\infty} A_n \int_0^a \sin \frac{2\pi n}{a} x \sin \frac{2\pi m}{a} x \, dx$$

$$+ \sum_{n=0}^{\infty} B_n \int_0^a \cos \frac{2\pi n}{a} x \sin \frac{2\pi m}{a} x \, dx$$

$$= A_m \frac{a}{2}$$

We have then

$$A_m = \frac{2}{a} \int_0^a f(x) \sin \frac{2\pi m}{a} x \, dx \tag{6-8-2}$$

Similarly

$$B_m = \frac{2}{a} \int_0^a f(x) \cos \frac{2\pi m}{a} x \, dx \tag{6-8-3}$$

As the range of x over which we wish to consider the function $f(x)$ becomes infinite, our series naturally becomes an integral. We write

$$f(x) = \int_0^{\infty} A(k) \sin kx \, dk + \int_0^{\infty} B(k) \cos kx \, dk \tag{6-8-4}$$

where $A(k) \, dk$ is the contribution to our "sum" from the interval between k and $k + dk$. We remember though that

$$\cos kx = \frac{e^{ikx} + e^{-ikx}}{2} \qquad \text{and} \qquad \sin kx = \frac{e^{ikx} - e^{-ikx}}{2i}$$

Substituting, we have

$$f(x) = \frac{1}{2} \int_0^{\infty} [B(k) - iA(k)] e^{ikx} \, dk + \frac{1}{2} \int_0^{\infty} [B(k) + iA(k)] e^{-ikx} \, dk$$

Both $A(k)$ and $B(k)$ must be real if $f(x)$ is to be real. Let $C(k) = \sqrt{\pi/2}$ $[B(k) - iA(k)]$. Then we have

$$f(x) = \frac{1}{\sqrt{2\pi}} \int_0^{\infty} C(k) e^{ikx} \, dk + \frac{1}{\sqrt{2\pi}} \int_0^{\infty} C^*(k) e^{-ikx} \, dk$$

If we change variables in the second integral, from k to $-k$, and define $C(-k) \equiv C^*(k)$, we can combine our terms to obtain

$$f(x) = \frac{1}{\sqrt{2\pi}} \int_{-\infty}^{\infty} C(k) e^{ikx} \, dk \tag{6-8-5}$$

To obtain $C(k)$ if we are given $f(x)$ we proceed in a manner analogous to that used to obtain A_n and B_n in the case of Fourier series. We multiply both sides by $e^{-ik'x}$ and integrate over x from $-\infty$ to ∞.

$$\int_{-\infty}^{\infty} f(x)e^{-ik'x}\,dx = \frac{1}{\sqrt{2\pi}}\int_{-\infty}^{\infty} C(k)\,dk \int_{-\infty}^{\infty} e^{i(k-k')x}\,dx \qquad (6\text{-}8\text{-}6)$$

We now wish to evaluate the integral

$$\int_{-\infty}^{\infty} e^{i(k-k')x}\,dx = \lim_{a\to\infty}\int_{-a}^{a} e^{i(k-k')x}\,dx$$

$$= \lim_{a\to\infty}\frac{1}{i(k-k')}\left(e^{i(k-k')a} - e^{-i(k-k')a}\right)$$

$$= \lim_{a\to\infty}\frac{2\sin(k-k')a}{k-k'} \qquad (6\text{-}8\text{-}7)$$

Let us have a look at the function $[2\sin(k-k')a]/(k-k')$ as we let a get very large. In Figs. 6-5 and 6-6 we plot the function for $a=1$ and $a=10$. As is apparent, the function becomes more and more peaked at $k=k'$ as we let a increase. However, the total area in the peak remains constant and has a value of about $\frac{1}{2}2a(2\pi/a) \cong 2\pi$, as can be seen from examining the graphs. (The actual area is in fact exactly 2π.) Now, as we take a larger and larger, we can begin to ignore the function outside the region $k=k'$, inasmuch as it will average to zero in any integral. We define the function

$$2\pi\delta(k-k') = \lim_{a\to\infty}\frac{2\sin(k-k')a}{k-k'}$$

(δ is called the **Dirac δ function**.) And we have then, for any integral including the peak,

$$\int_{k'-\mathscr{E}}^{k'+\mathscr{E}} \delta(k-k')\,dk = 1$$

We return then to evaluate

$$\int_{-\infty}^{\infty} f(x)e^{ik'x}\,dx = \frac{1}{\sqrt{2\pi}}\int_{-\infty}^{\infty} C(k)2\pi\delta(k-k')\,dk$$

$$= \sqrt{2\pi}\,C(k')$$

Fig. 6-5 A graph of the function $[2\sin(k-k')a]/(k-k')$ for $a=1$.

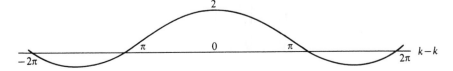

Hence we have

$$C(k') = \frac{1}{\sqrt{2\pi}} \int_{-\infty}^{\infty} f(x)e^{-ik'x}\,dx \tag{6-8-8}$$

The functions $C(k)$ and $f(x)$ are called **Fourier transforms** of each other.

Fig. 6-6 A graph of the function $[2\sin(k - k')a]/(k - k')$ for $a = 10$.

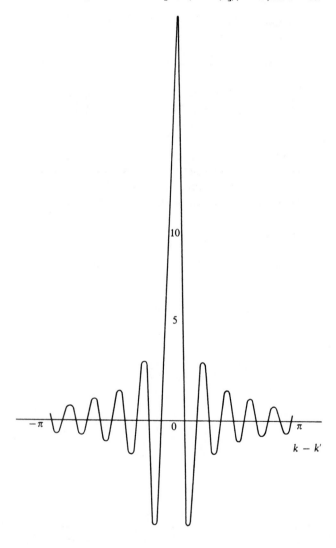

6-9 THE INTERACTION OF RADIATION WITH A CHARGE IN A HARMONIC POTENTIAL

We consider now in detail the "scattering" of radiation by a charge q in a harmonic potential. We will assume that the incident radiation is traveling in the $+z$ direction and is polarized with its electric vector along the x direction. This latter assumption involves no loss of generality for, as we shall see, it will be possible to superimpose the effects due to a superposition of incident radiation. We begin with a very simple incoming plane wave

$$E_x = E_0 \cos(kz - \omega t) \qquad \text{where } \omega = kc$$

At $z = 0$, we can write

$$E_x = \frac{E_0}{2} (e^{i\omega t} + e^{-i\omega t}) \tag{6-9-1}$$

We assume the restoring force on the charge to be equal to $m\omega_0^2$. In addition we assume a resistive force proportional to the velocity of the charge. Hence our differential equation for x, the displacement of the charge from its equilibrium position, becomes

$$\frac{d^2x}{dt^2} + \gamma \frac{dx}{dt} + \omega_0^2 x = \frac{qE_0}{2m} (e^{i\omega t} + e^{-i\omega t}) \tag{6-9-2}$$

We assume x to have the form $x = Ae^{i\omega t} + Be^{-i\omega t}$ and substitute back into our differential equation.

$$\left[A(-\omega^2 + i\gamma\omega + \omega_0^2) - \frac{qE_0}{2m} \right] e^{i\omega t}$$

$$+ \left[B(-\omega^2 - i\gamma\omega + \omega_0^2) - \frac{qE_0}{2m} \right] e^{-i\omega t} = 0$$

For this to be true at any time we must have

$$A = \frac{qE_0}{2m(\omega_0^2 - \omega^2 + i\gamma\omega)} \qquad B = \frac{qE_0}{2m(\omega_0^2 - \omega^2 - i\gamma\omega)} \tag{6-9-3}$$

We have then

$$x = \frac{qE_0}{2m(\omega_0^2 - \omega^2 + i\gamma\omega)} e^{i\omega t} + \frac{qE_0}{2m(\omega_0^2 - \omega^2 - i\gamma\omega)} e^{-i\omega t}$$

$$= \frac{qE_0}{m\sqrt{(\omega_0^2 - \omega^2)^2 + \gamma^2\omega^2}} \frac{e^{i(\omega t - \varphi)} + e^{-i(\omega t - \varphi)}}{2}$$

$$= \frac{qE_0}{m\sqrt{(\omega_0^2 - \omega^2)^2 + \gamma^2\omega^2}} \cos(\omega t - \varphi) \tag{6-9-4}$$

where

$$\tan \varphi = \frac{\gamma \omega}{\omega_0{}^2 - \omega^2} \tag{6-9-5}$$

Now, we are not quite finished, since we can add to x any function x' which is a solution to the differential equation

$$\frac{d^2 x'}{dt^2} + \gamma \frac{dx'}{dt} + \omega_0{}^2 x' = 0 \tag{6-9-6}$$

The sum $x + x'$ will also be a solution to our original equation. The solution for x' is well known. It is

$$x' = e^{-(\gamma/2)t} \left(A \sin \sqrt{\omega_0{}^2 - \frac{\gamma^2}{4}} t + B \cos \sqrt{\omega_0{}^2 - \frac{\gamma^2}{4}} t \right) \tag{6-9-7}$$

where A and B are arbitrary. Inasmuch as this solution is exponentially damped with time constant $2/\gamma$, we can effectively ignore it in most cases.

Now we wish to find the fraction of incident energy per unit area which is reradiated or scattered by the bound charge. Again we make use of the Larmor formula for the instantaneous rate at which energy is radiated.

$$P_{\text{rad}} = \frac{2}{3} \frac{q^2 a^2}{c^3} = \frac{2}{3} \frac{q^2 (d^2 x / dt^2)^2}{c^3} \tag{6-9-8}$$

But

$$\frac{d^2 x}{dt^2} = \frac{-q E_0 \omega^2}{m \sqrt{(\omega_0{}^2 - \omega^2) + \gamma^2 \omega^2}} \cos^2 (\omega t - \varphi) \tag{6-9-9}$$

Hence

$$P_{\text{rad}} = \frac{2}{3} \frac{q^4 E_0{}^2 \omega^4 \cos^2 (\omega t - \varphi)}{m^2 c^3 [(\omega_0{}^2 - \omega^2)^2 + \gamma^2 \omega^2]} \tag{6-9-10}$$

The average radiated power [remembering that the average value of $\cos^2 (\omega t - \varphi)$ is $\frac{1}{2}$] is just

$$\bar{P}_{\text{rad}} = \frac{1}{3} \frac{q^4 E_0{}^2 \omega^4}{m^2 c^3 [(\omega_0{}^2 - \omega^2)^2 + \gamma^2 \omega^2]} \tag{6-9-11}$$

The average incident power per unit area is just

$$|\bar{S}| = \frac{c}{4\pi} \frac{E_0{}^2}{2}$$

We have then, remembering that $r_0 = q^2/mc^2$,

$$\sigma_{\text{scattering}} = \frac{\bar{P}_{\text{rad}}}{|\bar{S}|} = \frac{8\pi}{3} r_0^2 \frac{\omega^4}{(\omega_0^2 - \omega^2)^2 + \gamma^2\omega^2} \tag{6-9-12}$$

Now, we might ask, what if the incoming radiation were not just a pure harmonic function of one frequency? We would like to make use of the power of Fourier analysis to achieve a solution to the problem. Let us assume again that our charge is at $z = 0$ and that the electric field at $z = 0$ is in the x direction. Let $E_x(t)$ be the field at the charge. Then we can write

$$E_x(t) = \frac{1}{\sqrt{2\pi}} \int_{-\infty}^{\infty} \mathscr{E}(\omega)e^{i\omega t} \, d\omega \tag{6-9-13}$$

We wish to solve the differential equation

$$\frac{d^2x}{dt^2} + \gamma \frac{dx}{dt} + \omega_0^2 x = \frac{qE_x(t)}{m} \tag{6-9-14}$$

If we let

$$x = \frac{1}{\sqrt{2\pi}} \int_{-\infty}^{\infty} G(\omega)e^{i\omega t} \, d\omega \tag{6-9-15}$$

then we obtain

$$\frac{1}{\sqrt{2\pi}} \int_{-\infty}^{\infty} \left\{ G(\omega)[-\omega^2 + i\gamma\omega + \omega_0^2] - q\frac{\mathscr{E}(\omega)}{m} \right\} e^{i\omega t} \, d\omega = 0 \tag{6-9-16}$$

If we wish this equation to be true for any value of t, we must have

$$G(\omega) = \frac{q\mathscr{E}(\omega)}{m(\omega_0^2 - \omega^2 + i\gamma\omega)} \tag{6-9-17}$$

We are thus in possession of the Fourier transform of x, which, upon integration, yields x directly:

$$x = \frac{q}{m\sqrt{2\pi}} \int_{-\infty}^{\infty} \frac{\mathscr{E}(\omega)}{(\omega_0^2 - \omega^2) + i\gamma\omega} e^{i\omega t} \, d\omega \tag{6-9-18}$$

To complete our solution we need only differentiate Eq. (6-9-18) twice and insert the result into the Larmor formula.

PROBLEMS

6-1. Consider a classical electron going in a circular orbit around a proton with an initial radius equal to r_0.

(a) Derive an expression for the radius r of the electron as a function of time.

(b) Calculate the time it would take for the electron to spiral into the proton if it started out at a radius of 10^{-8} cm. (This is actually typical of the time it takes for an electron to go from an excited state to the ground state.)

6-2. A nonrelativistic charged particle is brought to rest from an initial velocity v_0 by means of a constant acceleration a_0. That is,

$$v_x = \begin{cases} v_0 & \text{for } t \leq 0 \\ v_0 - a_0 t & \text{for } 0 \leq t \leq v_0/a_0 \\ 0 & \text{for } t \geq v_0/a_0 \end{cases}$$

What is the frequency distribution of the radiation seen by an observer at a great distance from the particle?

6-3. A relativistic particle is slowed down by means of a constant acceleration **a** directed opposite to its direction of motion. Plot the laboratory intensity per unit solid angle of the radiated energy as a function of $\cos \theta$ for $\beta = 0.6$, 0.8, and 0.95. The angle θ lies between the direction of motion of the particle and the direction of radiation in the laboratory.

6-4. Derive expressions for the electromagnetic fields **E** and **B** corresponding to a point charge moving with constant velocity **v** by transforming from the rest system of the charge into the laboratory frame of reference. Compare your results with Eqs. (6-3-9) and (6-3-11) as evaluated in the case where $\mathbf{a}' = 0$.

6-5. Consider a charge on the end of a spring moving according to the differential equation

$$\frac{d^2 x}{dt^2} + \gamma \frac{dx}{dt} + \omega_0^2 x = 0$$

where $\omega_0 \gg \gamma/2$. It starts out with some amplitude which then damps to zero exponentially.

(a) Find the frequency spectrum which characterizes the emitted electric and magnetic fields.

(b) Plot the intensity of radiation as a function of frequency in the neighborhood of ω_0. How does the width of the frequency distribution relate to the dissipation constant γ?

(c) What is the ratio of the mechanical energy stored in the system to the mechanical energy lost per cycle? (This number is called the Q of the system.)

6-6. Suppose that an oscillating charge is emitting electromagnetic radiation at a frequency ω_0. We turn on a receiver and investigate this radiation for a time period T. If we Fourier analyze the intensity of the received radiation, we will find that it is largely confined to a frequency range $\Delta\omega$ in the vicinity of ω_0. Carry out the Fourier transform of the observed radiation intensity and then estimate $\Delta\omega$.

6-7. A pair of equal and opposite point charges with electric dipole moment **p** rotate with angular frequency ω about an axis perpendicular to the line joining them.

Show that the energy radiated per unit time can be approximated by the expression

$$\frac{dU}{dt} = -\frac{2}{3}\frac{p^2\omega^4}{c^3}$$

in the event that the separation is much less than the wavelength of the radiation. Compare this with the sum of the energy which would be radiated were the two charges traveling separately and independently.

6-8. Suppose for a moment that magnetic monopoles existed. What are the radiation fields that would result from the acceleration of a point magnetic monopole $q^{(m)}$?

6-9. A magnetic dipole μ is perpendicular to an applied magnetic field and precesses about it with frequency ω. How much energy is radiated per unit time as a result of this precession if the wavelength corresponding to ω is much greater than any of the physical dimensions of the dipole? [Hint: Make use of the results of Probs. 6-7 and 6-8.]

7

The Interaction of Radiation with Matter

Having understood the interaction between radiation and a simple charge, be it free or be it bound in a harmonic potential, we are ready to take the next step and go to an assembly of charges. Inasmuch as matter is made up of just such an assembly, we would hope to be able to demonstrate all the well-known principles of optics by following the electromagnetic radiation in detail. The kinds of questions we will answer are the ones which have troubled almost anyone who has studied optics—how does the light get turned around in a mirror, and why does the light appear to go "slower" in glass than in air? Just as always, our approach will be the physical one of saying that currents produce fields which shake charges which produce more fields which shake other charges, ad infinitum. The beauty will lie in the rather simple way in which everything will fall together.

Now the actual situation within material at the microscopic level is quite complicated, and in the end we will need the power of quantum mechanics to deal with it exactly. There are numerous things we would have need to worry about including the absorption and reemission of radiation by atoms, the scattering of radiation by "free" electrons, the magnetic polarization of the atom, and even the electron spin. Our approach will be to solve several simple and idealized examples exactly and then develop a formalism so that when we have finally solved the microscopic problem exactly, we will be able to find the macroscopic properties of matter directly. The general expressions that we will derive will have far-reaching applicability (even into the realm of elementary particle physics), and even the simple models we take will come remarkably close to reality.

7-1 THE ABSORPTION AND REFLECTION OF RADIATION BY AN IDEALIZED CONDUCTING SHEET WITH NO MAGNETIZATION

We begin with the simplest of systems, radiation striking normally onto an idealized conducting sheet with conductivity σ. The sheet is assumed to have a magnetic permeability of unity; that is, no magnetic dipole moment is present anywhere. Our experience tells us that very little, if any, of the radiation is transmitted through the sheet. We also know that a large part of the radiation may be reflected back into the direction from which it came. We would like to understand these results in some detail.

As before we will assume that the origin of the incoming radiation is some accelerating charge a very long distance away. We can then consider that our incoming wave is a *plane wave*; the loci of points of equal field form planes perpendicular to the direction of propagation. As we just learned, we can subdivide the incoming wave into harmonic components and superimpose the results later; this permits us to use a simple incoming wave of the form[1]

$$E_x(z,t) = E_0 e^{-i(kz - \omega t)}$$
$$B_y(z,t) = E_0 e^{-i(kz - \omega t)}$$

$$(7\text{-}1\text{-}1)$$

where $\omega = kc$. We will, for convenience, take $E_y = B_x = 0$.

Now to begin our treatment of the problem, we place a sheet of conductor of infinitesimal thickness in the path of the radiation, as shown

---

[1] Our work will be immeasurably simplified by making use of the exponential in this way. If the true incoming wave has the form $E_0 \cos(kz - \omega t)$, then we need consider only the real parts of Eq. (7-1-1) and of all equations for the electromagnetic waves which derive from it.

in Fig. 7-1. We choose the thickness δ to be much less than $2\pi/k$, the wavelength of the radiation, and small enough so that the amount of radiation produced by the sheet is infinitesimal when compared with the incoming wave. We further assume that the current density induced in the conductor at any point is directly proportional to the electric field at that point. (Needless to say the proportionality constant may be frequency dependent. However at any given frequency, the assumption of proportionality between current density and electric field is a reasonable one.) Hence we write, at $z = 0$,

$$j_x = \sigma E_0 e^{i\omega t}$$
$$j_y = j_z = 0 \tag{7-1-2}$$

We now find the *induced* vector potential due to j_x at some point on the positive z axis at a distance z to the right of the sheet. (By **induced** we mean not present in the original incoming field but produced through the mediation of the current in the conducting sheet.) Using cylindrical coordinates, we write

$$A_x{}^{\text{ind}}(0,0,z,t) = \delta \int_0^{2\pi} d\theta \int_0^{r_{\text{max}}(\theta)} \frac{\sigma E_0 e^{i\omega(t - R/c)}}{R} r \, dr \tag{7-1-3}$$

Fig. 7-1 A thin sheet of conductor of thickness $\delta(\delta \ll 2\pi/k)$ is placed perpendicular to incoming radiation. Its conductivity is σ.

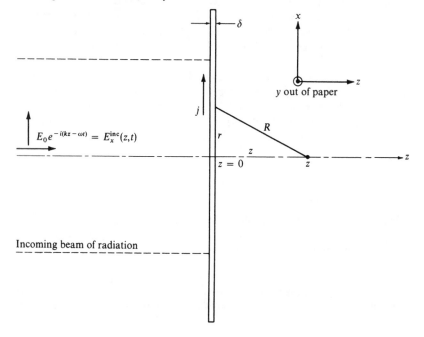

However, $r^2 + z^2 = R^2$, and, since z is fixed, we can write

$$2r \, dr = 2R \, dR \tag{7-1-4}$$

Substituting back into Eq. (7-1-3), we obtain

$$A_x^{\text{ind}} = \delta\sigma E_0 e^{i\omega t} \int_0^{2\pi} d\theta \int_z^{R_{\text{max}}(\theta)} e^{-ikR} \, dR$$

$$= \frac{\delta\sigma E_0 e^{i\omega t}}{ik} \int_0^{2\pi} d\theta (e^{-ikR_{\text{max}}(\theta)} - e^{-ikz}) \tag{7-1-5}$$

Now the term $e^{-ikR_{\text{max}}(\theta)}$ varies extremely rapidly with $R_{\text{max}}(\theta)$, taking on every possible phase if $R_{\text{max}}(\theta)$ varies as much as a wavelength. Hence we can ignore this term; it will average out to zero. This leads us to the result

$$A_x^{\text{ind}}(z,t) = \frac{2\pi\delta\sigma E_0}{ik} e^{-i(kz - \omega t)} \qquad \text{for } z > 0 \tag{7-1-6}$$

When we began, we chose our z axis arbitrarily, requiring only that it lie within our incoming beam of radiation. Hence the result we obtained in Eq. (7-1-6) is true for any point that lies a distance z to the right of the conducting sheet, provided that it is within the original beam. If we choose a point which is outside the original beam of radiation, we would have to replace z in Eq. (7-1-5) by $R_{\text{min}}(\theta)$ and the exponential would average out to zero. Hence there is no induced vector potential outside the original beam.

Now we can have a look at the induced vector potential to the left of the conducting sheet ($z < 0$). The complete symmetry of the situation requires that

$$A_x(-z) = A_x(z) \tag{7-1-7}$$

Hence we can write

$$A_x(z,t) = \frac{2\pi\delta\sigma E_0 e^{i(kz + \omega t)}}{ik} \qquad \text{for } z < 0 \tag{7-1-8}$$

We now proceed to find the induced electric and magnetic fields.

$$E_x^{\text{ind}}(z,t) = -\frac{1}{c} \frac{\partial A_x^{\text{ind}}}{\partial t}$$

$$= \begin{cases} -2\pi\delta\sigma E_0 e^{-i(kz - \omega t)} & \text{for } z > 0 \\ -2\pi\delta\sigma E_0 e^{i(kz + \omega t)} & \text{for } z < 0 \end{cases} \tag{7-1-9}$$

(We have made use of the fact that $k = \omega/c$ in obtaining the above expressions.)

$$B_y^{\text{ind}}(z,t) = (\nabla \times \mathbf{A}^{\text{ind}})_y$$

$$= \frac{\partial A_x^{\text{ind}}}{\partial z}$$

$$= \begin{cases} -2\pi\delta\sigma E_0 e^{-i(kz - \omega t)} & \text{for } z > 0 \\ 2\pi\delta\sigma E_0 e^{i(kz + \omega t)} & \text{for } z < 0 \end{cases} \qquad (7\text{-}1\text{-}10)$$

We should make a number of observations about the induced fields. First, going off to the right of the sheet is an induced plane wave proportional to and exactly $180°$ out of phase with the incident wave. When added to the incident wave, it diminishes its amplitude by a factor $1 - 2\pi\sigma\delta$. Obviously, $\mathbf{E}$ and $\mathbf{B}$ remain mutually perpendicular, and the Poynting vector is still in the $+z$ direction. Second, going off to the left of the sheet is a reflected plane wave. The electric field in this induced wave at position $-z$ is identical with the electric field in the induced wave at position $+z$. The induced magnetic fields, on the other band, are equal and opposite at equal distances to the left and right of the sheet. The Poynting vector associated with the reflected plane wave points naturally in the $-z$ direction. Finally we note that the amplitudes of the induced fields are independent of frequency (provided, of course, that $\delta \ll \lambda = 2\pi/k$). Any frequency dependence which we discover when dealing with a thick conducting plate must result then from the interference between the fields generated by layers at various depths below the surface.

Before proceeding, let us have a look at the energy balance within the thin sheet. We would like to show that the energy coming in per unit time minus the energy leaving per unit time is just equal to the ohmic heating of the sheet. Immediately to the right of the sheet we have

$$E_x(+0,t) = E_x^{\text{inc}}(0,t) + E_x^{\text{ind}}(+0,t) \qquad (7\text{-}1\text{-}11)$$

where $+0$ refers to "0" approached from the right. Using Eq. (7-1-9), we find

$$E_x(+0,t) = E_z^{\text{inc}}(0,t)(1 - 2\pi\delta\sigma) \qquad (7\text{-}1\text{-}12)$$

Similarly, using Eq. (7-1-10), we obtain

$$B_y(+0,t) = B_y^{\text{inc}}(0,t)(1 - 2\pi\delta\sigma) \qquad (7\text{-}1\text{-}13)$$

Combining these results, we derive the Poynting vector just to the right of the sheet.

$$S_z(+0,t) = \frac{c}{4\pi} E_x^{\text{inc}}(0,t) B_y^{\text{inc}}(0,t) (1 - 4\pi\delta\sigma + 4\pi^2\delta^2\sigma^2)$$

Ignoring terms of order δ^2, we have then

$$S_z(+0,t) = S_z^{\text{inc}}(0,t)(1 - 4\pi\delta\sigma)$$

$$= \frac{c}{4\pi} (1 - 4\pi\delta\sigma)[E_x{}^{\text{inc}}(0,t)]^2$$

= energy leaving unit area of sheet, to the right, per unit time

(7-1-14)

The amount of energy leaving to the left per unit time per unit area is proportional to δ^2 and can thus be ignored. The rate at which energy is being delivered to the sheet per unit area is then

$$\frac{dU}{dt} = c\delta\sigma \, [E_x{}^{\text{inc}}(0,t)]^2 \tag{7-1-15}$$

This is just equal to the rate of ohmic heating for a volume δ of material [see Eq. (4-1-6)].

Now that we know how to find the radiation from an infinitesimally thin slice of conducting sheet all of which has the same current density j_x, we can proceed to the case of a thick conducting plate being irradiated normal to one face by a plane wave of frequency ω. For convenience we will first allow the conductor to extend from $z = 0$ to $z = \infty$ (see Fig. 7-2). As before, the incoming radiation will have the form

$$E_x{}^{\text{inc}}(z,t) = E_0 e^{-i(kz - \omega t)}$$
$$B_y{}^{\text{inc}}(z,t) = E_0 e^{-i(kz - \omega t)}$$

(7-1-16)

Fig. 7-2 Incoming radiation is normally incident upon a semi-infinite conducting sheet with conductivity σ.

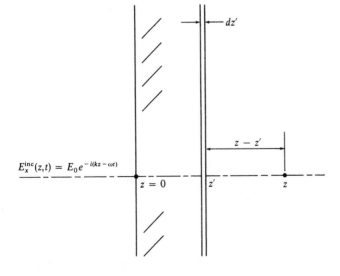

Naturally, we can slice our conductor into an infinitude of individual, infinitesimally thick slices and add together each of their contributions. A typical such slice, of thickness dz', is shown in Fig. 7-2. We are interested in finding the fields at a distance z from the face of the conductor, and our typical slice is taken at a distance z' from that face.

Unfortunately, however, the field which is shaking the charges at z' is precisely the field we would like to find. We can no longer make the approximation that this is equal to the incoming field. Hence we allow the function $E_x(z,t)$ to be an unknown. Inasmuch as our incoming plane wave has frequency ω, we shall assume that

$$E_x(z,t) = E_x(z)e^{i\omega t} \tag{7-1-17}$$

Now if our typical slice dz' at z' is to the left of z ($z \geqq z'$), then its contribution to the field at z is just given by [see Eq. (7-1-9)]

$$dE_x^{\text{ind}}(z,t) = -2\pi\sigma E_x(z')e^{-i[k(z-z')-\omega t]}\,dz' \tag{7-1-18}$$

On the other hand, if $z' > z$, then the contribution of dz' to the field at z is

$$dE_x^{\text{ind}}(z,t) = -2\pi\sigma E_x(z')e^{i[k(z-z')+\omega t]}\,dz' \tag{7-1-19}$$

The *total field* at z is given by a sum of three terms, the incident field, the contributions from slices to the left of z, and the contributions from slices to the right of z. Thus we have

$$E_x(z)e^{i\omega t} = E_0 e^{-ikz}e^{i\omega t} - 2\pi\sigma e^{i\omega t}\int_0^z E_x(z')e^{-ik(z-z')}\,dz'$$

$$-2\pi\sigma e^{i\omega t}\int_z^\infty E_x(z')e^{ik(z-z')}dz'$$

Canceling out $e^{i\omega t}$ and rewriting slightly, we have the basic integral equation for $E_x(z)$:

$$E_x(z) = E_0 e^{-ikz} - 2\pi\sigma e^{-ikz}\int_0^z E_x(z')e^{ikz'}\,dz'$$

$$-2\pi\sigma e^{ikz}\int_z^\infty E_x(z')e^{-ikz'}\,dz' \tag{7-1-20}$$

Similarly, we find the basic integral equation for $B_y(z)$ by adding together the three parts, the incoming plane wave, the contribution from slices to the left of z, and the contribution from slices to the right of z [(see Eq. (7-1-10)].

$$B_y(z) = E_0 e^{-ikz} - 2\pi\sigma e^{-ikz}\int_0^z E_x(z')e^{ikz'}\,dz'$$

$$+2\pi\sigma e^{ikz}\int_z^\infty E_x(z')e^{-ikz'}\,dz' \tag{7-1-21}$$

Now we address ourselves to finding the solution to Eqs. (7-1-20) and (7-1-21). The simplest way to proceed is to differentiate $E_x(z)$ twice with respect to z, obtaining a differential equation.

$$\frac{dE_x}{dz} = -ikE_0 e^{-ikz} + 2\pi\sigma ik e^{-ikz} \int_0^z E_x(z') e^{ikz'}\, dz'$$

$$- 2\pi\sigma ik e^{ikz} \int_z^\infty E_x(z') e^{-ikz'}\, dz'$$

$$\frac{d^2 E_x}{dz^2} = -k^2 E_0 e^{-ikz} + 2\pi\sigma k^2 e^{-ikz} \int_0^z E_x(z') e^{ikz'}\, dz'$$

$$+ 2\pi\sigma k^2 e^{ikz} \int_z^\infty E_x(z') e^{-ikz'}\, dz' + 4\pi\sigma ik E_x(z)$$

or

$$\frac{d^2 E_x}{dz^2} = (-k^2 + 4\pi\sigma ik) E_x(z) \tag{7-1-22}$$

We also note that

$$\frac{dE_x(z)}{dz} = -ikB_y \tag{7-1-23}$$

Differentiating $B_y(z)$ we will find that it satisfies the same differential equation as $E_x(z)$. In any case it will be much simpler to just calculate dE_x/dz and divide by $-ik$ to find B_y.

Before proceeding, we will make a slight digression. The sophisticated student will observe immediately that we could have obtained these differential equations much more rapidly if we had started with Maxwell's equations. We would however have lost all our beautiful insight into the origin of the fields in terms of flowing currents. Nevertheless we shall do so now if for no other reason than to provide the student with an easily remembered way of rapidly obtaining the appropriate coefficients in the differential equation. We start with

$$\nabla \times \mathbf{B} = 4\pi\mathbf{j} + \frac{1}{c}\frac{\partial \mathbf{E}}{\partial t} = 4\pi\sigma\mathbf{E} + \frac{1}{c}\frac{\partial \mathbf{E}}{\partial t}$$

$$\nabla \times \mathbf{E} = -\frac{1}{c}\frac{\partial \mathbf{B}}{\partial t}$$

Taking the curl of the second equation and remembering that in this case $\nabla \cdot \mathbf{E} = 4\pi\rho = 0$, we have

$$\nabla \times (\nabla \times \mathbf{E}) = \nabla(\nabla \cdot \mathbf{E}) - \nabla^2 \mathbf{E}$$

$$= -\nabla^2 \mathbf{E} = -\frac{1}{c}\frac{\partial}{\partial t}\nabla \times \mathbf{B}$$

$$= -\frac{1}{c}\left(4\pi\sigma \frac{\partial \mathbf{E}}{\partial t} + \frac{1}{c}\frac{\partial^2 \mathbf{E}}{\partial t^2}\right) \qquad (7\text{-}1\text{-}24)$$

If we search for a solution of the form

$$E_x(z,t) = E_x(z)e^{i\omega t} \qquad \text{and} \qquad E_y = E_z = 0$$

we obtain for our differential equation

$$\frac{d^2 E_x}{dz^2} = \frac{4\pi\sigma}{c}\, i\omega E_x - \frac{\omega^2}{c^2} E_x$$

$$= (4\pi\sigma ik - k^2)E_x$$

just as before. Taking the curl of $\nabla \times \mathbf{B}$ will obtain for us the same equation for B_y.

We return to the problem at hand, the solution of which is now apparent. In fact there are two possible solutions:

$$E_x(z) = \begin{cases} \exp\left(-\sqrt{4\pi\sigma ik - k^2}z\right) & (7\text{-}1\text{-}25) \\ \exp\left(+\sqrt{4\pi\sigma ik - k^2}z\right) & (7\text{-}1\text{-}26) \end{cases}$$

Our first problem is how do we evaluate $\sqrt{4\pi\sigma ik - k^2}$? The answer will be immediately apparent if we just write $4\pi\sigma ik - k^2 = \sqrt{k^4 + (4\pi\sigma k)^2}\, e^{i\varphi}$. Then

$$\sqrt{4\pi\rho ik - k^2} = [k^4 + (4\pi\sigma k)^2]^{\frac{1}{4}}e^{i\varphi/2} \qquad (7\text{-}1\text{-}27)$$

The important thing to note here is that $\varphi/2$ is always in the first quadrant, and hence $\sqrt{4\pi\sigma ik - k^2}$ has a *positive real part* if σ has any value at all other than zero (see Fig. 7-3). The solution given by Eq. (7-1-26) would *diverge* as we approached infinity and hence must be rejected. We are left then with

$$E_x(z) = A \exp\left(-\sqrt{4\pi\sigma ik - k^2}z\right) \qquad (7\text{-}1\text{-}28)$$

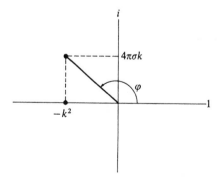

Fig. 7-3   The expression $-k^2 + 4\pi\sigma ik$ plotted in the complex plane.

We wish to find A to complete our solution to the problem. To do this we need only substitute back into our integral equation.

$$A \exp\left(-\sqrt{4\pi\sigma ik - k^2}z\right) = E_0 e^{-ikz}$$

$$- 2\pi\sigma A e^{-ikz} \int_0^z \exp\left[(ik - \sqrt{4\pi\sigma ik - k^2})z'\right] dz'$$

$$- 2\pi\sigma A e^{ikz} \int_z^\infty \exp\left[(-ik - \sqrt{4\pi\sigma ik - k^2})z'\right] dz' \qquad (7\text{-}1\text{-}29)$$

Carrying out the integrals indicated, we find

$$\left(E_0 + \frac{2\pi\sigma A}{ik - \sqrt{4\pi\sigma ik - k^2}}\right) e^{-ikz} = 0 \qquad (7\text{-}1\text{-}30)$$

This leads to the result

$$A = \frac{\sqrt{4\pi\sigma ik - k^2} - ik}{2\pi\sigma} E_0 \qquad (7\text{-}1\text{-}31)$$

We have then for the field within the conductor

$$E_x(z,t) = \frac{\sqrt{4\pi\sigma ik - k^2} - ik}{2\pi\sigma} E_0 \exp\left(-\sqrt{4\pi\sigma ik - k^2}z\right) e^{i\omega t} \qquad (7\text{-}1\text{-}32)$$

To find $B_y(z,t)$ we just differentiate E_x with respect to z and divide by $-ik$.

$$B_y(z,t) = \sqrt{4\pi\sigma ik - k^2} \frac{\sqrt{4\pi\sigma ik - k^2} - ik}{2\pi\sigma ik}$$

$$E_0 \exp\left(-\sqrt{4\pi\sigma ik - k^2}z\right) e^{i\omega t} \qquad (7\text{-}1\text{-}33)$$

Now these expressions look quite hopeless except for the fact that, for most metals over a large frequency range, we can ignore k by comparison with $4\pi\sigma$. In that frequency range we have

$$E_x(z,t) \cong \left[\sqrt{\frac{k}{2\pi\sigma}} + i\left(\sqrt{\frac{k}{2\pi\sigma}} - \frac{k}{2\pi\sigma}\right)\right] E_0 \exp\left(-\sqrt{2\pi\sigma kz}\right)$$

$$\exp\left[-i(\sqrt{2\pi\sigma kz} - \omega t)\right] \qquad (7\text{-}1\text{-}34)$$

$$B_y(z,t) \cong 2\left[\left(1 - \sqrt{\frac{k}{8\pi\sigma}}\right) - i\sqrt{\frac{k}{8\pi\sigma}}\right]$$

$$E_0 \exp\left(-\sqrt{2\pi\sigma kz}\right) \exp\left[-i(\sqrt{2\pi\sigma kz} - \omega t)\right] \qquad (7\text{-}1\text{-}35)$$

As the ratio of k to σ becomes smaller and smaller (either better conductivity or lower frequency), we can ignore $k/2\pi\sigma$ by comparison with $\sqrt{k/2\pi\sigma}$ and $\sqrt{k/8\pi\sigma}$ by comparison with 1. In that case we can further approximate E and B as follows:

$$E_x(z,t) \cong \sqrt{\frac{k}{\pi\sigma}} E_0 \exp(-\sqrt{2\pi\sigma k}z)$$

$$\exp\left[-i\left(\sqrt{2\pi\sigma k}z - \omega t - \frac{\pi}{4}\right)\right] \quad (7\text{-}1\text{-}36)$$

$$B_y(z,t) \cong 2E_0 \exp(-\sqrt{2\pi\sigma k}z) \exp[-i(\sqrt{2\pi\sigma k}z - \omega t)] \quad (7\text{-}1\text{-}37)$$

The magnetic field in the conductor thus has a magnitude that is much larger than that of the electric field. It also lags the electric field by about an eighth of a cycle in the case where $k/\sigma \ll 1$. Both fields drop off in magnitude as $\exp(-\sqrt{2\pi\sigma k}z)$. The term $1/\sqrt{2\pi\sigma k}$, called the **skin depth,** is the distance within which the fields drop by a factor of e.

Now to round out our evaluation of the electromagnetic fields in this simple example, let us calculate the reflected electric and magnetic fields. We make use of our integral equations again and note that

$$E_x{}^{\text{refl}}(z,t) = -2\pi\sigma e^{i(kz+\omega t)} A \int_0^\infty \exp\left[-(\sqrt{4\pi\sigma ik - k^2} + ik)z'\right] dz'$$

$$= -E_0 \frac{\sqrt{4\pi\sigma ik - k^2} - ik}{\sqrt{4\pi\sigma ik - k^2} + ik} e^{i(kz+\omega t)} \quad (7\text{-}1\text{-}38)$$

We obtain $B_y{}^{\text{refl}}(z,t)$ through the equation

$$B_y{}^{\text{refl}}(z,t) = -E_x{}^{\text{refl}}(z,t) \quad (7\text{-}1\text{-}39)$$

It is interesting to note one other rather important point which is apparent from looking at the integral equations for $E_x(z)$ and $B_y(z)$. Both $E_x(z)$ and $B_y(z)$ are continuous at $z = 0$. In general the tangential components of the electric and magnetic fields are both continuous when going across a nonmagnetic boundary. This can be ascertained through an inspection of Maxwell's equations, as follows.

Faraday's law states that

$$\oint \mathbf{E} \cdot d\mathbf{l} = -\frac{1}{c} \frac{\partial \Phi}{\partial t}$$

If we take our curve of integration as shown in Fig. 7-4, the leg δ can be taken arbitrarily small and hence Φ can be made arbitrarily small. Thus $E_T(\text{region } 1) = E_T(\text{region } 2)$. Similarly, unless there is a finite surface current within a layer of zero thickness, the equation

$$\nabla \times \mathbf{B} = 4\pi\mathbf{j} + \frac{1}{c} \frac{\partial \mathbf{E}}{\partial t}$$

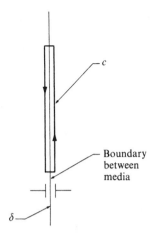

Fig. 7-4 The tangential components of E and B are continuous across the boundary of our nonmagnetic conductor.

ensures that the tangential component of **B** is continuous across the boundary. Obviously, if we are dealing with a piece of magnetic material, this continuity no longer holds because the sharp charge in magnetization is equivalent to a finite surface current.

These boundary conditions along with the solution to the differential equation for **E** and **B** in the conductor can serve to fix **E** and **B** completely without having recourse to the integral equation. Before we leave our conducting sheets, let us see what happens if we remove the restriction that they be of infinite thickness. We begin by treating one sheet of finite thickness and then extend our technique to a succession of any number of such sheets of varying conductivity. We take our sheet to lie between $z = 0$ and $z = a$ (see Fig. 7-5). For some point x within the sheet ($0 \leqq x \leqq a$), we can again express $E_z(x)$ in terms of a sum of the incoming field, the field contributed by currents on the left, and the field contributed by currents

Fig. 7-5 Electromagnetic radiation strikes normal to a conducting sheet of thickness a and conductivity σ.

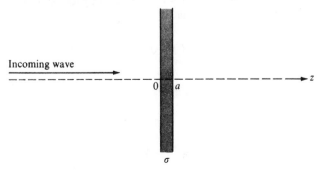

on the right. The equation is identical to Eq. (7-1-20) except for the fact that ∞ is replaced by a.

$$E_x(z) = E_0 e^{-ikz} - 2\pi\sigma e^{-ikz} \int_0^z E_x(z')e^{ikz'}\,dz'$$

$$- 2\pi\sigma e^{ikz} \int_z^a E_x(z')e^{-ikz'}\,dz' \quad (7\text{-}1\text{-}40)$$

Again $E_x(z)$ satisfies the differential equation $d^2E_x/dz^2 = (-k^2 + 4\pi\sigma ik)E_x$, and again we can write, in general,

$$E_x = A\exp\left(-\sqrt{4\pi\sigma ik - k^2}z\right) + B\exp\left(+\sqrt{4\pi\sigma ik - k^2}z\right) \quad (7\text{-}1\text{-}41)$$

This time, however, we have no right to remove the second solution $[\exp(\sqrt{4\pi\sigma ik - k^2}z)]$ because we are constrained to the region $0 \leq z \leq a$. If we substitute back into our integral equation, we find A and B by insisting that the equation hold at any two points $z = 0$ and $z = a$, for example. (The equation will then automatically hold at all other points in the conductor.) We obtain

$$A + B = E_0 - 2\pi\sigma \int_0^a \left[A\exp\left(-\sqrt{4\pi\sigma ik - k^2}z'\right)\right.$$

$$\left. + B\exp\left(+\sqrt{4\pi\sigma ik - k^2}z'\right)\right]e^{-ikz'}\,dz' \quad (7\text{-}1\text{-}42)$$

and

$$A\exp\left(-\sqrt{4\pi\sigma ik - k^2}a\right) + B\exp\left(+\sqrt{4\pi\sigma ik - k^2}a\right) = E_0 e^{-ika}$$

$$- 2\pi\sigma e^{-ika} \int_0^a \left[A\exp\left(-\sqrt{4\pi\sigma ik - k^2}z'\right)\right.$$

$$\left. + B\exp\left(+\sqrt{4\pi\sigma ik - k^2}z'\right)\right]e^{ikz'}\,dz' \quad (7\text{-}1\text{-}43)$$

As you can easily see, the solution to the problem, while messy, is quite straightforward. There are two independent equations for A and B in terms of E_0, σ, k, and a. Having found A and B, we can proceed to find the reflected and transmitted waves just as before:

$$E_x^{\text{refl}}(z,t) = -2\pi\sigma e^{i(kz + \omega t)}\left[\int_0^a A\exp\left(-\sqrt{4\pi\sigma ik - k^2}z'\right)e^{-ikz'}\,dz'\right.$$

$$\left. + \int_0^a B\exp\left(+\sqrt{4\pi\sigma ik - k^2}z'\right)e^{-ikz'}\,dz'\right] \quad (7\text{-}1\text{-}44)$$

$$E_x^{\text{trans}}(z,t) = E_0 e^{-i(kz - \omega t)} - 2\pi\sigma e^{-i(kz - \omega t)}$$

$$\times \left[\int_0^a A\exp\left(-\sqrt{4\pi\sigma ik - k^2}z'\right)e^{ikz'}\,dz'\right.$$

$$+ \int_0^a B \exp \left(+\sqrt{4\pi\sigma ik - k^2} z' \right) e^{ikz'} \, dz' \right] \quad (7\text{-}1\text{-}45)$$

Suppose now that instead of having to deal with only one sheet we had two sheets, as shown in Fig. 7-6, the first with conductivity σ_1 and the second with conductivity σ_2. If we let $E_{x,1}(z,t)$ and $E_{x,2}(z,t)$ be the solutions inside the two conductors, respectively, we can write, in general,

$$E_{x,1}(z,t) = E_{x,1}(z)e^{i\omega t}$$

$$= [A \exp \left(-\sqrt{4\pi\sigma_1 ik - k^2} z \right)$$

$$+ B \exp \left(+\sqrt{4\pi\sigma_1 ik - k^2} z \right)]e^{i\omega t} \quad (7\text{-}1\text{-}46)$$

$$E_{x,2}(z,t) = E_{x,2}(z)e^{i\omega t}$$

$$= [C \exp \left(-\sqrt{4\pi\sigma_2 ik - k^2} z \right)$$

$$+ D \exp \left(+\sqrt{4\pi\sigma_2 ik - k^2} z \right)]e^{i\omega t} \quad (7\text{-}1\text{-}47)$$

We can write down two integral equations

$$E_{x,1}(z) = E_0 e^{-ikz} - 2\pi\sigma_1 e^{-ikz} \int_0^z E_{x,1}(z')e^{+ikz'} \, dz'$$

$$- 2\pi\sigma_1 e^{+ikz} \int_z^a E_{x,1}(z')e^{-ikz'} \, dz'$$

$$- 2\pi\sigma_2 e^{ikz} \int_b^c E_{x,2}(z')e^{-ikz'} \, dz' \quad (7\text{-}1\text{-}48)$$

$$E_{x,2}(z) = E_0 e^{-ikz} - 2\pi\sigma_1 e^{-ikz} \int_0^a E_{x,1}(z')e^{ikz'} \, dz'$$

Fig. 7-6 Electromagnetic radiation strikes normal to two conducting sheets.

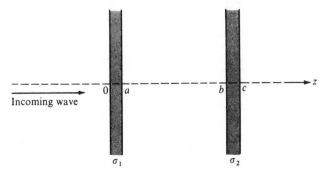

$$- 2\pi\sigma_2 e^{-ikz} \int_c^z E_{x,2}(z')e^{ikz'}\,dz' - 2\pi\sigma_2 e^{ikz} \int_z^d E_{x,2}(z')e^{-ikz'}\,dz'$$

$$(7\text{-}1\text{-}49)$$

If we insert Eqs. (7-1-46) and (7-1-47) for $E_{x,1}(z)$ and $E_{x,2}(z)$ into our integrals and evaluate Eq. (7-1-48) at $z = 0,a$ and Eq. (7-1-49) at $z = c,d$, we will have four simultaneous equations for A,B,C,D. Thus our solution to $E_x(z)$ will be complete at least within the conductors. The extension to the regions outside of the conductors is trivial and is left to the reader.

We might add here that any number of parallel conducting sheets can be treated in this manner. For N sheets we will need $2N$ coefficients for the solutions, and the N integral equations will give them to us.

7-2   WE ALLOW THE CONDUCTOR TO HAVE MAGNETIC PERMEABILITY μ

We will now generalize slightly by allowing our conductor to have uniform magnetic permeability μ. We can of course retrace all our steps and ask what a thin sheet of magnetic-moment distribution will yield for $\mathbf{A}$ and then for $\mathbf{E}$ and $\mathbf{B}$. It is simpler however to just make use of Maxwell's equations in the same manner as we did on page 241. We now have

$$\nabla \times \left(\frac{\mathbf{B}}{\mu}\right) = 4\pi\sigma\mathbf{E} + \frac{1}{c}\,\frac{\partial \mathbf{E}}{\partial t}$$

$$\nabla \times \mathbf{E} = \frac{-1}{c}\,\frac{\partial \mathbf{B}}{\partial t}$$

$$(7\text{-}2\text{-}1)$$

This leads to the equations

$$\nabla^2\mathbf{E} = \frac{4\pi\sigma\mu}{c}\,\frac{\partial \mathbf{E}}{\partial t} + \frac{\mu}{c^2}\,\frac{\partial^2 \mathbf{E}}{\partial t^2}$$

$$\nabla^2\mathbf{B} = \frac{4\pi\sigma\mu}{c}\,\frac{\partial \mathbf{B}}{\partial t} + \frac{\mu}{c^2}\,\frac{\partial^2 \mathbf{B}}{\partial t^2}$$

$$(7\text{-}2\text{-}2)$$

Again we search for solutions of the form

$$E_x(z,t) = E_x(z)e^{i\omega t}$$

$$E_y = E_x = 0$$

$$(7\text{-}2\text{-}3)$$

We find the differential equation for E_x to be

$$\frac{d^2 E_x}{dz^2} = (4\pi\sigma\mu ik - \mu k^2)E_x$$

$$(7\text{-}2\text{-}4)$$

The result is then

$$E_x(z) = \begin{cases} \exp\left[-\sqrt{\mu(4\pi\sigma ik - k^2)z}\right] & \text{(7-2-5)} \\ \exp\left[\sqrt{\mu(4\pi\sigma ik - k^2)z}\right] & \text{(7-2-6)} \end{cases}$$

Let us now find the complete set of solutions for the case of the semi-infinite slab of conductor extending from $z = 0$ to $z = \infty$. As before, we can reject solution Eq. (7-2-6) as diverging at infinity. This yields

$$E_x(z,t) = A\exp\left[-\sqrt{\mu(4\pi\sigma ik - k^2)z}\right]e^{i\omega t} \qquad \text{for } z > 0 \qquad \text{(7-2-7)}$$

The magnetic field B_y can be obtained from Maxwell's equation:

$$\frac{\partial E_x}{\partial z} = -\frac{1}{c}\frac{\partial B_y}{\partial t} = -ikB_y \qquad \text{(7-2-8)}$$

Hence

$$B_y(z,t) = -\frac{1}{ik}\frac{\partial E_x}{\partial z}$$

$$= \frac{A\sqrt{\mu(4\pi\sigma ik - k^2)}}{ik}\exp\left[-\sqrt{\mu(4\pi\sigma ik - k^2)z}\right]e^{i\omega t}$$

$$\text{for } z > 0 \quad \text{(7-2-9)}$$

The incident wave has the form

$$E_x^{\text{inc}}(z,t) = E_0 e^{-i(kz - \omega t)} \qquad \text{(7-2-10)}$$

$$B_y^{\text{inc}}(z,t) = E_0 e^{-i(kz - \omega t)} \qquad \text{(7-2-11)}$$

The reflected wave has the form

$$E_x^{\text{refl}}(z,t) = De^{i(kz + \omega t)} \qquad \text{(7-2-12)}$$

$$B_y^{\text{refl}}(z,t) = -De^{i(kz + \omega t)} \qquad \text{(7-2-13)}$$

where D is a constant, to be determined from the boundary conditions.

Our job is now reduced to finding A and D. We do this by requiring that the tangential components of **H** and **E** be continuous across the boundary. This yields

$$E_0 + D = A$$

$$E_0 - D = \frac{A\sqrt{\mu(4\pi\sigma ik - k^2)}}{ik\mu} \qquad \text{(7-2-14)}$$

The solutions for A and D are

$$A = \frac{2ik\mu E_0}{\sqrt{\mu(4\pi\sigma ik - k^2)} + ik\mu} \tag{7-2-15}$$

$$D = \frac{ik\mu - \sqrt{\mu(4\pi\sigma ik - k^2)}}{ik\mu + \sqrt{\mu(4\pi\sigma ik - k^2)}} E_0 \tag{7-2-16}$$

Again, it is instructive to look at these solutions in the limit of large σ. We have then

$$A \cong \frac{2ik\mu}{\sqrt{4\pi\sigma ik\mu}} E_0 = \sqrt{\frac{ik\mu}{\pi\sigma}} E_0 \tag{7-2-17}$$

The fields in the conductor become

$$E_x(z,t) = \sqrt{\frac{k\mu}{\pi\sigma}} E_0 \exp(-\sqrt{2\pi\sigma k\mu}\, z)$$

$$\exp\left[-i\left(\sqrt{2\pi\sigma k\mu}\, z - \omega t - \frac{\pi}{4}\right)\right] \tag{7-2-18}$$

$$B_y(z,t) = 2\mu E_0 \exp(-\sqrt{2\pi\sigma k\mu}\, z)\exp\left[-i(\sqrt{2\pi\sigma k\mu}\, z - \omega t)\right] \tag{7-2-19}$$

We compare with Eqs. (7-1-36) and (7-1-37) to note the approximate effect of the permeability μ.

1. The skin depth is decreased by the factor $\sqrt{\mu}$.
2. The magnetic field B_y just inside the conductor's surface is increased by a factor of μ.
3. The electric field just inside the conductors is increased by a factor of $\sqrt{\mu}$.

7-3 THE PHYSICAL ORIGIN OF THE REFRACTIVE INDEX

We now find ourselves in the beautiful position of having developed a machine so powerful that we can just stand back and pile up our profits. When we carried out our analysis of the interaction of radiation with a conducting sheet we made only one simple assumption, that the current density $j_x(z)$ was proportional to the electric field at that point $E_x(z)$. We called the proportionality constant σ and identified it with the conductivity of the material. However, nothing that we did after that implied that the constant had to be real (as it is in the case of free electrons in a conductor) or even that it had to be frequency independent. The physical arguments which we followed for adding together the effects of all the currents at various values of z would have been the same if σ had been a complex number with some frequency dependence. Thus we open up a wide range of applicability for the treatment we have just completed. In particular, we can deal with all situations where the local induced currents are proportional to the applied electric field, provided, of course, that the medium in question is sufficiently uniform and dense so that the variation in the number of scattering centers within a volume of order λ^3 (λ is the wavelength of the radiation) is negligible. (It is the violation of this last proviso that causes the sky to be blue. Why?)

We will now begin with a rather simple model of a dielectric as a collection of charges, each of which is trapped within a harmonic oscillator potential. Using this model, we will show that the currents which flow are indeed proportional to the applied fields. By carrying over the expressions we have derived for a conductor, we will discover that dielectrics appear to have a classical index of refraction, and we will derive an expression for this index in terms of the frequency of the incident radiation, the natural frequency of the oscillations, and the density of electrons. We will then show how it is possible to generalize our results to permit us to treat a dielectric exactly, given a description of the microscopic *forward-scattering amplitude*.

Consider first an individual electron of charge q and mass m in a harmonic oscillator potential with spring constant $m\omega_0{}^2$. (Its resonant frequency is ω_0.) We let the incident radiation on this electron be of the form

$$E_x(z,t) = E_0 e^{-i(kz - \omega t)} \tag{7-3-1}$$

If the electron is at $z = 0$ and it sees a resistive force equal to $m\gamma\, dx/dt$, then the differential equation describing its motion (see Sec. 6-9) is

$$\frac{d^2 x}{dt^2} + \gamma\, \frac{dx}{dt} + \omega_0{}^2 x = \frac{qE_0}{m} e^{i\omega t} \tag{7-3-2}$$

Letting $x = x_0 e^{i\omega t}$, we find

$$x_0 = \frac{qE_0}{m(\omega_0{}^2 - \omega^2 + i\gamma\omega)} \tag{7-3-3}$$

Suppose we now take a thin sheet of material made up of N such charges per unit volume. If we subject this sheet to an electric field $E_x = E_0 e^{i\omega t}$, then we will induce a current density

$$
\begin{aligned}
j_x &= N\left(\frac{q}{c}\right)\left(\frac{dx}{dt}\right) \\
&= N\left(\frac{q}{c}\right)(i\omega)\, \frac{qE_0 e^{i\omega t}}{m(\omega_0{}^2 - \omega^2 + i\gamma\omega)}
\end{aligned} \tag{7-3-4}
$$

(The factor c converts j_x into emu.) Hence it appears as though the material has an effective *conductivity* given by

$$\sigma_{\text{eff}} = \frac{iNq^2k}{m(\omega_0^2 - \omega^2 + i\gamma\omega)} \qquad (7\text{-}3\text{-}5)$$

From here on, we can make direct use of all the machinery we developed in Sec. 7-1. Wherever we find an expression involving σ, we can use σ_{eff} instead.

Let us return then to the semi-infinite slab of dielectric extending from $z = 0$ to $z = \infty$ (see Fig. 7-2). Again we can take the incoming radiation to have the form

$$E_x^{\text{inc}}(z,t) = E_0 e^{-i(kz - \omega t)} \qquad (7\text{-}3\text{-}6)$$

There are two possible solutions for the field in the dielectric. Using Eqs. (7-1-25) and (7-1-26) and substituting σ_{eff} for σ, we find

$$E_x(z) = \begin{cases} \exp\left[-\sqrt{-k^2 - \dfrac{4\pi q^2 k^2 N}{m(\omega_0^2 - \omega^2 + i\gamma\omega)}}\; z \right] & (7\text{-}3\text{-}7) \\[3mm] \exp\left[\sqrt{-k^2 - \dfrac{4\pi q^2 k^2 N}{m(\omega_0^2 - \omega^2 + i\gamma\omega)}}\; z \right] & (7\text{-}3\text{-}8) \end{cases}$$

Simple inspection will show that the radical will have a positive real part, and hence Eq. (7-3-8) can be removed as a possible solution. This leaves us with

$$E_x(z,t) = A e^{-i(knz - \omega t)} \qquad (7\text{-}3\text{-}9)$$

where

$$n = \sqrt{1 + \frac{4\pi q^2 N}{m(\omega_0^2 - \omega^2 + i\gamma\omega)}} \qquad (7\text{-}3\text{-}10)$$

The quantity n is called the **index of refraction.** We see that a given phase of the plane wave within the dielectric moves with an apparent velocity given by

$$v_{\text{phase}} = \frac{c}{n} \qquad (7\text{-}3\text{-}11)$$

Going back to Eq. (7-1-31), we find A:

$$A = \frac{\sqrt{4\pi\sigma_{\text{eff}} ik - k^2} - ik}{2\pi\sigma_{\text{eff}}} E_0$$

$$= \frac{ink - ik}{2\pi\sigma_{\text{eff}}} E_0 = \frac{2(n-1)}{n^2 - 1} E_0$$

$$= \frac{2E_0}{n+1} \qquad (7\text{-}3\text{-}12)$$

We have then our final solution for the electric field in the dielectric:

$$E_x(z,t) = \frac{2E_0}{n+1} e^{-i(knz - \omega t)} \qquad \text{for } z > 0 \qquad (7\text{-}3\text{-}13)$$

The magnetic field $B_y(z,t)$ can be obtained by differentiating E_x with respect to z and dividing by $-ik$.

$$B_y(z,t) = \frac{2nE_0}{n+1} e^{-i(knz - \omega t)} \qquad \text{for } z > 0 \qquad (7\text{-}3\text{-}14)$$

To obtain the reflected waves we make use of Eqs. (7-1-38) and (7-1-39):

$$E_x^{\text{refl}}(z,t) = -\frac{ink - ik}{ink + ik} E_0 e^{i(kz + \omega t)}$$

$$= \frac{1-n}{1+n} E_0 e^{i(kz + \omega t)} \qquad \text{for } z < 0 \qquad (7\text{-}3\text{-}15)$$

$$B_y^{\text{refl}}(z,t) = \frac{n-1}{n+1} E_0 e^{i(kz + \omega t)} \qquad \text{for } z < 0 \qquad (7\text{-}3\text{-}16)$$

7-4 WHAT HAPPENS WHEN $n < 1$? PHASE VELOCITY AND GROUP VELOCITY

We will now have a good hard look at our expression for index of refraction and see if we can understand something about the propagation of signals within the corresponding medium. To simplify our considerations we will set $\gamma = 0$. This is not a bad approximation for most dielectrics far from their resonant frequencies; a detailed treatment of the behavior of a dielectric in the vicinity of its resonances is beyond the scope of this book. Under this assumption we can write

$$n = \sqrt{1 + \frac{4\pi q^2 N}{m(\omega_0^2 - \omega^2)}} \qquad (7\text{-}4\text{-}1)$$

Again we remember that our wave travels within this medium as $e^{-i(knz - \omega t)}$. But what does this mean? Can it be that for $n < 1$ (which happens whenever $\omega > \omega_0$) we have a signal traveling faster than c? This would seem to be in patent violation of everything we have learned when we studied relativity. Furthermore, it seems absurd, since we never added together anything but waves traveling at velocity c.

The answer of course is very simple. Just because a wave appears to travel with velocity greater than c does not mean that information is moving with that same velocity. We will illustrate by making use of an elementary mechanical example.

Let us set out a long row of uncoupled springs as shown in Fig. 7-7. Each spring has a mass m hanging on it and has a spring constant equal to $m\omega_0^2$. (Hence the spring and mass will oscillate naturally at frequency ω_0.) We take the distance between springs to be δz.

Suppose we now pluck the first mass so as to set it into oscillation. A very short time δt later we pluck the second mass in exactly the same way. A time δt after that we pluck the third mass, and so on until all the masses are oscillating. As we watch the springs oscillate, it will look as though the peak of the oscillation is moving along with a *phase velocity* equal to $\delta z/\delta t$. Since δt can be made arbitrarily small, the phase velocity can be made arbitrarily large (see Fig. 7-7).

So far nothing bothers us at all because no information is being transferred from one spring to the next. They are all moving completely independently. Suppose we now wanted to transfer information along the chain. We would have to link the masses by some means and then we would have to give one mass an extra pluck so that it had a bit more energy than the others. This extra bit of energy would then be transferred down the line, and we would see the various oscillators gain amplitude and then lose amplitude again as the information passed through them. Needless to say, the velocity with which the peak amplitude moved down the line would not in general be the same as the phase velocity.

From looking at our mechanical model we now can extract the key idea. Information is transferred *only* when energy gets transferred, and hence there must somehow be a change in the world wide distribution of energy density. As long as we are dealing with nothing but a plane wave of fixed frequency extending from $-\infty$ to $+\infty$, there is no such change with time. Either we have to turn the light on at some time, in which case we would have to track the leading edge of the light beam, or we would have to produce fluctuations in the energy density of our beam and watch those fluctuations move along.

Consider first the problem of tracking the leading edge of a light

Fig. 7-7   We cause a wave to propagate with phase velocity equal to $\delta z/\delta t$ by plucking the masses sequentially with time interval δt between plucks (see text). No information is being transferred because the masses are all uncoupled.

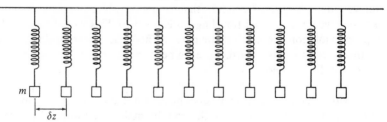

beam that has just been turned on. Initially the individual electrons as they were struck by this light beam would exhibit all the transient behavior which we have so far blithely ignored. We could solve the problem exactly by making our light beam up out of a superposition of different frequencies using Fourier analysis techniques. We could then include the transient terms and add everything together. Needless to say, the very leading edge would not move faster than c. (In fact, as seems intuitively reasonable, it will move with a velocity just equal to c.) In any case, to solve this problem exactly is rather tedious and somewhat beyond our scope.

How about producing fluctuations in the energy density of our electromagnetic wave and watching those fluctuations propagate? We can do this rather easily by adding together two waves with frequencies that lie very close to one another. We let $\kappa = kn$ and use frequencies ω and $\omega + d\omega$ which correspond, respectively, to κ and $\kappa + d\kappa$. We have then

$$
\begin{aligned}
E_x(z,t) &= E_0 e^{-i(\kappa z - \omega t)} + E_1 e^{-i[(\kappa + d\kappa)z - (\omega + d\omega)t]} \\
&= e^{-i(\kappa z - \omega t)}(E_0 + E_1 e^{-i(d\kappa z - d\omega t)})
\end{aligned}
\tag{7-4-2}
$$

As can be readily seen, the energy density corresponding to E_x varies with z and has an apparent wavelength equal to $2\pi/d\kappa$. The peaks in energy also move with velocity equal to $d\omega/d\kappa$. This velocity is the one we are interested in evaluating and is called the **group velocity** v_g.

$$
\begin{aligned}
v_g &= \frac{d\omega}{d\kappa} \\
&= \frac{cn}{1 + \dfrac{4\pi q^2 N \omega_0{}^2}{m(\omega^2 - \omega_0{}^2)^2}}
\end{aligned}
\tag{7-4-3}
$$

Inspection shows that v_g is always less than c. When $\omega > \omega_0$, the index n is less than 1 and the denominator is greater than 1. For $\omega < \omega_0$, the denominator is always greater than the numerator divided by c. In the limit that we can ignore ω^2 compared with $\omega_0{}^2$, we have

$$
v_g = \frac{cn}{n^2} = \frac{c}{n} = v_p = \text{phase velocity}
$$

Finally, before we leave these harmonic oscillators, let us see how reasonable our expression for index of refraction is. We would not have much use for it if it yielded a value of 1000 in the visible when everyone knows that for most transparent materials with a density of about 1 gram/cm^3 the index is about 1.5.

First we note that ω_0 is typically in the ultraviolet for a material like glass. Let us take $\omega_0 = 2 \times 10^{16}$ rad/sec and $\omega = 4 \times 10^{15}$ rad/sec.

Let us take $N \cong$ Avogadro's number $= 6 \times 10^{23}$. We know that

$$\frac{q^2}{m} = (c^2) \text{ (classical electron radius } r_0)$$

$$= (9 \times 10^{20})(2.8 \times 10^{-13})$$

We have then

$$n \cong \sqrt{1 + \frac{4\pi(6 \times 10^{23})(2.5 \times 10^8)}{4 \times 10^{32}}}$$

$$= \sqrt{5.5} \cong 2.3$$

Not bad for a crude guess with a crude model.

7-5 THE INDEX OF REFRACTION IN TERMS OF THE FORWARD-SCATTERING AMPLITUDE

Unfortunately, the real world is not made up of a collection of simple harmonic oscillators, and we would never be able to handle a real problem unless we developed some techniques which had broad practical applicability. It turns out to be quite simple to do so. We will assume that the microscopic problem of the interaction of radiation with a typical atom has been solved and show how we can go from there to an expression for the index of refraction of macroscopic matter made up of these atoms.

We will assume that the entire atom is coherent insofar as the frequencies that interest us are concerned. By this we mean that the atom is so much smaller than the wavelength of light that we need not pay any attention to its physical dimension. Now if we allow a plane wave of radiation to fall upon the atom, it will absorb and reradiate some of it in all directions. In particular, some of it will be reradiated in the same direction as the incoming radiation, and in interfering with the latter will lead to the effective index of refraction.

It is clear that our index of refraction should depend then on the number of atoms per unit volume, on the frequency of the incident light, and on the ratio of the amplitude of the forward-scattered radiation to the amplitude of the incoming radiation. To make the connection we will resort to a very simple trick. We will replace our atom with a classical accelerating charge that leads to exactly the same amplitude of forwardly scattered radiation. We will then make use of what we have learned to calculate the *effective conductivity* for an assembly of these classical charges. Having done this, we can immediately write down an expression for the index of refraction.

To begin with, let us define the so-called **forward-scattering amplitude** $f(0)$ for a single atom, as follows.

$$f(0) = \frac{\text{amplitude of forwardly emitted radiation}}{\text{at point 1 cm directly forward of scatterer}}$$
$$\frac{}{\text{amplitude that incident radiation would have}}$$
$$\text{at that same point in absence of scatterer}$$

(It is of course assumed that the incoming radiation and the forwardly emitted radiation have the same polarization. We shall shortly learn how to deal with the more general case where this assumption is removed.) Suppose now that we had an incident plane wave given by

$$E_x(z,t) = E_0 e^{-i(kz - \omega t)} \tag{7-5-1}$$

Imagine now that we place our classical charge at $z = 0$ and allow it to have an acceleration given by

$$a_x = a_0 e^{i\omega t} \tag{7-5-2}$$

At a distance z *in front* of the charge we would see a field due to this acceleration and equal to [see Eq. (6-4-2)]

$$E_x(\text{due to charge}) = \frac{-qa_0 e^{i\omega(t - z/c)}}{c^2 z}$$
$$= \frac{-qa_0 e^{-i(kz - \omega t)}}{c^2 z} \tag{7-5-3}$$

The forward-scattering amplitude will then be given by

$$f(0) = \frac{-qa_0}{c^2 E_0} \tag{7-5-4}$$

In other words, if we know $f(0)$, then we can replace our atom with a charge whose acceleration is just given by

$$a_x = \frac{-c^2 f(0) E_0}{q} e^{i\omega t} \tag{7-5-5}$$

The velocity of the charge is just

$$v_x = \frac{ic^2 f(0) E_0}{\omega q} e^{i\omega t} \tag{7-5-6}$$

If we make up a macroscopic sheet of these charges with N charges per unit volume, we have a current density given by

$$j_x = N \frac{q}{c} v_x = \frac{Nif(0) E_0 e^{i\omega t}}{k} \tag{7-5-7}$$

The effective conductivity is then

$$\sigma_{\text{eff}} = \frac{Nif(0)}{k} \tag{7-5-8}$$

Substituting back into Eq. (7-1-25) we find

$$n = \sqrt{1 + \frac{4\pi Nf(0)}{k^2}} \tag{7-5-9}$$

This expression relates the index of refraction to the atomic forward-scattering amplitude $f(0)$ and the number of atoms per unit volume, N. The expression is exact for wavelengths much larger than atomic dimensions and permits us to solve the macroscopic problem as soon as we have solved the microscopic problem.

7-6 THE FARADAY EFFECT

Of the various phenomena of physical optics none is more beautifully illustrative of the techniques and ideas we have developed than the Faraday effect. Let us apply an external magnetic field to a dielectric and then introduce plane-polarized radiation into it along the direction of the magnetic field. We will find that the polarization direction will rotate through an angle proportional to the distance we penetrate into the medium. The ratio of angle turned to distance traversed will depend approximately linearly on the magnetic field for reasonable fields. When evaluated for a unit magnetic field the ratio is called the **Verdet constant.** We will find in addition that this entire phenomenon can be understood in terms of a difference in the index of refraction for left- and right-hand circularly polarized light.

We begin as usual by considering a rather simple model, an electron of charge q and mass m bound in a harmonic oscillator potential. We will assume a restoring force equal to $-m\omega_0{}^2\mathbf{r}$ where $\mathbf{r}$ is the vector displacement from the equilibrium position. In addition we will assume an applied magnetic field $B\hat{\varepsilon}_3$ in the z direction and incoming radiation of the form

$$\mathbf{E}(z,t) = E_{0x}e^{-i(kz-\omega t)}\hat{\varepsilon}_1 + E_{0y}e^{-i(kz-\omega t)}\hat{\varepsilon}_2 \tag{7-6-1}$$

(We use $\hat{\varepsilon}_1$, $\hat{\varepsilon}_2$, and $\hat{\varepsilon}_3$ as the unit vectors in the x, y, z directions to avoid confusion with other quantities labeled by i, j, k.)

For convenience we will place the center of our harmonic oscillator at the origin of our coordinate system. We set the force on the charge equal to its mass times its acceleration

$$m\left(\frac{d^2x}{dt^2}\hat{\varepsilon}_1 + \frac{d^2y}{dt^2}\hat{\varepsilon}_2\right) = -m\omega_0{}^2(x\hat{\varepsilon}_1 + y\hat{\varepsilon}_2) + qE_{0x}e^{i\omega t}\hat{\varepsilon}_1$$
$$+ qE_{0y}e^{i\omega t}\hat{\varepsilon}_2 + \frac{qB}{c}\frac{dy}{dt}\hat{\varepsilon}_1 - \frac{qB}{c}\frac{dx}{dt}\hat{\varepsilon}_2 \tag{7-6-2}$$

Now, rearranging terms and letting ω_c = cyclotron frequency = qB/mc, we have two equations

$$\frac{d^2 x}{dt^2} + \omega_0^2 x - \omega_c \frac{dy}{dt} = \frac{qE_{0x}}{m} e^{i\omega t} \tag{7-6-3}$$

$$\frac{d^2 y}{dt^2} + \omega_0^2 y + \omega_c \frac{dx}{dt} = \frac{qE_{0y}}{m} e^{i\omega t} \tag{7-6-4}$$

Letting $x = x_0 e^{i\omega t}$ and $y = y_0 e^{i\omega t}$, we find

$$(\omega_0^2 - \omega^2)x_0 - i\omega\omega_c y_0 = \frac{qE_{0x}}{m} \tag{7-6-5}$$

$$(\omega_0^2 - \omega^2)y_0 + i\omega\omega_c x_0 = \frac{qE_{0y}}{m} \tag{7-6-6}$$

We solve for y_0 and x_0 and obtain

$$x_0 = \frac{q[(\omega_0^2 - \omega^2)E_{0x} + i\omega\omega_c E_{0y}]}{m[(\omega_0^2 - \omega^2)^2 - \omega^2\omega_c^2]} \tag{7-6-7}$$

and

$$y_0 = \frac{q[(\omega_0^2 - \omega^2)E_{0y} - i\omega\omega_c E_{0x}]}{m[(\omega_0^2 - \omega^2)^2 - \omega^2\omega_c^2]} \tag{7-6-8}$$

We can now proceed in one of two ways. We can calculate the currents in the x and y directions in terms of E_x and E_y, leading to effective conductivities (in this case, four numbers), and then put everything together as we have before. Alternatively, we can think a bit about the physical situation and by being clever come to a conclusion much more rapidly. We shall do both inasmuch as cleverness is not always a substitute for brute force in the world of real problems.

First, let us be clever. We ask what would happen if the incident radiation were circularly polarized? Certainly the trapped electron has no choice but to go in a circle since there is no preferred direction defined in the xy plane. The magnetic field then would either push the electron toward the center of the circle or pull it away from the center. Thus there would be two different radii, depending on whether the light coming in was left-hand circularly polarized or right-hand circularly polarized. Reversing the direction of the applied field would clearly interchange the radii corresponding to right- and left-hand circular polarization.

To test our intuition, we can now solve the problem exactly. Take first the case of left-hand circularly polarized light incident upon the charge. As we remember, this means that at a fixed value of z the electric field vector appears to be rotating counterclockwise as we look along the $+z$ axis. How shall we express E_{0x} and E_{0y} in order that the total electric field behave in this manner? Very simple. We remember that in the end we will be interested in only the real parts of the expressions we obtain for E_x and E_y. If we let $E_{0y} = iE_{0x}$, then we have

$$E_x = E_{0x}e^{i\omega t} \quad \text{and} \quad E_y = E_{0x}e^{i(\omega t + \pi/2)} \quad (7\text{-}6\text{-}9)$$

The real parts of E_x and E_y then rotate in the desired direction as a function of time.

Substituting this into our expressions for x_0 and y_0, we have

$$x_0 = \frac{qE_{0x}}{m(\omega_0{}^2 - \omega^2 + \omega\omega_c)} \quad (7\text{-}6\text{-}10)$$

$$y_0 = \frac{qE_{0y}}{m(\omega_0{}^2 - \omega^2 + \omega\omega_c)} \quad (7\text{-}6\text{-}11)$$

The charge q just follows the electric field, either in phase or $180°$ out of phase, depending on the sign of q and on the sign of $\omega_0{}^2 - \omega^2 + \omega\omega_c$.

In any case the rotating charge produces its own left-hand circularly polarized radiation in the forward direction. The forward-scattering amplitude, defined as before, is just

$$f_L(0) = \frac{-q}{c^2}(-\omega^2)\frac{q}{m(\omega_0{}^2 - \omega^2 + \omega\omega_c)} \quad (7\text{-}6\text{-}12)$$

If matter were made up of N such harmonic oscillators per unit volume, the index of refraction for left-hand circularly polarized light traveling in the direction of a magnetic field B would be

$$n_L = \sqrt{1 + \frac{4\pi Nq^2}{m(\omega_0{}^2 - \omega^2 + \omega\omega_c)}} \quad (7\text{-}6\text{-}13)$$

To go from left- to right-hand circularly polarized light is trivial. We remember that reversing the handedness of the polarization gave the same result as reversing the magnetic field B. Indeed, it is only the direction of B which establishes a handedness in the first place. Hence we can obtain our result for n_R by just letting ω_c go to $-\omega_c$.

$$n_R = \sqrt{1 + \frac{4\pi q^2 N}{m(\omega_0{}^2 - \omega^2 - \omega\omega_c)}} \quad (7\text{-}6\text{-}14)$$

(Alternatively we could let $E_{0y} = -iE_{0x}$ and substitute in the expressions for x_0 and y_0.) At this point it is reasonable to note that n_R and n_L differ very little from what they would be with no field. Typically ω_0 is about 2×10^{16} rad/sec and ω_c is 17.6×10^6 rad/sec for 1 gauss. Inasmuch as magnetic fields are not usually more than 10^4 to 10^5 gauss, we expect ω_c to be no more than 10^{11} to 10^{12} rad/sec. Typical values of ω, in the visible, are in the region of 4×10^{15} rad/sec. Hence ω_0^2 is larger than $\omega\omega_c$ by a factor of about 10^5 to 10^6. We are then entitled to make a linear approximation for n_L and n_R.

$$n_L \cong n + \left(\frac{dn_L}{d\omega_c}\right)_0 \omega_c$$

$$= n - \frac{n^2 - 1}{2n}\frac{\omega\omega_c}{\omega_0^2 - \omega^2} \tag{7-6-15}$$

$$n_R = n + \frac{n^2 - 1}{2n}\frac{\omega\omega_c}{\omega_0^2 - \omega^2} \tag{7-6-16}$$

Now that we have shown that right- and left-hand circularly polarized light have different indices of refraction, we would like to see what would happen if we began with plane-polarized light. As we have seen earlier, the essential characteristic of left-hand circularly polarized light propagating in the z direction is that the amplitude of the y component is i times the amplitude of the x component. Remembering that left-hand circularly polarized light propagates with index of refraction n_L, we write

$$\mathbf{E}_L(z,t) = Ae^{-i(kn_Lz - \omega t)}\hat{\pmb{\varepsilon}}_1 + iAe^{-i(kn_Lz - \omega t)}\hat{\pmb{\varepsilon}}_2 \tag{7-6-17}$$

where A is an arbitrary constant to be determined to fit given boundary conditions. Similarly

$$\mathbf{E}_R(z,t) = Be^{-i(kn_Rz - \omega t)}\hat{\pmb{\varepsilon}}_1 - iBe^{-i(kn_Rz - \omega t)}\hat{\pmb{\varepsilon}}_2 \tag{7-6-18}$$

We now ask can we superimpose right-hand circularly polarized light and left-hand circularly polarized light so as to produce light which is linearly polarized in the x direction at $z = 0$? That is to say, we search for values of A and B so as to produce a field of the form

$$\mathbf{E}(0,t) = E_0 e^{i\omega t}\hat{\pmb{\varepsilon}}_1 \tag{7-6-19}$$

at $z = 0$. Obviously this can be done by choosing $A = B = E_0/2$ and adding together the two circularly polarized plane waves. We have then

$$\mathbf{E}(z,t) = \frac{E_0}{2}e^{i\omega t}\big[(e^{-ikn_Lz} + e^{-ikn_Rz})\hat{\pmb{\varepsilon}}_1$$
$$+ i(e^{-ikn_Lz} - e^{-ikn_Rz})\hat{\pmb{\varepsilon}}_2\big] \tag{7-6-20}$$

If we make the approximation

$$n_R = n + \Delta$$
$$n_L = n - \Delta \tag{7-6-21}$$

then we obtain a simplified expression for $\mathbf{E}(z,t)$:

$$\mathbf{E}(z,t) = E_0 e^{-i(knz - \omega t)}\big[(\cos k\Delta z)\hat{\pmb{\varepsilon}}_1 - (\sin k\Delta z)\hat{\pmb{\varepsilon}}_2\big] \tag{7-6-22}$$

The polarization thus rotates from the x direction toward the $-y$ direction (for positive Δ), going through an angle $k\Delta z$ in a distance z. The Verdet constant is the rate of change of this angle with respect to distance, per unit field.

$$V = \frac{k\Delta}{B} \tag{7-6-23}$$

For the particular model we have taken, we would have [see Eqs. (7-6-15) and (7-6-16)]

$$V = \frac{q}{mc^2} \frac{\omega^2}{\omega_0{}^2 - \omega^2} \frac{n^2 - 1}{2n} \qquad \text{rad/cm-gauss} \tag{7-6-24}$$

Now, what if we had not been so clever as to see that the natural solution to this problem makes use of circularly polarized radiation? Is there some way in which we can treat the problem in a "brute force" manner which would lead in the end to the "natural" states for a description of the system? Fortunately, the answer to this question is affirmative. Furthermore, the technique we will develop will be of such broad generality that we will be able to make use of it to deal with systems of substantial complexity, provided that we can solve the problem on the atomic level.

Before generalizing, however, let us go back to our solution to x_0 and y_0 in terms of E_{0x} and E_{0y} [see Eqs. (7-6-7) and (7-6-8)]. Having found x and y, we can determine the current densities j_x and j_y by differentiating with respect to time and multiplying by Nq/c. We have then

$$j_x = \frac{i\omega Nq}{c} x_0 e^{i\omega t} = \frac{ikNq^2}{m} \frac{(\omega_0{}^2 - \omega^2)E_{0x} + i\omega\omega_c E_{0y}}{(\omega_0{}^2 - \omega^2)^2 - \omega^2\omega_c{}^2} e^{i\omega t} \tag{7-6-25}$$

$$j_y = \frac{i\omega Nq}{c} y_0 e^{i\omega t} = \frac{ikNq^2}{m} \frac{(\omega_0{}^2 - \omega^2)E_{0y} - i\omega\omega_c E_{0x}}{(\omega_0{}^2 - \omega^2)^2 - \omega^2\omega_c{}^2} e^{i\omega t} \tag{7-6-26}$$

We notice immediately that these equations have the general form

$$\begin{aligned}
j_x &= \sigma_{xx}E_x + \sigma_{xy}E_y \\
j_y &= \sigma_{yx}E_x + \sigma_{yy}E_y
\end{aligned} \tag{7-6-27}$$

Indeed, this form is sufficient to describe the most general linear *index of refraction* problem where transverse fields are propagating only in the z direction. Let us proceed to solve this general problem; the solution to our specific problem of the Faraday effect will then fall right out.

Again we make use of integral equations for the case of a semi-infinite slab to give us the physical insight we need to derive the differential equations. As before, we break up the contributions to the fields E_x and E_y at some position z into those arising from the incoming fields, the currents to the left of z, and the currents to the right of z. The only difference is that both x and y currents can be caused by either x or y fields. We have then, canceling out the $e^{i\omega t}$ term from both sides,

$$E_x(z) = E_{0x}e^{-ikz} - 2\pi e^{-ikz} \int_0^z [\sigma_{xx}E_x(z') + \sigma_{xy}E_y(z')]e^{ikz'} \, dz'$$

$$- 2\pi e^{ikz} \int_z^\infty [\sigma_{xx}E_x(z') + \sigma_{xy}E_y(z')]e^{-ikz'} \, dz' \quad (7\text{-}6\text{-}28)$$

$$E_y(z) = E_{0y}e^{-ikz} - 2\pi e^{-ikz} \int_0^z [\sigma_{yx}E_x(z') + \sigma_{yy}E_y(z')]e^{ikz'} \, dz'$$

$$- 2\pi e^{ikz} \int_z^\infty [\sigma_{yx}E_x(z') + \sigma_{yy}E_y(z')]e^{-ikz'} \, dz' \quad (7\text{-}6\text{-}29)$$

If we differentiate twice with respect to x, we obtain the coupled differential equations

$$\frac{d^2 E_x}{dz^2} = M_{xx}E_x + M_{xy}E_y$$

$$\frac{d^2 E_y}{dz^2} = M_{yx}E_x + M_{yy}E_y \qquad (7\text{-}6\text{-}30)$$

where

$$M_{xx} = -k^2 + 4\pi i k \sigma_{xx}$$

$$M_{xy} = 4\pi i k \sigma_{xy}$$

$$M_{yx} = 4\pi i k \sigma_{yx} \qquad (7\text{-}6\text{-}31)$$

$$M_{yy} = -k^2 + 4\pi i k \sigma_{yy}$$

We note that if we let ψ be the state given by $\psi = E_x \hat{\varepsilon}_1 + E_y \hat{\varepsilon}_2$, then we can write

$$\frac{d^2 \psi}{dz^2} = M\psi$$

where

$$M = \begin{bmatrix} M_{xx} & M_{xy} \\ M_{yx} & M_{yy} \end{bmatrix} \qquad (7\text{-}6\text{-}32)$$

(M is a tensor of the second rank in this two-dimensional space.)

Let us see if we can find some ψ for which $M\psi = \lambda\psi$ and λ is a constant scalar. That is, we look for a polarization state which goes into itself when operated upon by M. ψ is then called an *eigenstate* and λ is its eigenvalue. We desire that

$$(M_{xx} - \lambda)E_x + M_{xy}E_y = 0$$
$$M_{yx}E + (M_{yy} - \lambda)E_y = 0 \qquad (7\text{-}6\text{-}33)$$

This requires that the determinant

$$\begin{Vmatrix} M_{xx} - \lambda & M_{xy} \\ M_{yx} & M_{yy} - \lambda \end{Vmatrix} = 0 \qquad (7\text{-}6\text{-}34)$$

We have then, for λ, the equation

$$\lambda^2 - \lambda(M_{xx} + M_{yy}) + (M_{xx}M_{yy} - M_{xy}M_{yx}) = 0 \qquad (7\text{-}6\text{-}35)$$

There are two solutions for λ which we call λ_+ and λ_-, respectively.

$$\lambda_{+,-} = \frac{M_{xx} + M_{yy}}{2} \pm \tfrac{1}{2}\sqrt{(M_{xx} - M_{yy})^2 + 4M_{xy}M_{yx}} \qquad (7\text{-}6\text{-}36)$$

The two solutions corresponding to $\lambda_{+,-}$ are $\psi_{+,-}$, given, respectively, by

$$\frac{E_y}{E_x} = \frac{\lambda_\pm - M_{xx}}{M_{xy}} \qquad (7\text{-}6\text{-}37)$$

Now ψ_+ obeys the equation

$$\frac{d^2\psi_+}{dz^2} = \lambda_+\psi_+ \qquad (7\text{-}6\text{-}38)$$

and hence

$$\psi_+(z) = \psi_+(0)\exp\left(\pm i\sqrt{-\lambda_+}\,z\right) \qquad (7\text{-}6\text{-}39)$$

Similarly

$$\psi_-(z) = \psi_-(0)\exp\left(\pm i\sqrt{-\lambda_-}\,z\right) \qquad (7\text{-}6\text{-}40)$$

Now for any given initial polarization $\psi(0) = E_{0x}\hat{\varepsilon}_1 + E_{0y}\hat{\varepsilon}_2$, we can decompose $\psi(0)$ into $\psi_+(0)$ and $\psi_-(0)$ components.

$$\psi(0) = A\psi_+(0) + B\psi_-(0) \qquad (7\text{-}6\text{-}41)$$

At some other value of z,t we would have then, for waves propagating in the $+z$ direction,

$$\psi(z,t) = A\psi_+(0)\exp\left[-i(\sqrt{-\lambda_+}\,z - \omega t)\right]$$
$$+ B\psi_-(0)\exp\left[-i(\sqrt{-\lambda_-}\,z - \omega t)\right] \qquad (7\text{-}6\text{-}42)$$

In this way we obtain the solution to the general problem of radiation propagating within a medium. Incidentally, the indices of refraction corresponding to the two values of λ are just

$$n_{+,-} = \frac{\sqrt{-\lambda_{+,-}}}{k} \tag{7-6-43}$$

We now apply this general method to the solution of our model for the Faraday effect. We have in that case

$$M_{xx} = M_{yy} = -k^2 + 4\pi i k \frac{ikNq^2}{m} \frac{\omega_0^2 - \omega^2}{(\omega_0^2 - \omega^2)^2 - \omega^2\omega_c^2} \tag{7-6-44}$$

$$M_{xy} = -M_{yx} = \frac{4\pi i k}{m} \frac{-kNq^2\omega\omega_c}{(\omega_0^2 - \omega^2)^2 - \omega^2\omega_c^2} \tag{7-6-45}$$

Thus

$$
\begin{aligned}
\lambda_{+,-} &= -k^2 - \frac{4\pi Nk^2q^2(\omega_0^2 - \omega^2)}{m\left[(\omega_0^2 - \omega^2)^2 - \omega^2\omega_c^2\right]} \pm \frac{4\pi Nk^2q^2\omega\omega_c}{\left[(\omega_0^2 - \omega^2)^2 - \omega^2\omega_c^2\right]m} \\
&= -k^2\left[1 + \frac{4\pi Nq^2}{m\left[(\omega_0^2 - \omega^2) \pm \omega\omega_c\right]}\right]
\end{aligned} \tag{7-6-46}
$$

The two indices of refraction are

$$n_{+,-} = \sqrt{1 + \frac{4\pi Nq^2}{\left[(\omega_0^2 - \omega^2) \pm \omega\omega_c\right]m}} \tag{7-6-47}$$

The eigenstate corresponding to λ_+ has

$$\frac{E_y}{E_x} = \frac{\lambda_+ - M_{xx}}{M_{xy}} = i \qquad \text{(left-hand circular polarization)} \tag{7-6-48}$$

The eigenstate corresponding to λ_- has

$$\frac{E_y}{E_x} = \frac{\lambda_- - M_{xx}}{M_{xy}} = -i \qquad \text{(right-hand circular polarization)} \tag{7-6-49}$$

We have thus reproduced our earlier result in this completely general manner.

7-7 WE REMOVE THE REQUIREMENT OF NORMAL INCIDENCE; FRESNEL'S EQUATIONS; TOTAL INTERNAL REFLECTION

We have heretofore concerned ourselves only with situations where radiation was incident normally onto the interface between the materials. We will now remove this restriction and examine in detail what happens when the radiation has an angle of incidence equal to θ_i. The reflected radiation will then have an angle of reflection θ_r and the transmitted radiation an angle of refraction θ_t (see Fig. 7-8).

We first ask how one can write down a plane wave traveling in a direction given by a unit vector $\hat{\varepsilon}_a$. The plane wave will have two possible polarizations given by $\hat{\varepsilon}_b$ and $\hat{\varepsilon}_c$, where we take

$$\hat{\varepsilon}_a \times \hat{\varepsilon}_b = \hat{\varepsilon}_c$$

$$\hat{\varepsilon}_b \times \hat{\varepsilon}_c = \hat{\varepsilon}_a$$

$$\hat{\varepsilon}_c \times \hat{\varepsilon}_a = \hat{\varepsilon}_b$$

As usual, we let ω be the frequency of the radiation and define $k \equiv \omega/c$. Then the wavelength is just given by $2\pi/nk$ where n is the index of refraction. The phase of our plane wave must then change by 2π if we proceed along $\hat{\varepsilon}_a$ by a distance equal to $2\pi/nk$. Hence we can write, in general,

$$\mathbf{E}(\mathbf{r},t) = (E_{0b}\hat{\varepsilon}_b + E_{0c}\hat{\varepsilon}_c) \exp\left[-i(nk\hat{\varepsilon}_a \cdot \mathbf{r} - \omega t)\right] \qquad (7\text{-}7\text{-}1)$$

Fig. 7-8 Radiation is incident upon an interface between two media, having indices of refraction n_1 and n_2, respectively. The angle of incidence is θ_i.

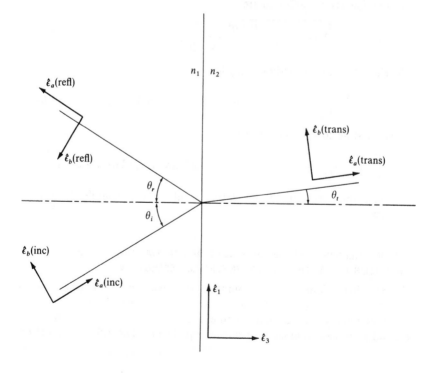

We also remember that the magnetic field within a medium is n times the electric field and is at right angles to it so that $\mathbf{E} \times \mathbf{B}$ points in the propagation direction. The magnetic field corresponding to Eq. (7-7-1) is thus

$$\mathbf{B}(\mathbf{r},t) = -n(E_{0c}\hat{\varepsilon}_b - E_{0b}\hat{\varepsilon}_c) \exp\left[-i(nk\hat{\varepsilon}_a \cdot \mathbf{r} - \omega t)\right] \quad (7\text{-}7\text{-}2)$$

We now return to Fig. 7-8 and set up a coordinate system at some point on the interface. We let the z axis represented by unit vector $\hat{\varepsilon}_3$ be normal to the interface and point from medium n_1 to medium n_2. The incident radiation is taken to come from the n_1 side, and the incident direction is taken to lie in the xz plane. On general grounds of symmetry, the directions of the reflected and transmitted radiation must also lie in the xz plane.

It is now convenient to rewrite Eqs. (7-7-1) and (7-7-2) in terms of one standard coordinate system. We can take $\hat{\varepsilon}_c$ to be along the y direction for the incident, reflected, and transmitted radiation. Hence we will uniformly set $\hat{\varepsilon}_c = \hat{\varepsilon}_2$. The component of electric field in this direction, which was previously called E_{0c}, will now be called $E_{0\perp}$, inasmuch as it is perpendicular to the plane of incidence. As far as $\hat{\varepsilon}_b$ is concerned, we see that

$$\hat{\varepsilon}_b(\text{incident}) = \cos\theta_i\,\hat{\varepsilon}_1 - \sin\theta_i\,\hat{\varepsilon}_3$$

$$\hat{\varepsilon}_b(\text{reflected}) = -\cos\theta_r\,\hat{\varepsilon}_1 - \sin\theta_r\,\hat{\varepsilon}_3$$

$$\hat{\varepsilon}_b(\text{transmitted}) = \cos\theta_t\,\hat{\varepsilon}_1 - \sin\theta_t\,\hat{\varepsilon}_3$$

The amplitudes of electric field along these directions will be called $E_{0\|}$, rather than E_{0b}, to indicate that the field is parallel to the plane of incidence. The incident electric and magnetic fields then have the form

$$\mathbf{E}^{\text{inc}}(\mathbf{r},t) = \left[E_{0\perp}{}^{\text{inc}}\,\hat{\varepsilon}_2 + E_{0\|}{}^{\text{inc}}\left(\cos\theta_i\,\hat{\varepsilon}_1 - \sin\theta_i\,\hat{\varepsilon}_3\right)\right]$$
$$e^{-i[n_1k(z\cos\theta_i + x\sin\theta_i) - \omega t]} \quad (7\text{-}7\text{-}3)$$

$$\mathbf{B}^{\text{inc}}(\mathbf{r},t) = -n_1\left[E_{0\perp}{}^{\text{inc}}(\cos\theta_i\,\hat{\varepsilon}_1 - \sin\theta_i\,\hat{\varepsilon}_3) - E_{0\|}{}^{\text{inc}}\,\hat{\varepsilon}_2\right]$$
$$e^{-i[n_1k(z\cos\theta_i + x\sin\theta_i) - \omega t]} \quad (7\text{-}7\text{-}4)$$

Similarly, the reflected and transmitted plane waves have the form

$$\mathbf{E}^{\text{refl}}(\mathbf{r},t) = \left[E_{0\perp}{}^{\text{refl}}\,\hat{\varepsilon}_2 - E_{0\|}{}^{\text{refl}}\left(\cos\theta_r\,\hat{\varepsilon}_1 + \sin\theta_r\,\hat{\varepsilon}_3\right)\right.$$
$$e^{-i[n_1k(x\sin\theta_r - z\cos\theta_r) - \omega t]} \quad (7\text{-}7\text{-}5)$$

$$\mathbf{B}^{\text{refl}}(\mathbf{r},t) = n_1\left[E_{0\perp}{}^{\text{refl}}\left(\cos\theta_r\,\hat{\varepsilon}_1 + \sin\theta_r\,\hat{\varepsilon}_3\right) + E_{0\|}{}^{\text{refl}}\,\hat{\varepsilon}_2\right]$$
$$e^{-i[n_1k(x\sin\theta_r - z\cos\theta_r) - \omega t]} \quad (7\text{-}7\text{-}6)$$

$$\mathbf{E}^{\text{trans}}(\mathbf{r},t) = \left[E_{0\perp}{}^{\text{trans}}\,\hat{\varepsilon}_2 + E_{0\|}{}^{\text{trans}}\left(\cos\theta_t\,\hat{\varepsilon}_1 - \sin\theta_t\,\hat{\varepsilon}_3\right)\right]$$
$$e^{-i[n_2k(z\cos\theta_t + x\sin\theta_t) - \omega t]} \quad (7\text{-}7\text{-}7)$$

$$\mathbf{B}^{\text{trans}}(\mathbf{r},t) = -n_2[E_{0\perp}{}^{\text{trans}}(\cos\theta_t\,\hat{\varepsilon}_1 - \sin\theta_t\,\hat{\varepsilon}_3) - E_{0\parallel}{}^{\text{trans}}\,\hat{\varepsilon}_2]$$
$$e^{-i[n_1 k(z\cos\theta_t + x\sin\theta_t)-\omega t]} \quad (7\text{-}7\text{-}8)$$

We are now ready to apply the boundary conditions at the interface. We require that the tangential components of $\mathbf{E}$ and $\mathbf{B}$ be continuous. Since the tangential components are those along $\hat{\varepsilon}_1$ and $\hat{\varepsilon}_2$, we can write (remembering that $z = 0$ at the boundary and factoring out the $e^{i\omega t}$ term)

$$E_{0\perp}{}^{\text{inc}}e^{-in_1 kx\sin\theta_i} + E_{0\perp}{}^{\text{refl}}e^{-in_1 kx\sin\theta_r} = E_{0\perp}{}^{\text{trans}}e^{-in_2 kx\sin\theta_t} \quad (7\text{-}7\text{-}9)$$

$$E_{0\parallel}{}^{\text{inc}}\cos\theta_i\,e^{-in_1 kx\sin\theta_i} - E_{0\parallel}{}^{\text{refl}}\cos\theta_r\,e^{-in_1 kx\sin\theta_r}$$
$$= E_{0\parallel}{}^{\text{trans}}\cos\theta_t\,e^{-in_2 kx\sin\theta_t} \quad (7\text{-}7\text{-}10)$$

In order that these equations hold for any value of x, we must have

$$n_1 kx\sin\theta_i = n_1 kx\sin\theta_r = n_2 kx\sin\theta_t$$

Thus we learn that the angle of incidence and the angle of reflection are equal. Furthermore, we have derived Snell's law relating θ_t to θ_i.

$$n_1\sin\theta_i = n_2\sin\theta_t \quad (7\text{-}7\text{-}11)$$

Rewriting Eqs. (7-7-9) and (7-7-10) and adding the equations for the tangential components of $\mathbf{B}$, we have

$$E_{0\perp}{}^{\text{inc}} + E_{0\perp}{}^{\text{refl}} = E_{0\perp}{}^{\text{trans}}$$
$$(E_{0\parallel}{}^{\text{inc}} - E_{0\parallel}{}^{\text{refl}})\cos\theta_i = E_{0\parallel}{}^{\text{trans}}\cos\theta_t$$
$$n_1(E_{0\perp}{}^{\text{inc}} - E_{0\perp}{}^{\text{refl}})\cos\theta_i = n_2 E_{0\perp}{}^{\text{trans}}\cos\theta_t \quad (7\text{-}7\text{-}12)$$
$$n_1(E_{0\parallel}{}^{\text{inc}} + E_{0\parallel}{}^{\text{refl}}) = n_2 E_{0\parallel}{}^{\text{trans}}$$

Solving these, we obtain the well-known **Fresnel equations**

$$E_{0\perp}{}^{\text{refl}} = \frac{n_1\cos\theta_i - n_2\cos\theta_t}{n_1\cos\theta_i + n_2\cos\theta_t} E_{0\perp}{}^{\text{inc}} \quad (7\text{-}7\text{-}13)$$

$$E_{0\perp}{}^{\text{trans}} = \frac{2n_1\cos\theta_i\,E_{0\perp}{}^{\text{inc}}}{n_1\cos\theta_i + n_2\cos\theta_t} \quad (7\text{-}7\text{-}14)$$

$$E_{0\parallel}{}^{\text{refl}} = \frac{n_2\cos\theta_i - n_1\cos\theta_t}{n_1\cos\theta_t + n_2\cos\theta_i} E_{0\parallel}{}^{\text{inc}} \quad (7\text{-}7\text{-}15)$$

$$E_{0\parallel}{}^{\text{trans}} = \frac{2n_1\cos\theta_i\,E_{0\parallel}{}^{\text{inc}}}{n_1\cos\theta_t + n_2\cos\theta_i} \quad (7\text{-}7\text{-}16)$$

We notice immediately that $E_{0\parallel}{}^{\text{refl}}$ can be zero at a particular angle of incidence. This occurs whenever

$$n_2\cos\theta_i - n_1\cos\theta_t = 0$$

Inserting Snell's law into the above, we have, for no reflection,

$$\sin \theta_i \cos \theta_i = \sin \theta_t \cos \theta_t$$

or

$$\theta_i + \theta_t = \frac{\pi}{2} \tag{7-7-17}$$

If θ_i and θ_t satisfy this condition, then θ_i is called **Brewster's angle.**

Finally, let us consider what happens when we have total internal reflection. If n_1 is greater than n_2, then there certainly exists a range of incident angles for which

$$\frac{n_1}{n_2} \sin \theta_i = \sin \theta_t > 1 \tag{7-7-18}$$

A little bit of thought indicates that this is no problem if we allow θ_t to be a complex number. We find $\cos \theta_t$ by the usual relation

$$\cos \theta_t = \pm \sqrt{1 - \sin^2 \theta_t} = \pm i \sqrt{\sin^2 \theta_t - 1}$$

$$= \pm i \sqrt{\left(\frac{n_1}{n_2}\right)^2 \sin^2 \theta_i - 1} \tag{7-7-19}$$

To simplify our notation we let

$$\alpha \equiv \sqrt{\left(\frac{n_1}{n_2}\right)^2 \sin^2 \theta_i - 1} \tag{7-7-20}$$

We then note that the phase factor $e^{-in_2 kz \cos \theta_t}$ becomes

$$e^{-in_2 kz \cos \theta_t} = e^{\pm n_2 kz \alpha}$$

We must eliminate the positive exponential inasmuch as it diverges as we let z go to infinity. Thus we accept only the negative alternative in Eq. (7-7-19).

$$\cos \theta_t = -i\alpha \tag{7-7-21}$$

This gives us

$$E_{0\perp}{}^{\text{trans}} = \frac{2n_1 \cos \theta_i}{n_1 \cos \theta_i - in_2 \alpha} E_{0\perp}{}^{\text{inc}}$$

$$E_{0\perp}{}^{\text{refl}} = \frac{n_1 \cos \theta_i + in_2 \alpha}{n_1 \cos \theta_i - in_2 \alpha} E_{0\perp}{}^{\text{inc}}$$

$$E_{0\parallel}{}^{\text{trans}} = \frac{2n_1 \cos \theta_i}{n_2 \cos \theta_i - in_1 \alpha} E_{0\parallel}{}^{\text{inc}} \tag{7-7-22}$$

$$E_{0\parallel}{}^{\text{refl}} = \frac{n_2 \cos \theta_i + in_1 \alpha}{n_2 \cos \theta_i - in_1 \alpha} E_{0\parallel}{}^{\text{inc}}$$

As we anticipate, the magnitudes of the reflected waves are identical to the magnitudes of the corresponding incident waves but out of phase with them. The transmitted electric and magnetic fields are given by

$$\mathbf{E}^{trans}(\mathbf{r},t) = \left[E_{0\perp}{}^{trans}\hat{\varepsilon}_2 - E_{0\parallel}{}^{trans}\left(\frac{n_1}{n_2}\sin\theta_i\,\hat{\varepsilon}_3 + i\alpha\hat{\varepsilon}_1\right)\right]$$
$$e^{-n_2 k\alpha z - in_1 k \sin\theta_i\, x + i\omega t} \quad (7\text{-}7\text{-}23)$$

$$\mathbf{B}^{trans}(\mathbf{r},t) = n_2\left[E_{0\perp}{}^{trans}\left(i\alpha\hat{\varepsilon}_1 + \frac{n_1}{n_2}\sin\theta_i\,\hat{\varepsilon}_3\right) + E_{0\parallel}{}^{trans}\hat{\varepsilon}_2\right]$$
$$e^{-n_2 k\alpha z - in_1 k \sin\theta_i\, x + i\omega t} \quad (7\text{-}7\text{-}24)$$

A simple inspection of the Poynting vector will show that, averaged over time, no energy is flowing in the z direction. The only average energy flow is in the x direction, as expected. The characteristic distance in the z direction within which the amplitude decreases by a factor e is just

$$\delta = \frac{1}{n_2 k\alpha} = \frac{\lambda}{2\pi}\,\frac{1}{\sqrt{(n_1/n_2)^2\sin^2\theta_i - 1}} \quad (7\text{-}7\text{-}25)$$

We can obtain a feeling for the magnitude of δ by taking some typical values for n_1, n_2, and $\sin\theta_i$. We let $n_1 = 1.5$ and $n_2 = 1$ (corresponding to a glass-air interface) and take $\theta_i = 45°$.

$$\delta = \frac{\lambda}{2\pi}\,\frac{1}{\sqrt{\frac{9}{8} - 1}} \cong 0.45\lambda$$

Typically then the electromagnetic radiation can extend out for a wavelength or so from a totally reflecting surface.

PROBLEMS

7-1. Find the skin depth in copper, aluminum, nonmagnetic stainless steel, and sea water for radiation at $\omega = 10^6$ rad/sec, 10^8 rad/sec, and 10^{10} rad/sec.

$$\sigma_{copper} = 1.6 \times 10^7 \text{ abamperes/statvolt-cm}$$

$$\sigma_{aluminum} = 1.1 \times 10^7 \text{ abamperes/statvolt-cm}$$

$$\sigma_{stainless\ steel} = 0.3 \times 10^7 \text{ abamperes/statvolt-cm}$$

$$\sigma_{sea\ water} = 1.5 \text{ abamperes/statvolt-cm}$$

7-2. Referring to Fig. 7-2, calculate the time-averaged force per unit volume on the semi-infinite conductor by the magnetic field, as a function of z. Integrate from $z = 0$ to $z = \infty$ to find the average radiation pressure on the conductor. Compare this result with the change in momentum per unit area of the incoming and reflected radiation.

7-3. Radiation of frequency ω is normally incident upon a semi-infinite slab of index of refraction n_2, which has been coated with a thin layer of index n_1 (the thickness of the layer is δ).

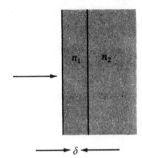

(a) Find a general expression for the reflected intensity as a function of ω, n_1, n_2, and δ.

(b) Given n_2, find a set of values for n_1 and δ for which the reflected intensity at frequency ω is zero.

7-4. Linearly polarized light of the form $E_x(z,t) = E_0 e^{-i(kz - \omega t)}$ is incident normally onto a material which has index of refraction n_R for right-hand circularly polarized light and n_L for left-hand circularly polarized light. Describe the reflected light quantitatively from the point of view of intensity and polarization.

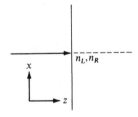

7-5. Electromagnetic radiation of frequency ω is incident on a thick conducting plate of conductivity σ at an angle of incidence θ_i. Find the intensity of the reflected

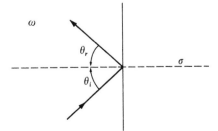

light for incident polarization parallel to and perpendicular to the plane of incidence.

7-6. A semi-infinite slab of matter is constituted of N charges per unit volume, each of which is held in place by an asymmetric harmonic oscillator potential. The restoring force on each charge for a displacement in the xy plane is

$$\mathbf{F} = -m\omega_{0x}^2 x \hat{\boldsymbol{\varepsilon}}_1 - m\omega_{0y}^2 y \hat{\boldsymbol{\varepsilon}}_2$$

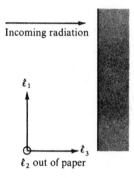

Incoming radiation

$\hat{\boldsymbol{\varepsilon}}_1$

$\hat{\boldsymbol{\varepsilon}}_3$

$\hat{\boldsymbol{\varepsilon}}_2$ out of paper

Left-hand circularly polarized light of frequency ω is normally incident upon the slab, as shown. The amplitude of the incident radiation is E_0.
(a) Find the electric field as a function of time at a distance z into the slab.
(b) Find an expression for the reflected radiation. What is its state of polarization?

7-7. A magnetic field B is now applied normal to the surface of the slab in Prob. 7-6. What are the two indices of refraction and what polarization states do they correspond to if we take $\omega_{0x}^2 = \frac{1}{4}\omega_{0y}^2$?

7-8. Explain how the Faraday effect can be used to determine the sign of the mobile charges in a block of dielectric.

8
Multipole Expansion of the Radiation Field; Some Further Considerations on the Interaction of Radiation with Matter; Interference and Diffraction

8-1 A GENERAL STATEMENT OF THE PROBLEM

When we studied electrostatics, we found it convenient to express the electro-static potential at a point r a large distance from a spatially limited charge distribution in terms of a power-series expansion. The expansion parameter was r'/r where the vector r' denoted the position of charge contributing to the potential. This sort of expansion is not particularly useful when we deal with radiation because as we have already learned, the radiation fields must decrease as $1/r$ in order that there be a net energy outflow from the radiator. There is however another parameter which is of prime importance, the size of the source as compared with the wavelength of the radiation. It is clear that if the size of the source is comparable to the radiation wavelength, then various parts of the source will interfere coherently.

We shall take then the point of view that r' is sufficiently less than r so that only electric and magnetic fields decreasing as $1/r$ must be considered. Subject to this assumption we will develop a rather general method for finding the fields and then make the further approximation that r' is much smaller than the wavelength of the radiation. In the event that the latter assumption is not true, we can in many cases subdivide the source to the point where it is true and add the results later. This technique will be used when we reexamine the index of refraction and when we study diffraction phenomena.

For convenience we will concern ourselves only with charge and current distributions which are harmonically periodic in their time dependence. (If the time dependence of our actual problem does not fit this criterion, we can decompose it into harmonic components by means of Fourier analysis.).

If ω is the angular frequency of the time dependence, then we write

$$\mathbf{j}(\mathbf{r}',t) = \mathbf{j}(\mathbf{r}')e^{i\omega t}$$
$$\rho(\mathbf{r}',t) = \rho(\mathbf{r}')e^{i\omega t}$$

(8-1-1)

Substituting into our well-known expressions for $\mathbf{A}$ and φ [see Eq. (5-6-3)], we have

$$\mathbf{A}(\mathbf{r},t) = e^{i\omega t} \int \frac{\mathbf{j}(\mathbf{r}') \exp(-ik|\mathbf{r} - \mathbf{r}'|)}{|\mathbf{r} - \mathbf{r}'|} dV'$$

(8-1-2)

$$\varphi(\mathbf{r},t) = e^{i\omega t} \int \frac{\rho(\mathbf{r}') \exp(-ik|\mathbf{r} - \mathbf{r}'|)}{|\mathbf{r} - \mathbf{r}'|} dV'$$

(8-1-3)

where, as usual, $k = \omega/c = 2\pi/\lambda$.

We have already agreed to consider r' to be infinitesimal compared with r. Hence $1/|\mathbf{r} - \mathbf{r}'|$ varies only negligibly as we vary $\mathbf{r}'$. We can replace it by $1/r$ and take it out of the integral. On the other hand, $\exp(-ik|\mathbf{r} - \mathbf{r}'|)$ can vary considerably because r' and λ are much more comparable. We note then that

$$|\mathbf{r} - \mathbf{r}'| = \sqrt{r^2 + r'^2 - 2\mathbf{r} \cdot \mathbf{r}'}$$

$$\cong r - \frac{\mathbf{r} \cdot \mathbf{r}'}{r}$$

(8-1-4)

If we let $\hat{\mathbf{n}} = \mathbf{r}/r$, then we can rewrite the potentials as

$$\mathbf{A}(\mathbf{r},t) = \frac{e^{-i(kr-\omega t)}}{r} \int \mathbf{j}(\mathbf{r}') \exp(ik\hat{\mathbf{n}} \cdot \mathbf{r}') \, dV' \qquad (8\text{-}1\text{-}5)$$

$$\varphi(\mathbf{r},t) = \frac{e^{-i(kr-\omega t)}}{r} \int \rho(\mathbf{r}') \exp(ik\hat{\mathbf{n}} \cdot \mathbf{r}') \, dV' \qquad (8\text{-}1\text{-}6)$$

So far we have only assumed that $r' \ll r$ and have not made any assumption about the magnitude of $2\pi r'/\lambda = kr'$. Indeed, in many cases we can carry out the above integration completely and never have any reason for further approximation. On the other hand, there are many situations where $kr' \ll 1$ and further approximation is useful. For example, an atom with a typical size of 10^{-8} cm will emit radiation with a wavelength of the order of 5×10^{-5} cm. For those situations, we are entitled to write

$$\exp(ik\hat{\mathbf{n}} \cdot \mathbf{r}') = 1 + ik\hat{\mathbf{n}} \cdot \mathbf{r}' - \tfrac{1}{2}k^2(\hat{\mathbf{n}} \cdot \mathbf{r}')^2 + \cdots \qquad (8\text{-}1\text{-}7)$$

Substituting into the above equations yields a power-series expansion for $\mathbf{A}$ and φ:

$$\mathbf{A} = \mathbf{A}_1 + \mathbf{A}_2 + \mathbf{A}_3 + \cdots$$

$$\varphi = \varphi_1 + \varphi_2 + \varphi_3 + \cdots$$

where

$$\mathbf{A}_1 = \frac{e^{-i(kr-\omega t)}}{r} \int \mathbf{j}(\mathbf{r}') \, dV'$$

$$\mathbf{A}_2 = \frac{e^{-i(kr-\omega t)}}{r} ik \int \mathbf{j}(\mathbf{r}')(\hat{\mathbf{n}} \cdot \mathbf{r}') \, dV'$$

and so forth.

We could now proceed to find the electric field by taking $-\nabla\varphi - \dfrac{1}{c} \dfrac{\partial \mathbf{A}}{\partial t}$ and the magnetic field by taking $\nabla \times \mathbf{A}$. It is much more convenient however to make the following observations:

$$\mathbf{E}(\mathbf{r},t) = \mathbf{E}(\mathbf{r})e^{i\omega t}$$
$$\mathbf{B}(\mathbf{r},t) = \mathbf{B}(\mathbf{r})e^{i\omega t} \qquad (8\text{-}1\text{-}8)$$

Hence, making use of Maxwell's equation $\nabla \times \mathbf{B} = \dfrac{1}{c} \dfrac{\partial \mathbf{E}}{\partial t}$, we have

$$\mathbf{E}(\mathbf{r}) = \frac{1}{ik} \nabla \times \mathbf{B}(\mathbf{r})$$

$$= \frac{1}{ik} \nabla \times [\nabla \times \mathbf{A}(\mathbf{r})] \qquad (8\text{-}1\text{-}9)$$

This obviates the necessity of knowing φ at all and simplifies our job considerably.

8-2 ELECTRIC DIPOLE RADIATION

As a first approximation we can ignore kr' completely. We then find

$$\mathbf{A}_1(\mathbf{r},t) = \frac{e^{-i(kr-\omega t)}}{r} \int \mathbf{j}(\mathbf{r}')\, dV' \tag{8-2-1}$$

We can cast this into a somewhat more elegant form by a bit of manipulation. We evaluate the x component of $\int \mathbf{j}(\mathbf{r}')\, dV'$:

$$\int j_x(\mathbf{r}')\, dV' = \int \mathbf{j}(\mathbf{r}') \cdot \hat{\boldsymbol{\varepsilon}}_1\, dV'$$

$$= \int \mathbf{j}(\mathbf{r}') \cdot \nabla' x'\, dV'$$

$$= \int \nabla' \cdot x' \mathbf{j}(\mathbf{r}')\, dV' - \int x' \nabla' \cdot \mathbf{j}(\mathbf{r}')\, dV' \tag{8-2-2}$$

Note that the first of the integrals on the right side of Eq. (8-2-2) can be turned into an integral over a surface enclosing the current distribution completely. On that surface $\mathbf{j} = 0$, and hence we conclude

$$\int j_x(\mathbf{r}')\, dV' = - \int x' \nabla' \cdot \mathbf{j}(\mathbf{r}')\, dV'$$

Generalizing to three dimensions, we obtain

$$\int \mathbf{j}(\mathbf{r}')\, dV' = - \int \mathbf{r}'[\nabla' \cdot \mathbf{j}(\mathbf{r}')]\, dV' \tag{8-2-3}$$

Next, we remember the conservation of charge equation:

$$\nabla' \cdot \mathbf{j}(\mathbf{r}',t) = -\frac{1}{c} \frac{\partial \rho(\mathbf{r}',t)}{\partial t}$$

Hence, using Eq. (8-1-1), we have

$$\nabla' \cdot \mathbf{j}(\mathbf{r}') = -ik\rho(\mathbf{r}') \tag{8-2-4}$$

Substituting back into Eqs. (8-2-3) and (8-2-1), we obtain our expression for $\mathbf{A}_1(\mathbf{r},t)$:

$$\mathbf{A}_1(\mathbf{r},t) = \frac{ike^{-i(kr-\omega t)}}{r} \int \mathbf{r}'\rho(\mathbf{r}')\, dV' \tag{8-2-5}$$

The integral now has a very familiar form. Consider the time-dependent electric dipole moment of the charge distribution.

$$\mathbf{p}(t) = \int \mathbf{r}' \rho(\mathbf{r}',t) \, dV'$$

$$= e^{i\omega t} \int \mathbf{r}' \rho(\mathbf{r}') \, dV'$$

$$= \mathbf{p}_0 e^{i\omega t} \tag{8-2-6}$$

Clearly then

$$\mathbf{A}_1(\mathbf{r},t) = \frac{ike^{-i(kr-\omega t)}}{r} \mathbf{p}_0 \tag{8-2-7}$$

Finally we must determine $\mathbf{B}$ and $\mathbf{E}$. We take the curl of $\mathbf{A}$ but ignore the term which goes as $1/r^2$.

$$\mathbf{B}_1 = \nabla \times \mathbf{A}_1$$

$$\cong \frac{ik\nabla e^{-i(kr-\omega t)}}{r} \times \mathbf{p}_0$$

$$= \frac{k^2 e^{-i(kr-\omega t)}}{r} (\hat{\mathbf{n}} \times \mathbf{p}_0) \tag{8-2-8}$$

To find $\mathbf{E}$ we take the curl of $\mathbf{B}$ and again ignore the $1/r^2$ term.

$$\mathbf{E}_1 = \frac{1}{ik} (\nabla \times \mathbf{B}_1)$$

$$= \frac{k^2}{ikr} \nabla e^{-i(kr-\omega t)} \times (\hat{\mathbf{n}} \times \mathbf{p}_0)$$

$$= \frac{-k^2 e^{-i(kr-\omega t)}}{r} \hat{\mathbf{n}} \times (\hat{\mathbf{n}} \times \mathbf{p}_0)$$

$$= \frac{k^2 e^{-i(kr-\omega t)}}{r} [\mathbf{p}_0 - (\mathbf{p}_0 \cdot \hat{\mathbf{n}})\hat{\mathbf{n}}] \tag{8-2-9}$$

A little bit of thought allows us to rewrite Eqs. (8-2-8) and (8-2-9) in slightly different form:

$$\mathbf{E}_1(\mathbf{r},t) = \frac{k^2}{r} \mathbf{p}_P\left(t - \frac{r}{c}\right)$$

$$\mathbf{B}_1(\mathbf{r},t) = \hat{\mathbf{n}} \times \mathbf{E}_1(\mathbf{r},t) \tag{8-2-10}$$

where $\mathbf{p}_P(t - r/c)$ is the component of the dipole moment perpendicular to

$\hat{n}$ and evaluated at the time $t - r/c$. Physically then, to find the electric field at r, we put our eye there and observe the charge distribution as it appears to us at that instant. (Naturally we are really looking at the charge distribution as it was a time r/c in the past.) We calculate the electric dipole moment of this *observed* distribution, take the component perpendicular to our line of sight, multiply by k^2, divide by r, and we have our answer.

We would like to compare this result with the one we found earlier for the case of an accelerating charge [see Eq. (6-4-2)]. Imagine that we have a charge q undergoing harmonic motion of the form

$$\mathbf{r} = \mathbf{r}_0 e^{i\omega t}$$

Then the acceleration is just

$$\mathbf{a} = -\omega^2 \mathbf{r}_0 e^{i\omega t}$$

On the other hand, the electric dipole moment is given by

$$\mathbf{p}(t) = q\mathbf{r}(t) = q\mathbf{r}_0 e^{i\omega t}$$

Hence

$$\mathbf{p}(t) = -\frac{q}{\omega^2} \mathbf{a}(t)$$

Using Eq. (6-4-2), we have

$$\mathbf{E}(r,t) = \frac{-q\mathbf{a}_P(t - r/c)}{c^2 r}$$

$$= \frac{\omega^2 \mathbf{p}_P(t - r/c)}{c^2 r}$$

$$= \frac{k^2 \mathbf{p}_P(t - r/c)}{r}$$

This confirms the general result of Eq. (8-2-10) for the simple case of a harmonically oscillating point charge.

The Poynting vector can now be evaluated and averaged over one cycle. This would usually require that we find the real parts of $\mathbf{E}$ and $\mathbf{B}$, respectively, then take their vector product and finally do the time averaging. We can save ourselves some effort however by making the following observations:

$$\langle S \rangle_{\text{time av}} = \frac{c}{4\pi} \left\langle \frac{\mathbf{E} + \mathbf{E}^*}{2} \times \frac{\mathbf{B} + \mathbf{B}^*}{2} \right\rangle_{\text{time av}}$$

$$= \frac{c}{16\pi} \langle \mathbf{E} \times \mathbf{B} + \mathbf{E} \times \mathbf{B}^* + \mathbf{E}^* \times \mathbf{B} + \mathbf{E}^* \times \mathbf{B}^* \rangle_{\text{time av}}$$

Now $\mathbf{E} \times \mathbf{B}$ is proportional to $e^{2i\omega t}$, which is zero when averaged over time. Similarly $\mathbf{E}^* \times \mathbf{B}^*$ is proportional to $e^{-2i\omega t}$, which also averages to zero. Hence

$$\langle \mathbf{S} \rangle_{\text{time av}} = \frac{c}{16\pi} \langle \mathbf{E} \times \mathbf{B}^* + \mathbf{E}^* \times \mathbf{B} \rangle_{\text{time av}}$$

$$= \frac{c}{8\pi} \operatorname{Re} \langle \mathbf{E} \times \mathbf{B}^* \rangle_{\text{time av}}$$

Since $\mathbf{E} \times \mathbf{B}^*$ has no time dependence we need not actually carry out any time averaging and we can write directly

$$\langle \mathbf{S} \rangle_{\text{time av}} = \frac{c}{8\pi} \operatorname{Re} (\mathbf{E} \times \mathbf{B}^*) \tag{8-2-11}$$

Applying this general result in the case of our dipole we have

$$\langle \mathbf{S} \cdot \hat{\mathbf{n}} \rangle_{\text{time av}} = \frac{c}{8\pi} \frac{k^4 |p_0|^2 \sin^2 \theta}{r^2} \tag{8-2-12}$$

where

$$\cos \theta = \frac{\hat{\mathbf{n}} \cdot \mathbf{p}_0}{p_0}$$

Integrating over a sphere, we find the average amount of energy radiated by the dipole per unit time:

$$\frac{du}{dt} = \frac{ck^4 |p_0|^2}{8\pi} \int_0^{2\pi} d\psi \int_0^{\pi} \sin^3 \theta \, d\theta$$

$$= \frac{ck^4 |p_0|^2}{3} \tag{8-2-13}$$

Lastly, let us apply what we have just learned and calculate the radiation field due to the simple short dipole antenna shown in Fig. 8-1. We assume that the overall length of the antenna is a and that $a \ll \lambda$.

The current in the antenna is taken to be

$$I(z', t) = \begin{cases} I_0 \left(1 - \dfrac{2z'}{a}\right) e^{i\omega t} & \text{for } z' > 0 \\[2mm] I_0 \left(1 + \dfrac{2z'}{a}\right) e^{i\omega t} & \text{for } z' < 0 \end{cases} \tag{8-2-14}$$

In this model then, the current is in the same direction in each half of the antenna and falls off linearly as we approach the ends.

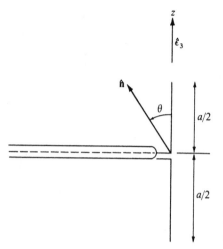

Fig. 8-1 We calculate the radiation field due to a short dipole antenna of length a. We assume that $a \ll \lambda$ where λ is the wavelength of the radiation.

To obtain the dipole moment, we can either find $\rho(z')$ or we can just make use of Eq. (8-2-1) directly. It is simpler to do the latter.

$$
\mathbf{p}_0 = \frac{1}{ik} \int \mathbf{j}(\mathbf{r}') \, dV'
$$

$$
= \frac{1}{ik} I_0 \left[\int_0^{a/2} \left(1 - \frac{2z'}{a} \right) dz' + \int_{-a/2}^{0} \left(1 + \frac{2z'}{a} \right) dz' \right] \hat{\mathbf{\epsilon}}_3
$$

$$
= \frac{I_0 a \hat{\mathbf{\epsilon}}_3}{2ik} \tag{8-2-15}
$$

The angular distribution of radiated power is just

$$
\left\langle \frac{d^2 u}{dt \, d\Omega} \right\rangle_{\text{time av}} = r^2 \langle \mathbf{S} \cdot \hat{\mathbf{n}} \rangle_{\text{time av}}
$$

$$
= \frac{c}{32\pi} (ka)^2 I_0^2 \sin^2 \theta \tag{8-2-16}
$$

Thus for a given maximum current, the average radiated power increases as the square of the frequency in the situation where $a \ll \lambda$.

8-3 MAGNETIC DIPOLE AND ELECTRIC QUADRUPOLE RADIATION

In the next order of approximation we consider the second term in the expansion of $\exp(ik\hat{\mathbf{n}} \cdot \mathbf{r}')$ [see Eq. (8-1-7)]. Substituting into Eq. (8-1-6), we obtain

$$A_2(\mathbf{r},t) = \frac{e^{-i(kr-\omega t)}}{r} ik \int \mathbf{j}(\mathbf{r}')(\hat{\mathbf{n}} \cdot \mathbf{r}') \, dV' \tag{8-3-1}$$

It is convenient to break the integrand up as follows:

$$\mathbf{j}(\hat{\mathbf{n}} \cdot \mathbf{r}') = \tfrac{1}{2}[\mathbf{j}(\hat{\mathbf{n}} \cdot \mathbf{r}') - \mathbf{r}'(\mathbf{j} \cdot \hat{\mathbf{n}})] + \tfrac{1}{2}[\mathbf{j}(\hat{\mathbf{n}} \cdot \mathbf{r}') + \mathbf{r}'(\mathbf{j} \cdot \hat{\mathbf{n}})]$$
$$= \tfrac{1}{2}[\hat{\mathbf{n}} \times (\mathbf{j} \times \mathbf{r}')] + \tfrac{1}{2}[\mathbf{j}(\hat{\mathbf{n}} \cdot \mathbf{r}') + \mathbf{r}'(\hat{\mathbf{n}} \cdot \mathbf{j})] \tag{8-3-2}$$

The part of the vector potential corresponding to the first term on the right of Eq. (8-3-2) will be called, for reasons which will become apparent shortly, the **magnetic dipole potential** A_{2m}. The part corresponding to the second term will be called the **electric quadrupole potential** A_{2q}. We examine the magnetic dipole potential first.

$$A_{2m}(\mathbf{r},t) = -ik \frac{e^{-i(kr-\omega t)}}{r} \hat{\mathbf{n}} \times \int \frac{[\mathbf{r}' \times \mathbf{j}(\mathbf{r}')] \, dV'}{2} \tag{8-3-3}$$

The integral on the right is nothing other than the magnetic moment of the current distribution. That is,

$$\mu(t) = \mu_0 e^{i\omega t} \tag{8-3-4}$$

where

$$\mu_0 = \tfrac{1}{2} \int \mathbf{r}' \times \mathbf{j}(\mathbf{r}') \, dV' \tag{8-3-5}$$

Hence

$$A_{2m}(\mathbf{r},t) = -ik \frac{e^{-i(kr-\omega t)}}{r} \hat{\mathbf{n}} \times \mu_0 \tag{8-3-6}$$

Taking the curl of A_m and ignoring all terms which go as $1/r^2$, we find the magnetic and electric fields.

$$B_{2m} = \nabla \times A_{2m}$$

$$\cong -k^2 \frac{e^{-i(kr-\omega t)}}{r} \hat{\mathbf{n}} \times (\hat{\mathbf{n}} \times \mu_0)$$

$$= k^2 \frac{e^{-i(kr-\omega t)}}{r} [\mu_0 - (\mu_0 \cdot \hat{\mathbf{n}})\hat{\mathbf{n}}]$$

$$= \frac{k^2}{r} \mu_P \left(t - \frac{r}{c} \right) \tag{8-3-7}$$

$$E_{2m} = \frac{1}{ik} \nabla \times B_{2m}$$

$$= -\hat{\mathbf{n}} \times B_{2m} \tag{8-3-8}$$

Note the amazing similarity to the expressions we obtained for electric dipole radiation. The magnetic field here has the same relation to the projected apparent magnetic dipole moment as the electric field has to the projected apparent electric dipole moment.

An example of such a magnetic dipole is the simple circular current loop shown in Fig. 8-2. We take the loop to have circumference a and let the current in the loop be given by

$$I(t) = I_0 e^{i\omega t} \tag{8-3-9}$$

(In this case the total length of wire involved in the antenna is the same as it is for our previous electric dipole.) We will again calculate the average energy radiated per unit time per unit solid angle and compare it directly with what was obtained in the case of the comparable electric dipole [see Eq. (8-2-15)]. The magnetic moment of the loop is just

$$\mu_0 = I_0 \frac{a^2}{4\pi} \hat{\varepsilon}_3 \tag{8-3-10}$$

Hence

$$\left\langle \frac{d^2 u}{dt\, d\Omega} \right\rangle_{\text{time av}} = r^2 \langle \mathbf{S} \cdot \hat{\mathbf{n}} \rangle_{\text{time av}}$$

$$= \frac{c}{8\pi} k^4 \mu_0^2 \sin^2 \theta$$

$$= \frac{c}{128\pi^3} (ka)^4 I_0^2 \sin^2 \theta \tag{8-3-11}$$

It is instructive then to compare this result with Eq. (8-2-16). We have, for comparable currents and a comparable size,

$$\frac{\left\langle \dfrac{d^2 u}{dt\, d\Omega} \right\rangle_{\text{mag dipole}}}{\left\langle \dfrac{d^2 u}{dt\, d\Omega} \right\rangle_{\text{elect dipole}}} = \frac{(ka)^2}{4\pi^2} = \left(\frac{a}{\lambda} \right)^2 \tag{8-3-12}$$

The intensity of magnetic dipole radiation is thus characteristically smaller by a factor of about $(a/\lambda)^2$ than that of electric dipole radiation.

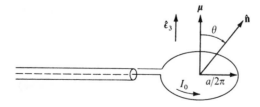

Fig. 8-2 A simple current loop with current $I(t) = I_0 e^{i\omega t}$ emits magnetic dipole radiation.

(This point is of considerable significance in analyzing the radiation of atomic states. When we get to study atomic transitions, we will find that some of them cannot take place through the medium of electric dipole radiation and hence are appropriately suppressed.)

We next return to the so-called electric quadrupole potential.

$$\mathbf{A}_{2q}(\mathbf{r},t) = ik \frac{e^{-i(kr - \omega t)}}{2r} \int \left[\mathbf{j}(\hat{\mathbf{n}} \cdot \mathbf{r}') + \mathbf{r}'(\hat{\mathbf{n}} \cdot \mathbf{j}) \right] dV' \tag{8-3-13}$$

Consider the first part of the integral. We will examine its x component.

$$\int j_x (\hat{\mathbf{n}} \cdot \mathbf{r}') \, dV' = \int (\hat{\mathbf{n}} \cdot \mathbf{r}') \mathbf{j} \cdot \hat{\boldsymbol{\varepsilon}}_1 \, dV'$$

$$= \int (\hat{\mathbf{n}} \cdot \mathbf{r}') \mathbf{j} \cdot \nabla' x' \, dV'$$

$$= \int \nabla' \cdot [x'(\hat{\mathbf{n}} \cdot \mathbf{r}') \mathbf{j}] \, dV' - \int x' \nabla' \cdot [(\hat{\mathbf{n}} \cdot \mathbf{r}') \mathbf{j}] \, dV'$$

We eliminate the first of these two integrals by converting it into a surface integral and by noting that $j = 0$ on the surface. Hence

$$\int j_x (\hat{\mathbf{n}} \cdot \mathbf{r}') \, dV' = - \int x' \nabla' \cdot [(\hat{\mathbf{n}} \cdot \mathbf{r}') \mathbf{j}] \, dV'$$

$$= - \int x' \nabla'(\hat{\mathbf{n}} \cdot \mathbf{r}') \cdot \mathbf{j} \, dV' - \int x'(\hat{\mathbf{n}} \cdot \mathbf{r}') \nabla' \cdot \mathbf{j} \, dV'$$

But

$$\nabla'(\hat{\mathbf{n}} \cdot \mathbf{r}') = \hat{\mathbf{n}}$$

and, remembering Eq. (8-2-4),

$$\nabla' \cdot \mathbf{j}(\mathbf{r}') = -ik\rho(\mathbf{r}')$$

We have thus

$$\int j_x (\hat{\mathbf{n}} \cdot \mathbf{r}') \, dV' = - \int x'(\hat{\mathbf{n}} \cdot \mathbf{j}) \, dV' + ik \int x'(\hat{\mathbf{n}} \cdot \mathbf{r}')\rho(\mathbf{r}') \, dV'$$

Generalizing to three dimensions, we conclude that

$$\int \mathbf{j}(\hat{\mathbf{n}} \cdot \mathbf{r}') \, dV' = - \int \mathbf{r}'(\hat{\mathbf{n}} \cdot \mathbf{j}) \, dV' + ik \int \mathbf{r}'(\hat{\mathbf{n}} \cdot \mathbf{r}')\rho(\mathbf{r}') \, dV' \tag{8-3-14}$$

Substituting back into Eq. (8-3-13), we find for $\mathbf{A}_{2q}$

$$\mathbf{A}_{2q}(\mathbf{r},t) = -k^2 \frac{e^{-i(kr - \omega t)}}{2r} \int \mathbf{r}'(\hat{\mathbf{n}} \cdot \mathbf{r}')\rho(\mathbf{r}') \, dV' \tag{8-3-15}$$

To proceed, we will take the curl of $\mathbf{A}_{2q}$ and find the magnetic field. Again we ignore terms of order $1/r^2$.

$$\mathbf{B}_{2q} = \nabla \times \mathbf{A}_{2q}$$

$$= \frac{ik^3 e^{-i(kr-\omega t)}}{2r} \, \hat{\mathbf{n}} \times \int \mathbf{r}'(\hat{\mathbf{n}} \cdot \mathbf{r}')\rho(\mathbf{r}') \, dV' \qquad (8\text{-}3\text{-}16)$$

We will add to the expression on the right of Eq. (8-3-16) a term which is zero but which helps us put the field in a somewhat more conventional form. We have

$$\mathbf{B}_{2q} = \frac{ik^3 e^{-i(kr-\omega t)}}{6r} \, \hat{\mathbf{n}} \times \int [3\mathbf{r}'(\hat{\mathbf{n}} \cdot \mathbf{r}') - \hat{\mathbf{n}} r'^2]\rho(\mathbf{r}') \, dV' \qquad (8\text{-}3\text{-}17)$$

The integral on the right side of Eq. (8-3-17) can now be written as the product of a second-rank tensor $\mathbf{Q}$ and the vector $\hat{\mathbf{n}}$:

$$\int [3\mathbf{r}'(\hat{\mathbf{n}} \cdot \mathbf{r}') - \hat{\mathbf{n}} r'^2]\rho(\mathbf{r}') \, dV' = \mathbf{Q}\hat{\mathbf{n}} \qquad (8\text{-}3\text{-}18)$$

where the elements of $\mathbf{Q}$ are

$$Q_{ij} = \int (3x_i' x_j' - r'^2 \, \delta_{ij})\rho(\mathbf{r}') \, dV' \qquad (8\text{-}3\text{-}19)$$

The magnetic field $\mathbf{B}_{2q}$ is thus

$$\mathbf{B}_{2q} = \frac{ik^3 e^{-i(kr-\omega t)}}{6r} \, \hat{\mathbf{n}} \times \mathbf{Q}\hat{\mathbf{n}} \qquad (8\text{-}3\text{-}20)$$

As usual the electric field is

$$\mathbf{E}_{2q} = -\hat{\mathbf{n}} \times \mathbf{B}_{2q} \qquad (8\text{-}3\text{-}21)$$

The tensor $\mathbf{Q}$ is called the **quadrupole tensor** of the charge distribution. In the event that the charge distribution is cylindrically symmetrical, we can calculate the elements Q_{ij} very easily. Letting Q_0 be the quadrupole moment of the charge distribution [see Eq. (2-14-20)],[1] we have

$$Q_{ij} = 0 \qquad \text{for } i \neq j$$

$$Q_{11} = Q_{22} = \int \left(\frac{r'^2}{2} - \frac{3z'^2}{2} \right) \rho(\mathbf{r}') \, dV'$$

$$= -Q_0 \qquad (8\text{-}3\text{-}22)$$

---

[1] Our definition of quadrupole moment differs by a factor of 2 from that occurring in many texts. Quite often, the quadrupole moment of a symmetrical distribution is directly defined as Q_{33}.

$$Q_{33} = \int (3z'^2 - r'^2)\rho(\mathbf{r}') \, dV'$$

$$= 2Q_0$$

We can calculate the distribution of radiated power as a function of position as follows:

$$\left\langle \frac{d^2u}{dt \, d\Omega} \right\rangle_{\text{time av}} = r^2 \langle \mathbf{S} \cdot \hat{\mathbf{n}} \rangle_{\text{time av}}$$

$$= \frac{ck^6}{288\pi} |\hat{\mathbf{n}} \times \mathbf{Q}\hat{\mathbf{n}}|^2 \tag{8-3-23}$$

Unfortunately this expression is rather complex and will not be evaluated in general. For the particular case of axial symmetry we have

$$\mathbf{Q}\hat{\mathbf{n}} = -Q_0 n_x \hat{\boldsymbol{\varepsilon}}_1 - Q_0 n_y \hat{\boldsymbol{\varepsilon}}_z + 2Q_0 n_z \hat{\boldsymbol{\varepsilon}}_3 \tag{8-3-24}$$

$$|\hat{\mathbf{n}} \times \mathbf{Q}\hat{\mathbf{n}}|^2 = 9n_z^2(n_x^2 + n_y^2)Q_0^2 = 9Q_0^2 \cos^2\theta \sin^2\theta \tag{8-3-25}$$

$$\left\langle \frac{d^2u}{dt \, d\Omega} \right\rangle_{\text{time av}} = \frac{ck^6 Q_0^2}{32\pi} \cos^2\theta \sin^2\theta \tag{8-3-26}$$

As a simple example of such a quadrupole radiator we consider the assembly of charges shown in Fig. 8-3. The charge $-2q$ is stationary at the origin of our coordinate system, and two positive charges oscillate harmonically, each with amplitude d, about the origin. The two positive

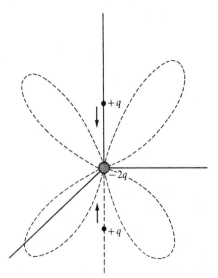

Fig. 8-3   As an example of a quadrupole radiator we consider the assembly of charges shown above. The negative charge is stationary at the origin of the coordinate system. The two positive charges oscillate harmonically with amplitude d along the z axis and about the origin. The radiation pattern is indicated.

charges are always on opposite sides of the origin and exchange positions every half-cycle.

To calculate the quadrupole moment Q_0 we assume that at $t = 0$ the charges are at their maximum amplitude. Then

$$Q_0 = \int \rho(\mathbf{r}')r'^2 P_2 (\cos \theta') \, dV'$$

$$= 2qd^2 \tag{8-3-27}$$

The distribution of radiation is gotten from Eq. (8-3-26).

$$\left\langle \frac{d^2u}{dt \, d\Omega} \right\rangle_{\text{time av}} = \frac{ck^6 d^4 q^2}{8\pi} \cos^2 \theta \sin^2 \theta \tag{8-3-28}$$

Integrating over the angle, we can find the total energy lost per unit time, first for the general symmetrical quadrupole and then for the special case we have just discussed.

$$\left\langle \frac{du}{dt} \right\rangle_{\text{time av}} = \frac{ck^6 Q_0^2}{60} \tag{8-3-29}$$

For our special example,

$$\left\langle \frac{du}{dt} \right\rangle_{\text{time av}} = \frac{ck^6 d^4 q^2}{15} \tag{8-3-30}$$

It is instructive to compare this with the dipole radiation which would result if one of the positive charges and half the negative charge were removed. We would then have a dipole moment $|p| = qd$, and, using Eq. (8-2-12), we would obtain

$$\left\langle \frac{du}{dt} \right\rangle_{\text{elect dipole}} = \frac{ck^4 d^2 q^2}{3} \tag{8-3-31}$$

Comparing directly, we have

$$\frac{\left\langle \dfrac{du}{dt} \right\rangle_{\text{elect quadrupole}}}{\left\langle \dfrac{du}{dt} \right\rangle_{\text{elect dipole}}} = \frac{(kd)^2}{5} \tag{8-3-32}$$

This result is very similar to Eq. (8-3-12), indicating, as anticipated, that electric quadrupole and magnetic dipole radiation have the same basic strength. Each is weaker than electric dipole radiation by a factor about equal to the ratio of the square of the characteristic size of the radiator to the square of the wavelength of the radiation.

Obtaining higher terms in the multipole expansion of the radiation field is beyond the scope of this book. The reader is referred to advanced texts[1] where more sophisticated mathematical methods are applied.

8-4   WE REEXAMINE THE PASSAGE OF RADIATION THROUGH MATTER

We will now make use of what we have learned about electric dipole radiation and reexamine some of the formalism we developed in Chap. 7 from a new point of view. We observe that an incoming electric field induces a dipole moment per unit volume in matter. This induced dipole moment, which has the same harmonic time dependence as the incident electric field, gives rise to a radiation field of its own. When we add together all contributions to the electric field at a point, we will again observe the characteristic interference which led us to the notion of refractive index.

We will begin by letting $P(r,t)$ be the polarization per unit volume of the matter in question. Then, since we will only be concerned with oscillations having frequency ω, we can write

$$P(r,t) = P(r)e^{i\omega t} \tag{8-4-1}$$

Consider then a thin sheet of material of uniform polarization and thickness δ (see Fig. 8-4). We wish to find the electric field at a point which lies a distance z from the sheet. To make matters simple we will assume that P

[1] For example, see J. D. Jackson, "Classical Electrodynamics," pp. 538–577, Wiley, New York, 1962.

Fig. 8-4   An incoming beam of radiation causes an induced dipole moment per unit volume $P(t) = Pe^{i\omega t}$ in a sheet of thickness δ. We are interested in finding the field due to this polarization at a distance z from the sheet.

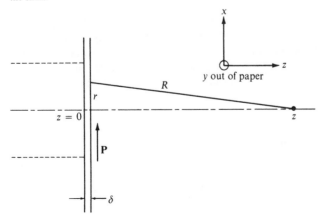

lies along the x axis. Later we can generalize somewhat to an arbitrary direction for **P** in the xy plane.

In carrying out the integration over the sheet, we will use the same technique as was applied in Sec. 7-1. We will consider the sheet to be limited in extent in order that we be able to consider the projected dipole moment to be equal to the full dipole moment. Making use of Eq. (8-2-10), we integrate to find the electric field resulting from the sheet.

$$
\begin{aligned}
E_x(z,t) &= k^2 P \int_{\text{sheet}} \frac{e^{-i(kR-\omega t)}}{R} \, dV \\
&= k^2 P \delta e^{i\omega t} \int_0^{2\pi} d\theta \int_0^{r_{\max}(\theta)} \frac{e^{-ikR} r \, dr}{R} \\
&= k^2 P \delta e^{i\omega t} \int_0^{2\pi} d\theta \int_z^{R_{\max}(\theta)} e^{-ikR} \, dR \\
&= ik P \delta e^{i\omega t} \int_0^{2\pi} d\theta (e^{-ikR_{\max}(\theta)} - e^{-ikz}) \quad (8\text{-}4\text{-}2)
\end{aligned}
$$

Just as we did when we dealt with Eq. (7-1-5), we ignore $e^{-ikR_{\max}(\theta)}$ since it will average to zero. This leaves us with the simple result

$$
E_x(z,t) = -2\pi i k P \delta e^{-i(kz-\omega t)} \quad (8\text{-}4\text{-}3)
$$

We can compare this result directly with Eq. (7-1-9). The equations are identical if we just identify ikP with σE_0. To see that this is reasonable, let us just go back to Eqs. (8-2-1) and (8-2-5) and consider a uniform current distribution and a unit volume. We have then, removing the integrals,

$$
ik\mathbf{P} = \mathbf{j} \quad (8\text{-}4\text{-}4)
$$

But $\mathbf{j} = \sigma \mathbf{E}$ and hence our result.

Now it often happens that **P** is related to the applied field by a proportionality constant χ, called the electric susceptibility. This proportionality constant which was introduced earlier for the static field will, in general, be a frequency-dependent second-rank tensor. In our simple case we have taken ω to be a constant number and we assume now that P is due to an incoming field.

$$
E_x^{\text{inc}}(z,t) = E_0 e^{-i(kz-\omega t)} \quad (8\text{-}4\text{-}5)
$$

Thus, taking the field at $z = 0$, we have

$$
P = \chi E_0 \quad (8\text{-}4\text{-}6)
$$

Substituting back into Eq. (8-4-3), we find

$$
E_x^{\text{ind}}(z,t) = -2\pi i k \chi \delta E_0 e^{-i(kz-\omega t)} \quad (8\text{-}4\text{-}7)
$$

Again, comparing with Eq. (7-1-9), we find

$$ik\chi = \sigma \tag{8-4-8}$$

This means that wherever we previously had σ we can now substitute $ik\chi$. In particular, going back to Eqs. (7-2-5) and (7-2-6), we find the general solutions for radiation traveling within this medium in the z direction.

$$E_x(z) = \begin{cases} \exp\left[-ik\sqrt{\mu(1 + 4\pi\chi)}z\right] \\ \exp\left[+ik\sqrt{\mu(1 + 4\pi\chi)}z\right] \end{cases} \tag{8-4-9}$$

We remember now that the expression $1 + 4\pi\chi$ was just equal to the dielectric constant ε in the case of electrostatics. We can define the dielectric constant at frequency ω in the same manner and thus observe that the refractive index can be put in terms of μ and ε.

$$n = \sqrt{\mu(1 + 4\pi\chi)} = \sqrt{\mu\varepsilon} \tag{8-4-10}$$

(Needless to say, the dielectric constant and the magnetic susceptibility that one must use here are not the values obtained from electrostatics and magnetostatics. The values we need here are to be obtained at the appropriate frequency ω.)

8-5 INTERFERENCE PHENOMENA FROM AN ARRAY OF DISCRETE DIPOLES; THE NOTION OF COHERENCE

With this section we begin a systematic study of the interference problems that result when we add together the radiation from a collection of coherent dipoles. Before we do much adding however it is wise to define precisely what it is that we mean by **coherence,** in simple physical terms.

Suppose for a moment that we have only two dipole oscillators operating at exactly the same frequency. If the phase between the oscillators remains absolutely fixed, then the amplitude of electric field at any point in space will remain constant in time. Wherever the electric fields from the two add constructively, we will have increased intensity. Wherever they add destructively, we will have decreased intensity. The main point is that the intensity pattern averaged over a cycle of the oscillation will not vary with time. These two oscillators are then completely coherent with respect to one another.

Suppose, on the other hand, that the two oscillators were going very fitfully in stops and starts so that every so often the phase between them would jump. If we looked at the combined electric field from the two oscillators at any given point in space, it would vary in amplitude each time such a jump in phase occurred. If the jumps took place in intervals which were short compared with the time constant of our sensing instruments, we would average over them and see no net constructive or destructive interference, just the sum of the independent intensities. If the phase jumps occurred in intervals which were long compared with the instrumental time constant, we would still see a complete interference pattern, but its nature would change at each jump. In any case the extent to which we will say that the sources are coherent will depend on the time constant of our instrument as compared with the time between phase jumps.

As an example of **incoherence** we might consider two independent monochromatic light sources. Since the light from any source is composed of the contribution from numerous atoms and since each atom only radiates for a very short time, about 10^{-8} sec, the interference pattern changes much too rapidly to follow with our instruments. On the other hand, if the atoms in the two oscillators can be driven together by having them each scatter the radiation from a third source, then they will act completely coherently. When we studied the origin of the refractive index, we observed exactly such a coherence among the scattering centers in our medium as they were driven by the incident plane wave.

As our first exercise then we examine the interference pattern at a great distance from a simple linear array of N parallel dipoles spaced a distance d apart (see Fig. 8-5). We will concern ourselves only with the electric field in the plane normal to the dipoles; for convenience we will call this the yz plane and take our dipoles in the x direction. The z axis is taken perpendicular to the line of dipoles.

We take the dipole moment of the nth dipole to be $p_n \hat{\varepsilon}_1$. The distance from the first dipole to the point at which we wish to evaluate the field is r. Since $r \gg d$, we can write

$$E_x(r,\theta,t) = \frac{k^2 e^{-i(kr-\omega t)}}{r} \sum_{n=0}^{N-1} p_n e^{iknd\sin\theta} \tag{8-5-1}$$

From here on, in principle, all we need to do is add together the complex numbers corresponding to each of the dipoles and take the real part of the sum. This procedure is particularly simple if all the dipoles are equal. In that case

$$E_x(r,\theta,t) = \frac{k^2 p e^{-i(kr-\omega t)}}{r} \sum_{n=0}^{N-1} e^{iknd\sin\theta} \tag{8-5-2}$$

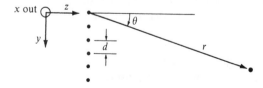

Fig. 8-5 We are interested in the radiation field at a position (r,θ) in the plane normal to an array of dipoles. We assume that r is much larger than d, the spacing between dipoles.

We can get a good idea of what takes place here by plotting the sum in the complex plane. If we let $\alpha = kd \sin \theta$, then the various terms in the sum are vectors of unit length, each of which is rotated at angle α relative to the previous term. For example, taking $n = 4$, we have plotted the sums in Fig. 8-6 for various values of α. It is clear that we have minima with $E_x = 0$ at $\alpha = \pi/2$, π, $3\pi/2$, $5\pi/2$, 3π, $7\pi/2$, etc., and maxima when α is any integral multiple of 2π. Generalizing to arbitrary N, we observe that the pattern

Fig. 8-6 We have plotted the sum $\sum_{n=0}^{3} e^{in\alpha}$ for various values of α. (The dashed line indicates the resultant.) We obtain maxima at $\alpha = 2m\pi$ and minima at $\alpha = \pi/2$, π, $3\pi/2$, $5\pi/2$, $\ldots$

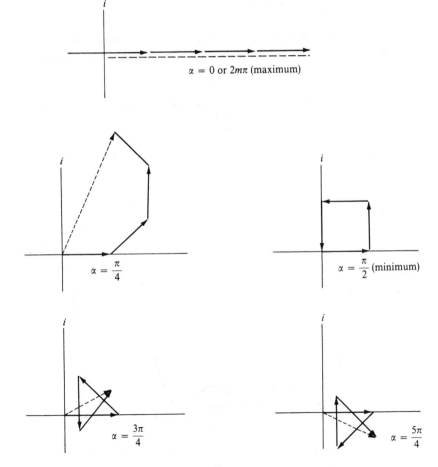

has a series of $N - 1$ nulls between large maxima. At these maxima,

$$\alpha = kd \sin \theta = 2m\pi \qquad (8\text{-}5\text{-}3)$$

where m is an integer. Since $k = 2\pi/\lambda$ where λ is the wavelength of the radiations, we can rewrite Eq. (8-5-3) into more conventional form:

$$m\lambda = d \sin \theta \qquad (8\text{-}5\text{-}4)$$

These peaks of constructive interference occur then when the path length difference between successive dipoles is just equal to an integral number of wavelengths.

We find the intensity by squaring the magnitude of E and multiplying by $c/8\pi$ [see Eq. (8-2-11)].

$$I(r,\theta) = \frac{ck^4|p|^2}{8\pi r^2} \left| \sum_{n=0}^{N-1} e^{ink\,d\sin\theta} \right|^2 \qquad (8\text{-}5\text{-}5)$$

The intensity pattern for a set of four dipoles is shown in Fig. 8-7. The potential significance of the analysis we have just completed is apparent.

Fig. 8-7 A plot of relative intensity versus $kd \sin \theta$ for a set of four synchronous dipole antennas spaced a distance d apart along a line (see Fig. 8-5).

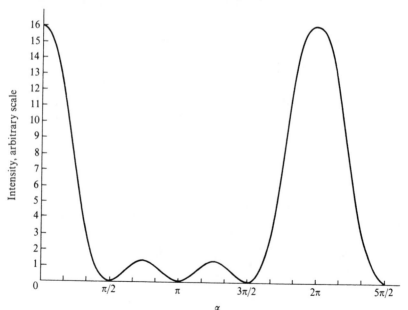

We have observed that radiation of a given frequency can be made more and more directional by increasing the number of phased antennas involved in the transmission. A bit of thought will indicate that by varying the phase among the antennas we can vary the direction in which the radiation is transmitted. Thus we see that a large field of phased antennas is the natural way of "beaming" broadcasts to specific areas around the world.

Incidentally, in a very analogous way we can consider the directionality of a set of receiving antennas. Suppose the dipoles of Fig. 8-5 had been set to receive rather than transmit radiation. If they were all connected to a common amplifier with no phase lag among them, then they would add completely constructively whenever the radiation was coming *from* a direction given by Eq. (8-5-4). By varying the relative phase lags from the antennas to the amplifier, the directions of maximum sensitivity can be varied. Much of radio astronomy these days is done with large fields of phased-antenna arrays.

8-6 FRAUENHOFER DIFFRACTION BY A SLIT; SCATTERING BY A DISK; THE DIFFRACTION GRATING

We are all qualitatively familiar with the fact that a plane wave of light passing through a small hole in a wall exhibits a remarkable interference pattern on the far side. This pattern is generally explained in terms of the so-called **Huygen's principle,** which tells us to consider each point on a wavefront as a new source of radiation and add the "radiation" from all of the new "sources" together. Physically this makes *no* sense at all. Light does not emit light; only accelerating charges emit light. Thus we will begin by throwing out Huygen's principle completely; later we will see that it actually does give the right answer for the wrong reasons.

What happens then as our radiation strikes a wall with a hole in it? Referring to Fig. 8-8 we see that the radiation to the right of the wall is a superposition of the incoming radiation and the radiation arising from the oscillating dipoles in the wall. If we were to fill in the hole so as to make the wall complete, then *nothing* would penetrate the wall. That is, the *complete* wall radiates just enough to completely cancel the incoming plane wave to the right of it. Hence the radiation which appears when the hole is open must be precisely canceled out by the radiation from the stopper as we cover up the hole. This makes our calculation of the diffraction pattern very simple. We need only calculate the radiation from the stopper itself. The radiation field to the right of the hole is equal and opposite in amplitude to that which would be emitted by the stopper if it were radiating all by itself with the same dipole moment per unit area as the rest of the screen.

Our job then consists of two parts. First, we must find out what the

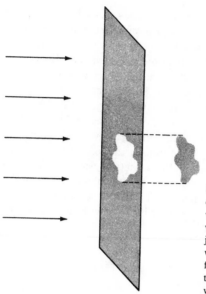

Fig. 8-8 An incoming plane wave strikes an opaque wall with a hole in it. If the "stopper" were inserted into the hole, then no radiation would penetrate. Hence the stopper radiates just enough to cancel the radiation which would be transmitted through the hole. We find the radiation pattern on the right by taking the negative of the electric field which would be radiated by the stopper.

dipole moment per unit area of our wall[1] is in terms of the incoming radiation. Second, we can replace the combination of incoming radiation and wall by the "equivalent" stopper alone. Giving this stopper a dipole moment per unit area which is equal and opposite to the rest of the wall, we calculate its radiation pattern and have the answer to our problem.

Let p be the dipole moment per unit area of our wall. If the wall were complete, then the field contributed by its oscillating charges [see Eq. (8-4-3)] would just be

$$E_x^{\text{wall}}(z,t) = -2\pi i k \, p \, e^{-i(kz - \omega t)}$$

if the incoming radiation had the form

$$E_x^{\text{inc}}(z,t) = E_0 e^{-i(kz - \omega t)}$$

Then in order that the total radiation to the right of the wall be zero, we must have

$$p = \frac{E_0}{2\pi i k} \tag{8-6-1}$$

---

[1] We assume throughout that the radiating dipole distribution in the wall has negligible thickness. In other words, the wall is made of exceedingly opaque material. We also ignore the displacement current at the edge of the hole or disk which results from the discontinuity in the current distribution. These edge effects are only important if the hole is of the same order of size as the wavelength.

We thus replace the incoming radiation and the wall with its hole by the stopper having dipole moment per unit area equal to $-E_0/2\pi ik$. If we let $\mathbf{r}'$ refer to a point on the aperture and $\mathbf{r}$ be the place at which we wish to know the electric field, then, using Eqs. (8-2-10) and (8-6-1), we obtain

$$E_x(\mathbf{r},t) \cong \frac{ikE_0}{2\pi} e^{i\omega t} \int_{\text{aperture}} \frac{\exp(-ik|\mathbf{r}-\mathbf{r}'|)}{|\mathbf{r}-\mathbf{r}'|} \, dA' \tag{8-6-2}$$

[We assume here that the diffraction angle is relatively small so that we need not worry about taking the projection of the dipole moment normal to the propagation direction. That is the reason for writing Eq. (8-6-2) as an approximation.]

As a simple exercise we can now calculate the diffraction pattern from a rectangular slit in the so-called **Frauenhofer limit,** where the distance from the slit to the point of interest $\mathbf{r}$ is much larger than any dimension of the slit itself. We set the origin of our coordinate system at the center of the aperture with the z axis normal to the wall and pointing in the direction of propagation of the incoming radiation. We take the aperture to extend from $x = -a/2$ to $x = +a/2$ and from $y = -b/2$ to $y = +b/2$. As before, the incoming radiation will be taken as polarized in the x direction. To simplify matters we will only consider relatively small angles where both x and y are quite a bit smaller than z, making Eq. (8-6-2) valid. We have then

$$E_x(x,y,z,t) = \frac{ikE_0}{2\pi} e^{i\omega t} \int_{-a/2}^{a/2} dx' \int_{-b/2}^{b/2} dy' \frac{\exp(-ik|\mathbf{r}-\mathbf{r}'|)}{|\mathbf{r}-\mathbf{r}'|} \tag{8-6-3}$$

We can take the denominator out of the integral, since it varies very little over the aperture. We can also approximate $|\mathbf{r}-\mathbf{r}'|$ using Eq. (8-1-4). We have then

$$E_x(x,y,z,t) \cong \frac{ikE_0 e^{-i(kr-\omega t)}}{2\pi r} \int_{-a/2}^{a/2} dx' \int_{-b/2}^{b/2} dy' \exp\left(ik\frac{\mathbf{r}\cdot\mathbf{r}'}{r}\right) \tag{8-6-4}$$

The integral splits very simply into the product of an integral over x' and one over y'. Thus we have

$$E_x(x,y,z,t) = \frac{ikE_0 e^{-i(kr-\omega t)}}{2\pi r} \left(\int_{-a/2}^{a/2} e^{ik(\sin\theta_x)x'} \, dx' \right)$$
$$\left(\int_{-b/2}^{b/2} e^{ik(\sin\theta_y)y'} \, dy' \right) \tag{8-6-5}$$

where

$$\sin\theta_x = \frac{x}{r} \quad \text{and} \quad \sin\theta_y = \frac{y}{r}$$

Integrating, we obtain

$$E_x(x,y,z,t) = \frac{iabkE_0 e^{-i(kr-\omega t)}}{2\pi r}$$

$$\frac{\sin\left(k\,\frac{a}{2}\sin\theta_x\right)}{k\,\frac{a}{2}\sin\theta_x}\,\frac{\sin\left(k\,\frac{b}{2}\sin\theta_y\right)}{k\,\frac{b}{2}\sin\theta_y} \quad (8\text{-}6\text{-}6)$$

Before we go on to calculate the intensity of the diffracted radiation, it is well to make some interesting observations. Note the phase of 90° between the directly forward-going diffracted field ($\theta_x = \theta_y = 0$) and the field that would be present there if the wall were removed. Note also that the amplitude of the directly forward-going diffracted field varies inversely as the wavelength of the incoming radiation. The longer the wavelength, the wider will be the diffraction pattern and hence the weaker will be the intensity at the center.

It is convenient to express the intensity of the diffraction pattern in terms of the amount of energy per unit solid angle at emission angles θ_x and θ_y. Since the Poynting vector determines the energy passing a given area per unit time, we can write

$$\left\langle\frac{d^2u}{dt\,d\Omega}\right\rangle_{\text{time av}} = r^2\left\langle\mathbf{S}\cdot\frac{\mathbf{r}}{r}\right\rangle_{\text{time av}}$$

$$= I_0\,\frac{\sin^2\left(k\,\frac{a}{2}\sin\theta_x\right)}{\left(k\,\frac{a}{2}\sin\theta_x\right)^2}\,\frac{\sin^2\left(k\,\frac{b}{2}\sin\theta_y\right)}{\left(k\,\frac{b}{2}\sin\theta_y\right)^2} \quad (8\text{-}6\text{-}7)$$

where

$I_0 =$ intensity per unit solid angle in forward direction

$$= \frac{ca^2b^2k^2E_0{}^2}{32\pi^3} \quad (8\text{-}6\text{-}8)$$

A plot of the intensity pattern versus $k\,\dfrac{a}{2}\sin\theta_x$ for $\theta_y = 0$ is shown in Fig. 8-9. Note that the minima occur at angles such that

$$\sin\theta_x = \frac{2m\pi}{ka} = \frac{m\lambda}{a} \quad\text{or}\quad \sin\theta_y = \frac{2n\pi}{kb} = \frac{n\lambda}{b} \quad (8\text{-}6\text{-}9)$$

where m and n are integers and λ is the wavelength of the radiation.

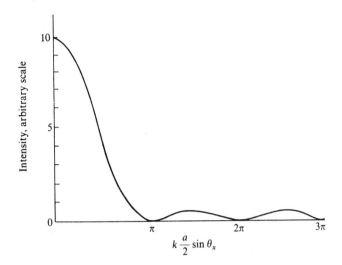

Fig. 8-9 A plot of intensity versus $(ka/2)\sin\theta_x$ for a slit of width a.

We will not go through the trouble of integrating the intensity over angle to find out how much of the incident energy makes it through the hole. Needless to say, *all the energy which strikes the hole* makes it through and contributes to the diffraction pattern. If we carried out the integral for any shape aperture, we would find

$$\int_{\substack{\text{all}\\ \text{angles}}} \left\langle \frac{d^2u}{d\Omega\, dt} \right\rangle_{\text{time av}} d\Omega = \langle \text{incident energy/unit area/unit time}\rangle_{\text{time av}}$$

$$\text{(area of aperture)} \quad (8\text{-}6\text{-}10)$$

Suppose that we replaced the wall with a hole in it (of arbitrary shape) by a disk which precisely matches the aperture (see Fig. 8-8). Now only a very small amount of energy is absorbed in the opaque disk and most of the radiation continues on as before in the plane wave.

To calculate the radiation pattern we need only ascribe to the disk a dipole moment per unit area equal to $E_0/2\pi ik$ [see Eq. (8-6-1)]. The total field is then obtained by summing the incident field and the field produced by the disk.

$$E_x(x,y,z,t) \cong E_0 e^{-i(kz-\omega t)} - \frac{ikE_0}{2\pi} e^{i\omega t} \int_{\text{disk}} \frac{\exp(-ik|\mathbf{r}-\mathbf{r}'|)}{|\mathbf{r}-\mathbf{r}'|}\, dA'$$

$$\cong E_0 e^{-i(kz-\omega t)} - \frac{ikE_0 e^{-i(kr-\omega t)}}{2\pi r} \int_{\text{disk}} \exp\left(ik\frac{\mathbf{r}\cdot\mathbf{r}'}{r}\right) dA'$$

$$(8\text{-}6\text{-}11)$$

It is interesting at this point to determine the forward-scattering amplitude of the disk. Back on page 257 we defined the forward-scattering amplitude $f(0)$ for a single scatterer. The definition here is identical; we take the ratio of the amplitude of forwardly emitted radiation at $z = 1$ cm to the amplitude that the incident radiation would have had at the same point if there were no scatterer. Hence we have in this case

$$f(0) = -\frac{ik}{2\pi} \int_{\text{disk}} dA'$$

$$= -\frac{ik}{2\pi} \text{(area of disk)} \qquad (8\text{-}6\text{-}12)$$

Taking another look at Eq. (8-6-11), we see that the scattered radiation is identical except for a minus sign to that which we obtained when we had a hole rather than its complementing disk [see Eq. (8-6-2)]. Hence, making use of Eq. (8-6-10), we conclude that

$$\sigma_{\text{elastic}} = \text{elastic scattering cross section}^1 \text{ of disk}$$

$$= \text{area of disk} \qquad (8\text{-}6\text{-}13)$$

Since the disk absorbs all the energy which is incident on it, we can also write

$$\sigma_{\text{abs}} = \text{area of disk} \qquad (8\text{-}6\text{-}14)$$

Summing the two, we have the total cross section.

$$\sigma_{\text{total}} = 2 \text{(area of disk)} \qquad (8\text{-}6\text{-}15)$$

We can rewrite Eq. (8-6-12) in terms of the total cross section, in a somewhat weaker form than we have it. As we shall see, this will be a result which will prove to be independent of the opacity of the disk.

$$\text{Im} f(0) = -\frac{k}{4\pi} \sigma_{\text{total}} \qquad (8\text{-}6\text{-}16)$$

Summarizing then, we have shown that the elastic scattering cross section and the absorption cross section of an opaque disk are identical. Furthermore, we have related the total cross section to the imaginary part of the forward-scattering amplitude. (In the case of an opaque disk the forward-scattering amplitude turned out to be completely imaginary.) These results will turn out to be of substantial importance when we study the scattering and absorption of elementary particles quantum mechanically. Indeed, the relation between the imaginary part of the forward-scattering

---

[1] For the definition of cross section see page 223.

amplitude and the total cross section goes under the name of the **optical theorem** in many quantum-mechanics texts. (The sign is often reversed because the sign convention for i is taken differently.)

What, we might ask, would be the situation if the disk had not been completely opaque? To answer that we first go back to our complete wall which is radiating the plane wave represented by Eq. (8-4-3). Again, for convenience, we let p be the dipole moment per unit area ($p = P\delta$). In the case of an opaque wall we found that $p = E_0/2\pi i k$ [see Eq. (8-6-1)]. We will modify this result by letting

$$p = \frac{\eta E_0}{2\pi i k} \tag{8-6-17}$$

The electric field to the right of the wall (see Fig. 8-4) would then be

$$E_x(z,t) = E_0(1 - \eta)e^{-i(kz - \omega t)} \qquad \text{for } z > 0 \tag{8-6-18}$$

The fraction $\mathscr{F}$ of the incoming energy which is transmitted through the wall is just given by

$$\mathscr{F} = |1 - \eta|^2 \tag{8-6-19}$$

If we now make up a disk having area A of this nonopaque material, we would anticipate its absorption cross section σ_{abs} to be given by

$$\sigma_{abs} = (1 - \mathscr{F})A \tag{8-6-20}$$

The outgoing radiation to the right of our disk would be

$$E_x(x,y,z,t) \cong E_0 e^{-i(kz - \omega t)} - \frac{\eta i k E_0}{2\pi r} e^{-i(kr - \omega t)} \int_{\text{disk}} \exp\left(ik\,\frac{\mathbf{r} \cdot \mathbf{r}'}{r}\right) dA' \tag{8-6-21}$$

The forward-scattering amplitude would be

$$f(0) = -\frac{\eta i k}{2\pi} A \tag{8-6-22}$$

Finally, the elastic cross section would be $|\eta|^2$ times the elastic cross section for an opaque disk of the same area. Hence, referring to Eq. (8-6-13), we have

$$\sigma_{\text{elastic}} = |\eta|^2 A \tag{8-6-23}$$

Summing the elastic and absorption cross section to find the total cross section, we obtain

$$\sigma_{\text{total}} = (1 - |1 - \eta|^2 + |\eta|^2)A$$

$$= 2A \operatorname{Re} \eta \tag{8-6-24}$$

Hence, the so-called *optical theorem* works for a nonopaque disk:

$$\text{Im} f(0) = -\frac{kA}{2\pi} \text{Re}\, \eta$$

$$= -\frac{k\sigma_{\text{total}}}{4\pi} \tag{8-6-25}$$

This general result was derived subject to the small-angle approximation which we have been making all along. It happens however to be independent of this approximation and is completely true in general. As mentioned before it plays a very important role in studying the quantum-mechanical scattering of elementary particles.

We now return to our wall with its aperture and replace our single slit with a series of N parallel slits each of height a and width b. The arrangement of the slits will be as shown in Fig. 8-10. We take the origin of our coordinate system at the center of the first slit. All slits extend from $x = -a/2$ to $x = +a/2$. The first slit extends from $y = -b/2$ to $y = b/2$. The nth slit then extends from $y = (n-1)d - b/2$ to $y = (n-1)d + b/2$. We calculate the electric field exactly as we did in the case of a single slit except

Fig. 8-10 We investigate the diffraction-interference pattern caused by a series of N slits each of width b that are spaced a distance d apart.

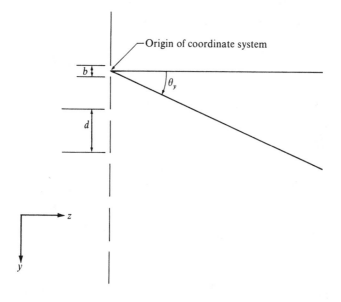

that the integral over y' consists of N different parts.

$$E_x(x,y,z,t) \cong \frac{ikE_0 e^{-i(kr-\omega t)}}{2\pi r} \left(\int_{-a/2}^{a/2} e^{ik(\sin\theta_x)x'} dx' \right)$$
$$\left(\sum_{n=1}^{N} \int_{(n-1)d-b/2}^{(n-1)d+b/2} e^{ik(\sin\theta_y)y'} dy' \right) \quad (8\text{-}6\text{-}26)$$

Our expression can be simplified somewhat by noting that

$$\int_{(n-1)d-b/2}^{(n-1)d+b/2} e^{ik(\sin\theta_y)y'} dy' = e^{ik(n-1)d\sin\theta_y} \int_{-b/2}^{b/2} e^{ik(\sin\theta_y)u} du \quad (8\text{-}6\text{-}27)$$

Substituting back and integrating, we find an expression for E_x.

$$E_x(x,y,z,t) = \frac{iabE_0 e^{-i(kr-\omega t)}}{2\pi r} \frac{\sin\left(k\frac{a}{2}\sin\theta_x\right)}{k\frac{a}{2}\sin\theta_x} \frac{\sin\left(k\frac{b}{2}\sin\theta_y\right)}{k\frac{b}{2}\sin\theta_y}$$
$$\cdot \sum_{n=1}^{N} e^{ik(n-1)d\sin\theta_y} \quad (8\text{-}6\text{-}28)$$

If we go back and compare this expression with Eqs. (8-6-2) and (8-6-6), we come to a very interesting conclusion. The *intensity* pattern arising from a set of N slits spaced a distance d apart can be arrived at by multiplying the intensity pattern for a single slit by the intensity pattern for N dipole antennas with the same spacing.

PROBLEMS

8-1. A dipole radiator (see Fig. 8-1) is set parallel to a perfectly conducting wall. The distance from the dipole to the wall is $\lambda/2$ where λ is the wavelength of the emitted radiation. What is the intensity distribution as a function of θ in a plane perpendicular to the dipole at a long distance away from it?

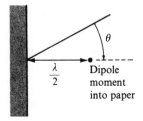

$\frac{\lambda}{2}$ Dipole moment into paper

θ

8-2. Suppose the electric dipole radiator of Prob. 8-1 were replaced with a magnetic dipole radiator (see Fig. 8-2) whose moment was oriented in the same direction. How would the intensity distribution change, if at all?

8-3. Consider the dipoles of Probs. 8-1 and 8-2 turned so as to have their moments perpendicular to the conducting plane. Evaluate the radiated intensity as a function of θ at a long distance away for each of the two cases.

8-4. A *classical* atom is made up of two electrons traveling in a common circle of radius a about a helium nucleus. (The orbits of the two electrons coincide.) Assume the frequency of rotation to be ω where $\omega \ll 2\pi c/a$.

(a) Find the radiation pattern if the two electrons are always on opposite sides of the circle from one another and are following one another around. How much energy is radiated per unit time?

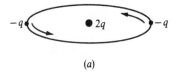

(a)

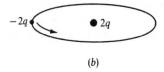

(b)

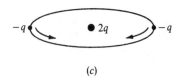

(c)

(b) How much energy would be radiated per unit time if the two electrons moved together as one doubly charged object?

(c) How much energy would be radiated per unit time if the two electrons traveled in opposite directions about the circle?

8-5. Calculate the Frauenhofer diffraction pattern that would result from a circular aperture of diameter D illuminated by radiation of wavelength λ. Sketch the pattern for the case where $D = \lambda$.

8-6. A beam of light of wavelength 5×10^{-5} cm is normally incident upon a square glass plate of thickness 10^{-4} cm and 10^{-3} cm on a side. The glass is ideal, completely nonabsorbing, and has refractive index 1.5.

   (a) What is the scattering cross section of the glass plate? That is, what fraction of the incident energy per square centimeter is removed from the beam and reradiated?

   (b) What is the angular distribution of the reradiated light?

   (c) What are the real and imaginary parts of the forward-scattering amplitude of the plate as a whole?

   (d) Suppose the phase of the scattered radiation could somehow be shifted relative to the incident radiation by $\pi/2$. What would the apparent absorption cross section of the plate be to an observer who was measuring its forward-scattering amplitude? Can you think of how to produce this shift? You would be reinventing the phase-contrast microscope.

8-7. As we fly at 30,000 ft, most of the atmosphere is below us. Why is it then that the air above us seems so much bluer than the air below us? Would the sky be blue if the density of the atmosphere were uniform throughout and the same as it is at sea level?

9
Waveguides and Cavities

One of the more useful applications of what we have just learned about the reflection of radiation at a conducting surface is in the field of waveguides and cavities. A waveguide is a rather simple device. Imagine that we have a long, hollow tube which is a perfect mirror on the inside. If we put a light bulb at one end, some light will obviously be carried to the other end. Hence the pipe or waveguide is a means of transferring radiant energy from one place to another with minimal attenuation.

Now in the pipe we have just described, we tacitly assumed that the pipe diameter was very much larger than the wavelength of the light. Hence we did not need to worry about any coherent interference effects as the light bounced back and forth across the pipe. If we were to take the wavelength to be comparable to the transverse dimension of the waveguide, we would observe a variety of such effects, and these indeed are largely the subjects of this chapter.

Now to make our work simple, we will limit our considerations to rectangular waveguides and cavities. (A cavity is a length of waveguide which is capped off at each end.) Without much physical difficulty, but with considerably mathematical difficulty, other shapes can be considered by the reader. We will begin by assuming the walls of our waveguide to be perfectly conducting. Later we will consider the effect of having only a finite conductivity.

9-1 THE PERFECTLY CONDUCTING, RECTANGULAR WAVEGUIDE

Consider then a straight pipe with rectangular cross section whose height and width are a and b, respectively. We will set up our coordinate system at some point in the waveguide, as shown in Fig. 9-1, with the z axis pointing in the direction in which we want our radiation to propagate. The inside of the waveguide extends from $x = 0$ to $x = a$ and from $y = 0$ to $y = b$.

Now we could begin by writing down a plane wave, letting it reflect each time it reaches a surface, and then adding everything together. It is however much simpler to start from scratch and look for solutions to Maxwell's equations subject to the boundary condition that the tangential component of **E** be zero at the surface. We have

$$\nabla \times \mathbf{B} = \frac{1}{c} \frac{\partial \mathbf{E}}{\partial t}$$

and

$$\nabla \times \mathbf{E} = -\frac{1}{c} \frac{\partial \mathbf{B}}{\partial t}$$

Taking the curl of the second equation, we obtain

$$\nabla \times (\nabla \times \mathbf{E}) = \nabla(\nabla \cdot \mathbf{E}) - \nabla^2 \mathbf{E}$$

$$= -\frac{1}{c} \frac{\partial}{\partial t}(\nabla \times \mathbf{B}) = -\frac{1}{c^2} \frac{\partial^2 \mathbf{E}}{\partial t^2}$$

Fig. 9-1 Segment of a rectangular waveguide of cross section $a \times b$.

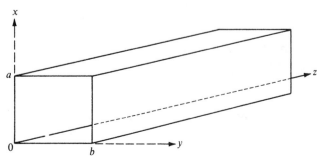

But $\nabla \cdot \mathbf{E} = 0$. Hence

$$\nabla^2 \mathbf{E} - \frac{1}{c^2} \frac{\partial^2 \mathbf{E}}{\partial t^2} = 0 \tag{9-1-1}$$

Similarly

$$\nabla^2 \mathbf{B} - \frac{1}{c^2} \frac{\partial^2 \mathbf{B}}{\partial t^2} = 0 \tag{9-1-2}$$

These wave equations hold for each of the components of $\mathbf{E}$ and $\mathbf{B}$ independently. For example, we can write

$$\nabla^2 E_x(x,y,z,t) - \frac{1}{c^2} \frac{\partial^2 E_x(x,y,z,t)}{\partial t^2} = 0 \tag{9-1-3}$$

Now we can search for a solution to this equation of the form

$$E_x(x,y,z,t) = E_{x_1}(x) E_{x_2}(y) E_{x_3}(z) E_{x_4}(t)$$

(It is possible to show that the solutions of this form make up a complete set; any solution to the problem can be expressed in terms of them.) Substituting back, we have

$$\frac{1}{E_{x_1}(x)} \frac{d^2 E_{x_1}}{dx^2} + \frac{1}{E_{x_2}(y)} \frac{d^2 E_{x_2}}{dy^2} + \frac{1}{E_{x_3}(z)} \frac{d^2 E_{x_3}}{dz^2}$$

$$- \frac{1}{c^2 E_{x_4}(t)} \frac{d^2 E_{x_4}}{dt^2} = 0 \tag{9-1-4}$$

Since each term in this sum involves a different and independent variable, we can set each equal to a constant.

$$\frac{d^2 E_{x_1}}{dx^2} = -k_{11}{}^2 E_{x_1} \qquad \frac{d^2 E_{x_3}}{dz^2} = -k_{31}{}^2 E_{x_3}$$

$$\frac{d^2 E_{x_2}}{dy^2} = -k_{21}{}^2 E_{x_2} \qquad \frac{d^2 E_{x_4}}{dt^2} = -c^2 k^2 E_{x_4} \tag{9-1-5}$$

As usual we define $\omega \equiv ck$. In order that Eq. (9-1-4) hold, we must have

$$k_{11}{}^2 + k_{21}{}^2 + k_{31}{}^2 = k^2 \tag{9-1-6}$$

Now the solutions for E_{x_1} and E_{x_2} are of the form

$$E_{x_1} = A_{11} \cos k_{11}x + B_{11} \sin k_{11}x$$

$$E_{x_2} = A_{21} \cos k_{21}y + B_{21} \sin k_{21}y \tag{9-1-7}$$

In order that the radiation be traveling in the direction of increasing z, we need only make use of those products of E_{x_3} and E_{x_4} which combine

$k_{31}z$ with $-\omega t$ and leave out the ones which combine $k_{31}z$ with $+\omega t$. That is to say, we keep only the terms leading to $\sin(k_{31}z - \omega t)$ and $\cos(k_{31}z - \omega t)$ among the various products of $\sin k_{31}z$, $\cos k_{31}z$, $\sin \omega t$, and $\cos \omega t$. We throw away the alternative terms $\sin(k_{31}z + \omega t)$ and $\cos(k_{31}z + \omega t)$. Now we can represent an arbitrary combination of $\sin (k_{31}z - \omega t)$ and $\cos(k_{31} - \omega t)$ by taking the real part of $E_{0x}e^{-i(k_{31}z - \omega t)}$ and allowing E_{0x} to be a complex number. We have then

$$E_x(x,y,z,t) = E_{0x}(A_{11} \cos k_{11}x + B_{11} \sin k_{11}x)(A_{21} \cos k_{21}y$$
$$+ B_{21} \sin k_{21}y)e^{-i(k_{31}z - \omega t)} \quad (9\text{-}1\text{-}8)$$

Similarly

$$E_y(x,y,z,t) = E_{0y}(A_{12} \cos k_{12}x + B_{12} \sin k_{12}x)(A_{22} \cos k_{22}y$$
$$+ B_{22} \sin k_{22}y)e^{-i(k_{32}z - \omega t)} \quad (9\text{-}1\text{-}9)$$

$$E_z(x,y,z,t) = E_{0z}(A_{13} \cos k_{13}x + B_{13} \sin k_{13}x)(A_{23} \cos k_{23}y$$
$$+ B_{23} \sin k_{23}y)e^{-i(k_{33}z - \omega t)} \quad (9\text{-}1\text{-}10)$$

The boundary conditions on the tangential components of E at $x = y = 0$ tell us that

$$A_{21} = A_{12} = A_{13} = A_{23} = 0 \quad (9\text{-}1\text{-}11)$$

The requirement that $\nabla \cdot \mathbf{E} = 0$ everywhere within the waveguide tells us that

$$B_{11} = B_{22} = 0$$

and

$$k_{11} = k_{12} = k_{13} \quad (=k_1 \text{ by definition})$$
$$k_{21} = k_{22} = k_{23} \quad (=k_2 \text{ by definition})$$
$$k_{31} = k_{32} = k_{33} \quad (=k_3 \text{ by definition}).$$

In addition, if we incorporate the remaining arbitrary constants into the definitions of E_{0x}, E_{0y}, and E_{0z}, we have

$$k_1 E_{0x} + k_2 E_{0y} + ik_3 E_{0z} = 0 \quad (9\text{-}1\text{-}12)$$

where

$$E_x = E_{0x} \cos k_1 x \sin k_2 y \, e^{-i(k_3 z - \omega t)}$$
$$E_y = E_{0y} \sin k_1 x \cos k_2 y \, e^{-i(k_3 z - \omega t)} \quad (9\text{-}1\text{-}13)$$
$$E_z = E_{0z} \sin k_1 x \sin k_2 y \, e^{-i(k_3 z - \omega t)}$$

Finally the requirements that $E_x = E_z = 0$ at $y = b$ and that $E_y = E_z = 0$

at $x = a$ yield the restrictions

$$k_1 = \frac{m\pi}{a} \qquad k_2 = \frac{n\pi}{b} \tag{9-1-14}$$

where n and m are integers. To find the magnetic field components B_x, B_y, and B_z we take the curl of $\mathbf{E}$ and remember that

$$\nabla \times \mathbf{E} = -\frac{1}{c}\frac{\partial \mathbf{B}}{\partial t}$$

Integrating the curl of $\mathbf{E}$ with respect to time and multiplying by $-c$, we obtain

$$B_x = \frac{ic}{\omega}(k_2 E_{0z} + ik_3 E_{0y}) \sin k_1 x \cos k_2 y \, e^{-i(k_3 z - \omega t)}$$

$$B_y = -\frac{ic}{\omega}(k_1 E_{0z} + ik_3 E_{0x}) \cos k_1 x \sin k_2 y \, e^{-i(k_3 z - \omega t)} \tag{9-1-15}$$

$$B_z = \frac{ic}{\omega}(k_1 E_{0y} - k_2 E_{0x}) \cos k_1 x \cos k_2 y \, e^{-i(k_3 z - \omega t)}$$

We notice incidentally that $\nabla \cdot \mathbf{B} = 0$ as expected.

Now because of the constraint that $k_1 E_{0x} + k_2 E_{0y} + ik_3 E_{0z} = 0$, we cannot choose E_{0x}, E_{0y}, and E_{0z} arbitrarily. It is convenient to define two different classes of solutions for a given choice of the integers n and m. One, called the **transverse electric mode (TE)**, has $E_{0z} = 0$. The other, called the **transverse magnetic mode (TM)**, has $B_{0z} = 0$. Hence, for the latter, $k_1 E_{0y} - k_2 E_{0x} = 0$.

We can easily show that any possible solution to the problem for a given m,n, and ω can be written in terms of TE and TM modes. If $E_{0x}{}^{\text{TE}}$, $E_{0y}{}^{\text{TE}}$, $E_{0x}{}^{\text{TM}}$, $E_{0y}{}^{\text{TM}}$, $E_{0z}{}^{\text{TM}}$ represent the coefficients corresponding to those modes, we require only that

$$E_{0x} = E_{0x}{}^{\text{TE}} + E_{0x}{}^{\text{TM}}$$

$$E_{0y} = E_{0y}{}^{\text{TE}} + E_{0y}{}^{\text{TM}} \tag{9-1-16}$$

$$E_{0z} = E_{0z}{}^{\text{TM}}$$

In addition, we have the two conditions that

$$k_1 E_{0x}{}^{\text{TE}} + k_2 E_{0y}{}^{\text{TE}} = 0$$

$$k_1 E_{0y}{}^{\text{TM}} - k_2 E_{0x}{}^{\text{TM}} = 0 \tag{9-1-17}$$

Solving these five equations for the five unknowns, we have

$$E_{0x}{}^{\text{TE}} = \frac{k_2{}^2 E_{0x} - k_1 k_2 E_{0y}}{k_1{}^2 + k_2{}^2}$$

$$E_{0y}{}^{\text{TE}} = \frac{k_1{}^2 E_{0y} - k_1 k_2 E_{0x}}{k_1{}^2 + k_2{}^2}$$

$$E_{0x}{}^{\text{TM}} = \frac{k_1{}^2 E_{0x} + k_1 k_2 E_{0y}}{k_1{}^2 + k_2{}^2}$$

$$E_{0y}{}^{\text{TM}} = \frac{k_2{}^2 E_{0y} + k_1 k_2 E_{0x}}{k_1{}^2 + k_2{}^2}$$

$$E_{0z}{}^{\text{TM}} = E_{0z}$$

Incidentally the requirement that $E_{0x}{}^{\text{TE}}/E_{0y}{}^{\text{TE}} = -k_2/k_1$ establishes that $E_{0x}{}^{\text{TE}}$ and $E_{0y}{}^{\text{TE}}$ have the same complex phase. The same is true of $E_{0x}{}^{\text{TM}}$ and $E_{0y}{}^{\text{TM}}$. However $E_{0z}{}^{\text{TM}} = i(k_1 E_{0x}{}^{\text{TM}} + k_2 E_{0y}{}^{\text{TM}})/k_3$ and is 90° out of phase with either $E_{0x}{}^{\text{TM}}$ or $E_{0y}{}^{\text{TM}}$.

It is instructive at this point to have a look at some of the simpler modes. A mode will be written as $\text{TE}_{m,n}$ or $\text{TM}_{m,n}$, depending on its type and the values of the integers m and n. The simplest is clearly TE_{01}, for which we have

$$k_1 = 0 \qquad k_2 = \frac{\pi}{b}$$

The requirement that $k_1 E_{0x}{}^{\text{TE}} + k_2 E_{0y}{}^{\text{TE}} = 0$ permits us to conclude that in this case $E_{0y} = 0$. Hence

$$E_x = E_{0x} \sin \frac{\pi}{b} y\, e^{-i(k_3 z - \omega t)}$$

$$E_y = E_z = 0$$

$$B_x = 0$$

$$B_y = \frac{k_3}{k} E_{0x} \sin \frac{\pi}{b} y\, e^{-i(k_3 z - \omega t)}$$

$$B_z = -i \frac{k_2}{k} E_{0x} \sin \frac{\pi}{b} y\, e^{-i(k_3 z - \omega t)}$$

If we now assume E_{0x} to be real, we can find the fields as they actually exist within the waveguide by taking the real parts of these equations.

$$E_x = E_{0x} \sin \frac{\pi}{b} y \cos (k_3 z - \omega t)$$

$$B_y = \frac{k_3}{k} E_{0x} \sin \frac{\pi}{b} y \cos (k_3 z - \omega t) \qquad (9\text{-}1\text{-}19)$$

$$B_z = -\frac{k_2}{k} E_{0x} \cos \frac{\pi}{b} y \sin (k_3 z - \omega t)$$

We see that the electric field just runs across from the plane at $x = 0$ to the plane at $x = a$. The magnetic field, on the other hand, appears to go in closed loops centered about the points where the electric field is zero. We indicate the directions of the field lines in Fig. 9-2. Solid lines represent electric field and dashed lines represent magnetic field.

The logical mode to examine next would be the TM_{01}. We note however that this mode as well as all modes of the form TM_{m0} or TM_{0m} are zero by virtue of the fact that $k_1 E_{0y} = k_2 E_{0x} = 0$ for these modes (since either k_1 or k_2 is zero). This means that the lowest frequency which can pass through the waveguide is in the transverse electric TE_{01} (or TE_{10}) mode

$$
\omega_{min} \begin{cases} = \dfrac{c\pi}{a} & \text{for } a > b \\[2mm] = \dfrac{c\pi}{b} & \text{for } b < a \end{cases}
$$

This mode, called the **dominant mode,** is the important one in the transmission of microwave power. At any frequency below the minimum value, k_3 will be imaginary and hence the radiation will be attenuated exponentially in going down the waveguide.

When we examine the velocity at which radiation travels down the waveguide, our first reaction is that it appears to be ω/k_3, which is greater than c (since $k_3 < k$). To understand this it is convenient to think of the entire process as composed of the superposition of waves being reflected back and forth from the walls. We illustrate this process in Fig. 9-3 where we show only one set of wavefronts (ignoring the reflected ones to avoid confusion). We see that in one period τ the wavefront will move a distance $c\tau$ at right angles to itself. The observer will think that the wave has moved a distance $v_p\tau$ downstream along the waveguide, where $v_p = c/\sin\theta$. The

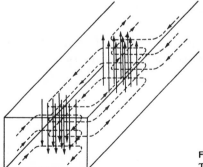

Fig. 9-2   Sketch of the field lines in the mode TE_{01}.

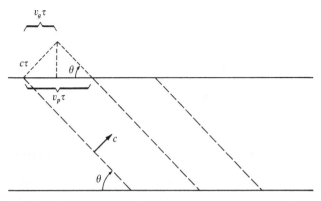

Fig. 9-3 Diagram of one set of wavefronts, illustrating the phase and group velocities and their relationship with c.

rate at which information moves down the waveguide is just $c \sin \theta = v_g$. Hence

$$v_p v_g = c^2 \tag{9-1-20}$$

We observe from our mathematical expressions for E_x, E_y, and E_z that $v_p = \omega/k_3$. The group velocity can also be obtained, as before, by differentiating ω with respect to k_3. We have

$$(k_1{}^2 + k_2{}^2 + k_3{}^2)c^2 = \omega^2$$

and hence

$$\frac{\partial \omega}{\partial k_3} = \frac{k_3}{\omega} c^2 = v_g$$

as expected. To see if all this makes sense we can transform into a frame of reference moving with the group velocity v_g down the waveguide. We should then have a situation where the solutions for E_x, E_y, and E_z should not depend upon z. We first examine the term $k_3 z - \omega t$. Our Lorentz transformation yields

$$z = \gamma(z' + \beta c t')$$

$$t = \gamma \left(t' + \frac{\beta}{c} z' \right)$$

In our case

$$\beta = \frac{k_3}{\omega} c \qquad \text{and} \qquad \gamma = \frac{\omega}{c\sqrt{k_1{}^2 + k_2{}^2}}$$

We have then

$$k_3 z = \gamma(k_3 z') + \frac{k_3{}^2 ct'}{\sqrt{k_1{}^2 + k_2{}^2}}$$

$$-\omega t = -\gamma(k_3 z') - \frac{\omega^2 ct'}{c^2\sqrt{k_1{}^2 + k_2{}^2}}$$

Hence

$$k_3 z - \omega t = \sqrt{k_1{}^2 + k_2{}^2}\, ct' \qquad (9\text{-}1\text{-}21)$$

We see then that the term $e^{-i(k_3 z - \omega t)}$ becomes $\exp(i\sqrt{k_1{}^2 + k_2{}^2}\, ct')$ and the z' dependence falls out. Since $x = x'$ and $y = y'$, the dependence on the x and y directions remains unchanged. The coefficients E_{0x}, E_{0y}, E_{0z}, B_{0x}, B_{0y}, B_{0z} are, of course, transformed like the appropriate elements of a second-rank tensor. Inasmuch as longitudinal fields transform into themselves under this transformation, the two sets of modes $TE_{n,m}$ and $TM_{n,m}$ retain their characters.

9-2   IDEAL RECTANGULAR CAVITIES

If we take a segment of waveguide of length d and cap it off with a sheet of conductor at each end, we have a rectangular cavity. We will let the cross-sectional dimensions of the waveguide segment be a and b as before and assume for the moment that the walls of the cavity are perfectly conducting.

Physically we can now think of the radiation as coming to one end of the cavity, reflecting back, coming to the other end, and reflecting again. If we choose the frequency right, the radiation will add constructively each time around, leading, as we shall see, to a standing wave. If the frequency is ever so slightly off resonance, then each reflection will throw the wave a bit more out of phase with the original wave, and the net result will be the destructive cancellation of the radiation by itself.

If we had considered a real cavity, on the other hand, rather than an ideal one, the radiation would be diminished in amplitude at each successive reflection. Hence, even if it was a bit out of phase after the reflection which brings it back to the original position, it still could never interfere completely destructively. We expect then that the resonance width, the frequency range over which the cavity resonates, will depend on the absorption of radiation by its walls.

At the moment though let us stick with our perfect walls and find the solutions to Maxwell's equations within the cavity subject to the same boundary conditions as before—the tangential components of the electric field are zero at the walls. As before we can search for solutions which are products of separate terms, each depending on one coordinate only. We follow the same procedures as we followed in dealing with waveguides and obtain

$$E_x = E_{0x} \cos k_1 x \sin k_2 y \sin k_3 z \, e^{i\omega t}$$

$$E_y = E_{0y} \sin k_1 x \cos k_2 y \sin k_3 z \, e^{i\omega t} \qquad (9\text{-}2\text{-}1)$$

$$E_z = E_{0z} \sin k_1 x \sin k_2 y \cos k_3 z \, e^{i\omega t}$$

with

$$k_1{}^2 + k_2{}^2 + k_3{}^2 = \frac{\omega^2}{c^2} = k^2 \qquad (9\text{-}2\text{-}2)$$

and

$$k_1 = \frac{l\pi}{a} \qquad k_2 = \frac{m\pi}{b} \qquad k_3 = \frac{n\pi}{d} \qquad (9\text{-}2\text{-}3)$$

where l, m, and n are integers. For convenience we can define the two vectors $\mathbf{k}$ and $\mathbf{E}_0$ as

$$\mathbf{k} = (k_1, k_2, k_3) \qquad (9\text{-}2\text{-}4)$$

$$\mathbf{E}_0 = (E_{0x}, E_{0y}, E_{0z}) \qquad (9\text{-}2\text{-}5)$$

The requirement that $\nabla \cdot \mathbf{E} = 0$ leads to the restriction on $\mathbf{E}_0$ given by

$$\mathbf{k} \cdot \mathbf{E}_0 = 0 \qquad (9\text{-}2\text{-}6)$$

We find the magnetic field as before by taking $\nabla \times \mathbf{E}$ and integrating with respect to time $\left(\nabla \times \mathbf{E} = -\dfrac{1}{c} \dfrac{\partial \mathbf{B}}{\partial t} \right)$.

$$B_x = \frac{ic}{\omega} (\mathbf{k} \times \mathbf{E}_0)_x \sin k_1 x \cos k_2 y \cos k_3 z \, e^{i\omega t}$$

$$B_y = \frac{ic}{\omega} (\mathbf{k} \times \mathbf{E}_0)_y \cos k_1 x \sin k_2 y \cos k_3 z \, e^{i\omega t} \qquad (9\text{-}2\text{-}7)$$

$$B_z = \frac{ic}{\omega} (\mathbf{k} \times \mathbf{E}_0)_z \cos k_1 x \cos k_1 y \sin k_3 z \, e^{i\omega t}$$

Now, since $\mathbf{E}_0$ must lie at right angles to $\mathbf{k}$ [see Eq. (9-2-6)], we can choose two standard vectors $\mathbf{E}_0{}^{\text{TM}}$ and $\mathbf{E}_0{}^{\text{TE}}$ and express any solution in terms of them. Again we define $\mathbf{E}_0{}^{\text{TE}}$ so as to have no z component. Thus

$$k_1 E_{0x}{}^{\text{TE}} + k_2 E_{0y}{}^{\text{TE}} = 0 \qquad (9\text{-}2\text{-}8)$$

We define $\mathbf{E}_0{}^{TM}$ so as to lead to no z component of magnetic field. That is,

$$(\mathbf{k} \times \mathbf{E}_0{}^{TM})_z = 0$$

or

$$k_1 E_{0y}{}^{TM} - k_2 E_{0x}{}^{TM} = 0 \tag{9-2-9}$$

We can see immediately that $\mathbf{E}_0{}^{TE}$ and $\mathbf{E}_0{}^{TM}$ are normal to each other:

$$E_{0x}{}^{TE}E_{0x}{}^{TM} + E_{0y}{}^{TE}E_{0y}{}^{TM} = E_{0x}{}^{TE}E_{0y}{}^{TM}$$

$$+ \left(-\frac{k_1}{k_2} E_{0x}{}^{TE}\right)\left(\frac{k_2}{k_1} E_{0x}{}^{TM}\right) = 0$$

Since $E_{0z}{}^{TE} = 0$, we have

$$\mathbf{E}_0{}^{TE} \cdot \mathbf{E}_0{}^{TM} = 0 \tag{9-2-10}$$

Incidentally our choice of the *normal modes* was somewhat arbitrary. We chose to give preference to the z axis, but we could just as easily have chosen the x or y axes. Obviously the normal modes obtained if we had used the x or y axes as our preferred direction can be expressed in terms of the normal modes we have just obtained. We see then that the system has two normal modes for each allowed frequency ω and an infinite number of possible solutions for ω.

Again we will label the modes as $TE_{l,m,n}$ and $TM_{l,m,n}$, respectively. Now for any mode we have either

$$k_1 E_{0x}{}^{TE} + k_2 E_{0y}{}^{TE} = 0 \qquad \text{or} \qquad k_1 E_{0y}{}^{TM} - k_2 E_{0x}{}^{TM} = 0$$

In either case E_{0x} and E_{0y} are in phase with each other because k_1 and k_2 are real numbers. In the case of a transverse magnetic mode we can write $k_1 E_{0x}{}^{TM} + k_2 E_{0y}{}^{TM} + k_3 E_{0z}{}^{TM} = 0$, and hence $E_{0z}{}^{TM}$ is also in phase with $E_{0x}{}^{TM}$ and $E_{0y}{}^{TM}$.

As a result we see that the electric and magnetic fields are 90° out of phase with respect to each other if the cavity is resonating in one of its modes. When the electric fields are at their maximum, the magnetic fields are zero, and vice versa. The energy stored in the fields thus is alternately electric and magnetic. To illustrate this, let us take the simplest mode corresponding to the lowest frequency. If we assume the dimensions a, b, and d are such that $a \leqq b \leqq d$, then the minimum frequency corresponds to $l = 0$, $m = 1$, $n = 1$. This frequency, ω_{011}, is just given by

$$\omega_{011} = c \sqrt{\left(\frac{\pi}{b}\right)^2 + \left(\frac{\pi}{d}\right)^2} \tag{9-2-11}$$

Only the TE_{011} mode exists for this frequency. If we take the real parts of

E_x, B_y, B_z, we have the following electromagnetic fields for this mode.

$$E_x = E_{0x} \sin \frac{\pi}{b} y \sin \frac{\pi}{d} z \cos \omega t$$

$$E_y = E_z = 0$$

$$B_x = 0 \tag{9-2-12}$$

$$B_y = -\frac{k_3}{k} E_{0x} \sin \frac{\pi}{b} y \cos \frac{\pi}{d} z \sin \omega t$$

$$B_z = \frac{k_2}{k} E_{0x} \cos \frac{\pi}{b} y \sin \frac{\pi}{d} z \sin \omega t$$

Figure 9-4 shows the electric fields at time $t = 0$ and the magnetic fields which follow a quarter cycle later. As we can see the charge will oscillate from one side of the box to the other, just as in the case of a capacitor shunted by a perfect inductance. As we expect, the magnetic field reaches its maximum at the time of maximum current flow, when the electric field is zero.

Other modes will, of course, correspond to different and more complicated charge distributions oscillating about the cavity. The principle is the same although the frequencies are higher.

9-3 LOSS IN THE CAVITY WALLS; THE NOTION OF Q IN GENERAL AND AS APPLIED TO OUR CAVITY

We begin by making some remarks about a resonant system in general. For any resonant system which is oscillating by itself, there is a given amount of energy loss per cycle which is often proportional to the energy stored. For example, in the case of our cavity, the energy loss per cycle is proportional to the average component of the Poynting vector normal to the cavity wall, which is in turn proportional to the energy density of the electromagnetic field. A mass vibrating at the end of a spring and being acted upon by a resistive force proportional to its velocity has an amplitude which

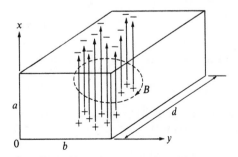

Fig. 9-4 The lowest mode TE_{011} in cavity of dimensions $a \times b \times d$ ($a < b < c$). Solid lines represent electric field. Dashed lines represent magnetic fields a quarter cycle later.

diminishes exponentially and hence loses energy at a rate proportional to its stored energy. For all systems of this sort we can define a parameter called the Q of the system as follows:

$$Q = \frac{\text{energy stored in system}}{\text{energy lost per cycle}} \tag{9-3-1}$$

If U is the energy stored in the system, we can write

$$\frac{dU}{dt} = -\frac{\omega_0}{2\pi Q} U \tag{9-3-2}$$

where ω_0 is the resonant frequency. Hence

$$U = U_0 e^{-(\omega_0/2\pi Q)t} \tag{9-3-3}$$

If the energy stored is proportional to the square of the amplitude of oscillation, as is most often the case, then the amplitude will diminish as $e^{-(\omega_0/4\pi Q)t}$. Thus, if A represents the coordinate which is oscillating (the electric field in the case of a cavity), we can write

$$A = A_0 e^{-(\omega_0/4\pi Q)t + i\omega_0 t} \tag{9-3-4}$$

We can ask now for the frequencies which are present if we Fourier analyze this oscillation. We have then (assuming the oscillation to begin at $t = 0$)

$$A(t) = \frac{1}{\sqrt{2\pi}} \int_{-\infty}^{\infty} g(\omega)e^{i\omega t} \, d\omega$$

$$g(\omega) = \frac{A_0}{\sqrt{2\pi}} \int_{0}^{\infty} e^{-(\omega_0/4\pi Q + i\omega_0 - i\omega)t} \, dt$$

$$= \frac{-A_0}{\sqrt{2\pi}\left[-\omega_0/4\pi Q + i(\omega_0 - \omega)\right]} \tag{9-3-5}$$

The energy density per unit frequency interval is proportional to $|g(\omega)|^2$, which turns out to be

$$|g(\omega)|^2 = \frac{A_0{}^2}{2\pi[(\omega_0 - \omega)^2 + (\omega_0/4\pi Q)^2]} \tag{9-3-6}$$

The full width at half-maximum of the *resonance curve* is just given by

$$\Delta\omega = \frac{\omega_0}{2\pi Q} \tag{9-3-7}$$

This is more or less what we anticipated on the basis of the arguments we made earlier about reflection and absorption within our cavity. The larger Q is, the sharper our resonance.

It remains for us to evaluate the Q of a cavity whose walls have conductivity σ. This is actually quite simple if we remember that the tangential components of both $\mathbf{E}$ and $\mathbf{B}$ are continuous as we cross into the metal. We remember also that $\mathbf{B}_\parallel$ (the tangential component of $\mathbf{B}$ at the surface) remains essentially unaltered as we reduce the conductivity from infinity to some finite value, as long as $\sqrt{k/8\pi\sigma}$ is much less than 1 [see page 243 and in particular Eqs. (7-1-36) and (7-1-37)]. On the other hand, $\mathbf{E}_\parallel$, which is zero for infinite conductivity, increases in proportion to $\sqrt{k/\pi\sigma}$ as we reduce the conductivity.

To find an exact expression for $\mathbf{E}_\parallel$ in terms of $\mathbf{B}_\parallel$, we can make use of Maxwell's equation for conducting material and insert the appropriate time dependence.

$$\nabla \times \mathbf{B} = 4\pi\sigma\mathbf{E} + \frac{1}{c}\frac{\partial\mathbf{E}}{\partial t} = (4\pi\sigma + ik)\mathbf{E} \tag{9-3-8}$$

We have already solved the problem however if we just go back to Eqs. (7-1-36) and (7-1-37). Letting $\hat{\mathbf{n}}$ be a unit normal vector pointing into the cavity wall, we can write

$$\mathbf{E}_\parallel = \frac{1}{2}\sqrt{\frac{k}{\pi\sigma}}(\mathbf{B}_\parallel \times \hat{\mathbf{n}})e^{i\pi/4} \tag{9-3-9}$$

To find the energy leaving the cavity per cycle we must evaluate the normal component of the Poynting vector along the cavity wall and average it over time. Referring back to Eq. (8-2-11), we have

$$\langle\mathbf{S}\rangle_{\text{time av}} = \frac{c}{8\pi}\operatorname{Re}(\mathbf{E} \times \mathbf{B}^*) \tag{8-2-11}$$

Applying this result to Eq. (9-3-9), we find that

$$\langle\mathbf{S}\cdot\hat{\mathbf{n}}\rangle_{\text{time av}} = \frac{c}{8\pi}\sqrt{\frac{k}{\pi\sigma}}\operatorname{Re}(|\mathbf{B}_\parallel|^2\, e^{i\pi/4})$$

$$= \frac{c}{8\pi}\sqrt{\frac{k}{2\pi\sigma}}|\mathbf{B}_\parallel|^2 \tag{9-3-10}$$

Let us return to Eq. (9-2-7) and evaluate the above expression for one wall at a time. For convenience we make the following notational changes:

$$B_{0x} = \frac{(\mathbf{k} \times \mathbf{E}_0)_x}{k}$$

$$B_{0y} = \frac{(\mathbf{k} \times \mathbf{E}_0)_y}{k} \tag{9-3-11}$$

$$B_{0z} = \frac{(\mathbf{k} \times \mathbf{E}_0)_z}{k}$$

For the wall at $x = 0$ we have then

$$B_x = 0$$

$$B_y = iB_{0y} \sin k_2 y \cos k_3 z \, e^{i\omega t}$$

$$B_z = iB_{0z} \cos k_2 y \sin k_3 z \, e^{i\omega t}$$

and

$$|\mathbf{B}_{\|}|^2 = |B_y|^2 + |B_z|^2$$

$$= B_{0y}{}^2 \sin^2 k_2 y \cos^2 k_3 z + B_{0z}{}^2 \cos^2 k_2 y \sin^2 k_3 z \qquad (9\text{-}3\text{-}12)$$

If we integrate the Poynting vector over the wall at $x = 0$, we find

$$\left(\frac{du}{dt}\right)_{\substack{\text{through wall} \\ \text{at } x = 0}} = -\frac{c}{32\pi} \sqrt{\frac{k}{2\pi\sigma}} \, (B_{0y}{}^2 + B_{0z}{}^2) \, bd \qquad (9\text{-}3\text{-}13)$$

The total energy lost per unit time (averaged over time) is thus

$$\left(\frac{du}{dt}\right)_{\substack{\text{through} \\ \text{all walls}}} = -\frac{c}{16\pi} \sqrt{\frac{k}{2\pi\sigma}} \, [(B_{0y}{}^2 + B_{0z}{}^2) \, bd + (B_{0x}{}^2 + B_{0y}{}^2) \, ab$$
$$+ (B_{0x}{}^2 + B_{0z}{}^2) \, ad] \qquad (9\text{-}3\text{-}14)$$

The energy lost per cycle is then

$$\frac{\Delta u}{\text{cycle}} = -\frac{1}{8\sqrt{2\pi\sigma k}} \, [aB_{0x}{}^2(b + d) + bB_{0y}{}^2(a + d)$$
$$+ dB_{0z}{}^2(a + b)] \qquad (9\text{-}3\text{-}15)$$

To complete our calculation of the Q of our rectangular cavity, we must know the total energy u stored within the cavity. We can best evaluate this energy at the time when the electric fields are all zero and the magnetic fields are at their maximum. At that moment we have

$$B_x = B_{0x} \sin k_1 x \cos k_2 y \cos k_3 z$$

$$B_y = B_{0y} \cos k_1 x \sin k_2 y \cos k_3 z \qquad (9\text{-}3\text{-}16)$$

$$B_z = B_{0z} \cos k_1 x \cos k_2 y \sin k_3 z$$

The energy stored at that moment per unit volume is

$$u = \frac{1}{8\pi} \int_0^a dx \int_0^b dy \int_0^d dz (B_x{}^2 + B_y{}^2 + B_z{}^2)$$

$$= \frac{abd}{64\pi} (B_{0x}{}^2 + B_{0y}{}^2 + B_{0z}{}^2) \qquad (9\text{-}3\text{-}17)$$

Finally dividing Eq. (8-3-18) by Eq. (8-3-16), we have

$$2\pi Q = \frac{abd(B_{0x}^2 + B_{0y}^2 + B_{0z}^2)}{4\delta[aB_{0x}^2(b + d) + bB_{0y}^2(a + d) + dB_{0z}^2(a + b)]} \qquad (9\text{-}3\text{-}18)$$

where δ = skin depth = $1/\sqrt{2\pi\sigma k}$. This simplifies considerably in the case of a cubical cavity where $a = b = d = L$. In that case

$$2\pi Q = \frac{L}{8\delta} \qquad (9\text{-}3\text{-}19)$$

The important thing to note is that $2\pi Q$ is approximately equal to the ratio of cavity volume to the volume corresponding to the skin depth. Over a wide range of frequencies this ratio increases as the square root of the frequency. Typical values of Q in the microwave region for silver are about 10^4.

PROBLEMS

9-1. A long perfectly conducting waveguide of cross-sectional area 5 cm by 10 cm carries microwave energy in the dominant TE_{01} mode. Keeping the maximum amplitude of electric field fixed, find and graph an expression for the transmitted power as a function of frequency. Does the asymptotic behavior as $\omega \to \infty$ make sense?

9-2. Suppose we now cap off a section of the waveguide described in Prob. 9-1 by means of a plate having finite conductivity σ. We introduce energy into the open end in the TE_{01} mode at frequency ω. What is the fraction of the incoming energy that is absorbed by the end plate?

9-3. Suppose the entire waveguide described in Prob. 9-1 were made of copper. How would radiation of frequency ω be attenuated in the TE_{01} mode as a function of distance along the waveguide? Suppose the frequency ω were sufficiently large so as to allow propagation in the TM_{11} mode. Would the attenuation be any different from that in the TE_{01} mode at the same frequency?

9-4. A rectangular cavity with perfectly conducting walls has dimensions of 2 cm by 4 cm by 8 cm.
 (a) What is the lowest resonant frequency at which this cavity will oscillate?
 (b) Suppose the cavity is opened to air. Estimate the change in its resonant frequency.
 (c) What would the Q of the cavity become if its 2 cm by 4 cm faces were made of copper?

9-5. A waveguide is made of two perfectly conducting coaxial cylinders with the radiation propagating in the space between them. Show that it is possible to have a mode in which both the electric and magnetic fields are perpendicular to the axis of the cylinder (**transverse electric and magnetic mode—TEM**).

Is there a cutoff frequency for this mode? What is the velocity of propagation of this mode?

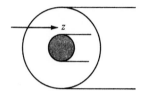

10
Electric and Magnetic Susceptibility

In earlier chapters we studied the effects of the macroscopic electric and magnetic properties of matter on the electric and magnetic field distributions. In this chapter we will try to explain, at least in a simple-minded sort of way, how it is that these properties exist. We will revert to a microscopic description of matter and see how the polarizability of individual molecules leads to a reasonable understanding of electric and magnetic susceptibilities. Needless to say, we are not prepared at this point for the full rigors of quantum theory and statistical mechanics; we will find though that we can go a long way with "seat-of-the-pants" reasoning and come up with results which are not too unreasonable.

Electric or magnetic polarization is the result of the average molecule having an electric or magnetic dipole moment in a given direction. This average moment can arise from two alternative sources or from a combination of them.

1. The molecule (or atom) can normally be free of any intrinsic moment. Applying an electric or magnetic field will however induce a moment. This, in general, will lead to a relatively small electric or magnetic susceptibility.

2. The molecule may already have an electric or magnetic dipole moment. In this case the applied field will try to align this moment. Clearly the average polarization is temperature dependent because collisions between molecules tend to disrupt the alignment. (It is also possible, as in the case of ferromagnetism, that a certain amount of long-range coherence exists among a fairly macroscopic group of molecules. The extent of this coherence will also be temperature dependent. Needless to say, coherence can lead to very large susceptibilities.)

We will begin by discussing the electric susceptibility of material whose molecules have *no* intrinsic electric dipole moments. This treatment is relatively simple because we have no need to worry about the disrupting influence of collisions. We will then introduce some notions which derive from statistical mechanics and which will permit us to deal with a large number of interacting molecules at a given temperature T. When we apply these notions to molecules having intrinsic electric dipole moments, we will be able to estimate the extent of their average polarization in the presence of an external electric field. The same types of treatment will be carried out for magnetism.

Finally, we will say a few words about ferromagnetism and its origin in the closely coupled behavior of neighboring atoms.

10-1 THE ELECTRIC POLARIZABILITY OF NONPOLAR MOLECULES HAVING SPHERICAL SYMMETRY

Naturally, we start with the simplest molecular system, an atom having complete spherical symmetry. We can think of the electrons' charge as being distributed over a sphere of about 0.5×10^{-8} cm. The nucleus is of negligible size ($\cong 10^{-13}$ cm) and can be thought of, for our purposes, as a point charge at the center of the electron cloud. We would like to find out how much of a dipole moment is induced if an external field E_a is applied (see Fig. 10-1).

To first approximation the nucleus will move relative to the electrons to the point where the restoring force from the electron cloud just equals the applied field. At a distance r from the center of a uniform charge distribution

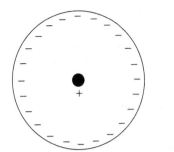

E_a

Fig. 10-1 An applied electric field causes the positive nucleus to be displaced relative to the center of the electron cloud, leading to a net electric dipole moment.

of radius R and total charge Q, the field is given by (see page 33)

$$E = \frac{Qr}{R^3} \tag{10-1-1}$$

In our case $Q = -Ze$ where $-e$ is the electron charge and Z is the atomic number. In order that E just precisely balance E_a, we must have

$$r = \frac{R^3}{Ze} E_a \tag{10-1-2}$$

The dipole moment **p** of the atom is then

$$\mathbf{p} = Ze \frac{R^3}{Ze} \mathbf{E}_a$$

$$= R^3 \mathbf{E}_a \tag{10-1-3}$$

In general, we can think of our atom as having an atomic polarizability α such that $\mathbf{p}_{ind} = \alpha \mathbf{E}_{app}$. In the case of the simple model we have taken above, the polarizability is just equal to R^3. This result is not in terrible disagreement with what we would have obtained had we done a complete quantum-mechanical calculation. For example, in the case of atomic hydrogen, α turns out to be equal to $\frac{9}{2}a_0{}^3$ where a_0 is the **Bohr radius** ($a_0 = 0.52 \times 10^{-8}$ cm).

Notice that the displacement of the nucleus is not very large for the sort of fields that are available in the laboratory. Typical high fields are in the hundreds of gauss (100 gauss = 30 kV/cm). Using our value of 0.5×10^{-8} cm for R and letting $Z = 1$, we find, for a field of 100 gauss, that $r = 2.5 \times 10^{-14}$ cm. This is a very small displacement indeed, smaller in fact than the radius of the nucleus itself.

On the whole we expect a certain amount of variation of atomic polarizability as we vary the details of the atomic structure. For example, we expect that noble gases with closed shells and no valence electrons might be harder to distort than alkali metals with a relatively free valence electron.

We also anticipate a modification of the elementary atomic polarizabilities when atoms are bound into molecules. These variations can be substantial and can run over several orders of magnitude. Hence precise measurements of polarizabilities can serve as important analytical tools in the study of chemical structures.

10-2 THE RELATION BETWEEN ATOMIC POLARIZABILITY AND ELECTRIC SUSCEPTIBILITY

Some time ago [see Eq. (2-6-22)] we defined the proportionality constant between electric field and dipole moment per unit volume to be the electric susceptibility χ_e. That is to say, $\mathbf{P} = \chi_e \mathbf{E}$. One would naively guess that the ratio between the electric susceptibility χ_e and the atomic polarizability (which we will call α_e) would just be the number of atoms per unit volume N. This is however not exactly the case; we must be careful to read the fine print in our definitions. When we defined polarizability, we related the induced dipole moment $\mathbf{p}$ to the *applied* electric field $\mathbf{E}_a$. That is to say, $\mathbf{p} = \alpha_e \mathbf{E}_a$. On the other hand, the susceptibility equation relates the average dipole moment per unit volume at some point in the dielectric to the total field at that point. Since the applied field on the dipole does *not* include any part of the field caused by the dipole itself, we must be careful to remove that field when evaluating the atomic polarizability in terms of χ_e.

The applied field $\mathbf{E}_a$ at the dipole is just equal to the total field $\mathbf{E}$ minus the field due to the dipole itself. We must average everything over the volume taken up by the dipole, which we will call V. We have then

$$\langle \mathbf{E}_a \rangle_V = \langle \mathbf{E} \rangle_V - \langle \mathbf{E}_{\text{atom}} \rangle_V \tag{10-2-1}$$

To carry out this averaging process rigorously, taking into account the complete charge distribution of the atom, is difficult. We will make the approximation that the distance between atoms is much larger than the atomic size. We will then let V be the average volume per dipole, or $1/N$. The assumption that V is much larger than the atomic size permits us to make use of only the leading dipole term [Eq. (2-6-7)] to obtain the field due to the atom itself within V. We have then

$$\langle \mathbf{E}_{\text{atom}} \rangle_V = \frac{1}{V} \int_V \mathbf{E}_{\text{atom}} \, dV$$

$$= \frac{1}{V} \int_V \nabla \left(\mathbf{p} \cdot \nabla \frac{1}{|\mathbf{r} - \mathbf{r}'|} \right) dV \tag{10-2-2}$$

where $\mathbf{r}'$ is the location of the atomic dipole.

We convert this to a surface integral, obtaining

$$\langle \mathbf{E}_{\text{atom}} \rangle_V = \frac{1}{V} \int_{\substack{\text{surface} \\ \text{of } V}} \left(\mathbf{p} \cdot \nabla \frac{1}{|\mathbf{r} - \mathbf{r}'|} \right) \hat{\mathbf{n}} \, dA \tag{10-2-3}$$

If we now simplify matters by letting V be a sphere of radius b about the point $\mathbf{r}'$, we find

$$\langle \mathbf{E}_{\text{atom}} \rangle_V = \frac{-1}{Vb^2} \int_{\substack{\text{surface} \\ \text{of } V}} (\mathbf{p} \cdot \hat{\mathbf{n}}) \hat{\mathbf{n}} \, dA$$

$$= \frac{-\frac{4}{3}\pi \mathbf{p}}{V} \tag{10-2-4}$$

(A little bit of thought will convince the reader that the shape of the volume V is irrelevent provided it is symmetrically situated about the position of the dipole.)

Taking $V = 1/N$, we have our result

$$\langle \mathbf{E}_{\text{atom}} \rangle_V = -\tfrac{4}{3}\pi N\mathbf{p} = -\tfrac{4}{3}\pi \mathbf{P} \tag{10-2-5}$$

Finally, going back to Eq. (10-2-1), we can write

$$\langle \mathbf{E}_a \rangle_V = \frac{\mathbf{P}}{N\alpha_e} = \langle \mathbf{E} \rangle_V + \tfrac{4}{3}\pi \mathbf{P}$$

$$= \left(\frac{1}{\chi_e} + \frac{4\pi}{3} \right) \mathbf{P}$$

This yields the equation for χ_e in terms of α_e:

$$\chi_e = \frac{N\alpha_e}{1 - 4\pi N\alpha_e/3} \tag{10-2-6}$$

In the limit where N becomes very small, χ_e reduces, as expected, to $N\alpha_e$. We might ask about the other limit, where N gets large. Is it really possible that χ_e becomes infinite for a particular value of N and then becomes negative for N greater than that value? We remember that α_e is about equal to the volume of the atomic charge distribution [see Eq. (10-1-3)]. When we began our derivation for χ_e, we assumed that the volume of the charge distribution was very small compared with the average volume per dipole. That is,

$$\alpha_e \ll \frac{1}{N}$$

or

$$N\alpha_e \ll 1 \tag{10-2-7}$$

Thus the domain of validity of Eq. (10-2-6) does *not* include the situation where $4\pi N\alpha_e/3$ is comparable with 1.

10-3   POLARIZABILITY AS A SECOND-RANK TENSOR

The most general linear relationship between the applied electric field $\mathbf{E}_a$ and the polarization $\mathbf{p}$ of a molecule is that of a second-rank tensor with components α_{ij}. That is,

$$p_i = \sum_{j=1}^{3} \alpha_{ij} E_j \qquad (10\text{-}3\text{-}1)$$

We will now demonstrate a rather important property of the tensor α, namely, its symmetry. The component α_{ij} is the same as the component α_{ji}. To see the physical meaning of this let us take our electric field, for example, in the y direction and of magnitude E_a. This would give rise to a polarization component in the x direction equal to $\alpha_{12} E_a$. If instead we had taken the same magnitude of electric field along the x direction, we would have found a y component of polarization equal to $\alpha_{21} E_a$. If $\alpha_{21} = \alpha_{12}$, we could conclude that these two resultant components are equal.

The demonstration of the symmetry of α rests upon the very simple observation that the electrostatic energy of a given charge distribution cannot depend upon the historical sequence of events by which it was built up. All that should matter when we determine the interaction energy of a given dipole $\mathbf{p}$ and the applied field $\mathbf{E}_a$ are the magnitude of $\mathbf{p}$ and $\mathbf{E}_a$ and the angle between them. Suppose we change $\mathbf{E}_a$ slightly and allow $\mathbf{p}$ to change accordingly by an amount $d\mathbf{p}$. The amount of work done on the charges in $\mathbf{p}$ by the field $\mathbf{E}_a$ is just

$$dW = \mathbf{E}_a \cdot d\mathbf{p} \qquad (10\text{-}3\text{-}2)$$

Suppose now that we allow the z component of $\mathbf{E}_a$ to be zero and let E_{a1} and E_{a2} be the x and y components, respectively. To arrive at these final values for the components of $\mathbf{E}_a$, we can proceed in one of two ways:

1. We can first build up the x component of $\mathbf{E}_a$ to its final value, allowing the y component of $\mathbf{E}_a$ to remain zero. We then raise the y component of $\mathbf{E}_a$ to its final value.
2. We can reverse the order, first bringing up the y component of $\mathbf{E}_a$ and then the x component of $\mathbf{E}_a$.

Let us begin first with procedure 1 and evaluate the work done by $\mathbf{E}_a$ on $\mathbf{p}$ as we bring it up by the indicated two-step process. In the first step we have

$$W_1(a) = \int_0^{E_{a1}} E_{a1}\, dp_1 = \int_0^{E_{a1}} \alpha_{11} E_{a1}\, dE_{a1}$$

$$= \tfrac{1}{2}\alpha_{11} E_{a1}{}^2$$

In the second step of procedure 1 we have

$$W_2(a) = \alpha_{12} E_{a1} E_{a2} + \tfrac{1}{2}\alpha_{22} E_{a2}{}^2$$

The total work for procedure 1 is thus

$$W(a) = \tfrac{1}{2}\alpha_{11} E_{a1}{}^2 + \tfrac{1}{2}\alpha_{22} E_{a2}{}^2 + \alpha_{12} E_{a1} E_{a2} \qquad (10\text{-}3\text{-}3)$$

Completely analogously, procedure 2 yields

$$W(b) = \tfrac{1}{2}\alpha_{11}E_{a1}^{2} + \tfrac{1}{2}\alpha_{22}E_{a2}^{2} + \alpha_{21}E_{a1}E_{a2} \qquad (10\text{-}3\text{-}4)$$

Letting $W(b) = W(a)$, we conclude that

$$\alpha_{21} = \alpha_{12}$$

Obviously, the same proof works for any pair of indices i and j. Hence we generalize for all i and j

$$\alpha_{ij} = \alpha_{ji} \qquad (10\text{-}3\text{-}5)$$

One very important consequence of this symmetry is the possibility of diagonalizing the polarizability tensor (see Sec. 1-7). We can find three mutually perpendicular directions in which $\mathbf{p}$ and $\mathbf{E}_a$ are colinear. These three axes are three fundamental axes of symmetry for the system. The remarkable point is that three such axes of symmetry exist for the polarizability even though the charge distribution itself may be totally asymmetrical and show no axis of symmetry.

10-4 THE POLARIZABILITY OF A POLAR MOLECULE

Quite commonly the molecules out of which our dielectric is composed have a built-in electric dipole moment. If the dielectric is a liquid or a gas, then the molecules will be relatively free to rotate and will tend to line themselves up with their dipole moments lying along the direction of an applied electric field. Were it not for the fact that collisions between molecules are continually upsetting this alignment, one might expect the molecules to eventually all point along the field. Our chore then in this section is to understand, first, how large the permanent dipole moment of a typical polar molecule is likely to be and, second, to calculate, making use of some basic ideas from statistical mechanics, a value for the polarizability of the molecule as a function of temperature.

Water is a very typical polar molecule. Referring to Fig. 10-2, we see that the water molecule is quite asymmetric. The two hydrogen atoms are

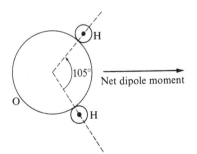

Fig. 10-2 The water molecule has a net dipole moment because of the fact that the electrons from the hydrogen spend a part of their time "going around" the oxygen nucleus.

attached to the oxygen atom so as to have the OH axes make an angle of about 105° with one another. The electrons from the hydrogen atoms then spend a good part of their time around the oxygen atom, leading to a net dipole moment, as shown. We can estimate its size very approximately by noting that each electron is displaced by about a Bohr radius ($a_0 = 0.5 \times 10^{-8}$ cm) on the average from its proton. This leads to a moment p_0 given by

$$p_0 \cong 2ea_0 \cos 52.5°$$

$$\cong 2(5 \times 10^{-10} \text{ esu})(0.5 \times 10^{-8} \text{ cm})(0.6)$$

$$\cong 3 \times 10^{-18} \text{ esu-cm}$$

In actual fact, the dipole moment of a water molecule is

$$p_0(\text{water}) = 1.84 \times 10^{-18} \text{ esu-cm} \tag{10-4-1}$$

We will next make use of the Boltzmann distribution to determine the polarizability of the average polar molecule in an applied electric field $\mathbf{E}_a$ at a temperature T. We shall not go through the details of deriving the Boltzmann distribution equation or justifying it. We will only assert that statistically the dipoles will be distributed according to the distribution function $e^{-W/kT}$ where W is the dipole's potential energy in the electric field and k is Boltzmann's constant. That is, the probability of finding the dipole pointing into a given solid angle $d\Omega$ will be proportional to $e^{-W/kT}$ and $d\Omega$. We can thus write

$$\frac{dN}{d\Omega} = Ae^{-W/kT} \tag{10-4-2}$$

where A is chosen to suitably normalize the distribution. If there are N_0 dipoles altogether, we have

$$N_0 = A \int e^{-W/kT} d\Omega \tag{10-4-3}$$

The potential energy of a dipole $\mathbf{p}_0$ in an applied field $\mathbf{E}_a$ is just

$$W = -\mathbf{p}_0 \cdot \mathbf{E}_a = -p_0 E_a \cos \theta \tag{10-4-4}$$

Substituting in Eq. (10-4-3), we obtain

$$N_0 = A \int_0^{2\pi} d\varphi \int_0^{\pi} e^{p_0 E_a \cos\theta/kT} \sin \theta \, d\theta$$

$$= 2\pi A \int_{-1}^{1} e^{p_0 E_a x/kT} dx$$

$$= \frac{2\pi A}{a} (e^a - e^{-a}) \tag{10-4-5}$$

where

$$a = \frac{p_0 E_a}{kT} \tag{10-4-6}$$

Solving for A, we now have

$$A = \frac{N_0 a}{2\pi(e^a - e^{-a})} \tag{10-4-7}$$

To find the average polarization we note that the only component which is not averaged out is that along E_a. If we let p be the component along E_a, we have

$$p = p_0 \cos \theta \tag{10-4-8}$$

and

$$\bar{p} = \frac{A}{N_0} \int p_0 \cos \theta \, e^{-W/kT} \, d\Omega$$

$$= \frac{2\pi p_0 A}{N_0} \int_{-1}^{1} x e^{-ax} \, dx$$

$$= p_0 \left(\frac{e^a + e^{-a}}{e^a - e^{-a}} - \frac{1}{a} \right) \tag{10-4-9}$$

The function on the right of Eq. (10-4-9), called the **Langevin function** $L(a)$, can be expanded for small values of a:

$$L(a) = \frac{e^a + e^{-a}}{e^a - e^{-a}} - \frac{1}{a}$$

$$\cong \tfrac{1}{3}a \qquad \text{for small } a \tag{10-4-10}$$

We have then our final result, relating $\bar{p}$ to E_a,

$$\bar{p} = \frac{1}{3} \frac{p_0^2 E_a}{kT} \tag{10-4-11}$$

The polarizability α_e is thus given by

$$\alpha_e = \frac{1}{3} \frac{p_0^2}{kT} \tag{10-4-12}$$

It is interesting to evaluate this for the water molecule at room temperature (300°K). We have then

$$\alpha_e(\text{water at } 300°\text{K}) = \frac{1}{3} \frac{(1.84)^2 \times 10^{-36}}{(1.4 \times 10^{-16})(300)}$$

$$= 2.7 \times 10^{-23} \text{ cm}^3 \tag{10-4-13}$$

Needless to say, this is an order of magnitude larger than the polarizability of a nonpolar molecule [see Eq. (10-1-3)].

10-5 DIAMAGNETISM

We are all quite familiar with the attractive force which the pole of a magnet exerts on a piece of iron, but very few of us realize that the same pole can repel a drop of water. This is because the forces related to diamagnetism, as displayed by the interaction of a magnetic field and water, are so weak on the scale we are accustomed to that we tend to ignore them. Nevertheless, diamagnetism represents an important example of the application of the laws of electrodynamics on the atomic scale and an excellent experimental tool for probing atomic structure.

Diamagnetism is most clearly observed in those atomic systems which have no intrinsic magnetic dipole moments. In the event that such a moment does exist, the applied field will try to align it, giving rise to paramagnetism, a phenomenon which generally masks diamagnetism at room temperature and which will be discussed shortly. To treat diamagnetism properly really requires solving the quantum-mechanical Schrödinger equation in the presence of a magnetic field; nevertheless, we can go a long way toward understanding it by means of a crude classical model. At the very least we will determine the order of magnitude for the effects we anticipate and their dependence on atomic size.

We will take as our simple model an electron with charge e traveling in a circular orbit of radius r_0 about a nucleus to which it is attracted by an electrostatic force F_0. The speed of the electron will be taken as v_0; its angular momentum about the nucleus has magnitude $L_0 = mv_0r_0$.

Naturally the above atomic system has a magnetic moment μ_0 which is parallel and opposite in direction to $\mathbf{L}_0$. Purely diamagnetic molecules can be produced by tying two of these atoms together in such a way that their angular momenta are forced to be in opposite directions. (This *anti-correlation* of the angular momenta of electrons arises out of the **Pauli exclusion principle,** a rather profound rule which governs the quantum-mechanical behavior of all particles having half-integer spin.) Alternatively, we can produce a purely diamagnetic medium by heating it sufficiently so as to keep the angular momenta of the molecules randomly oriented in direction. In any case, we will assume in our treatment that there is no preferential direction along which angular momenta are aligned, but rather that they are uniformly and isotropically distributed.

Incidentally, we recall from Eq. (4-10-1) that a simple relationship exists between μ_0 and $\mathbf{L}_0$:

$$\mu_0 = \frac{e\mathbf{L}_0}{2mc} \tag{10-5-1}$$

Let us see what happens when we apply a magnetic field **B** in the same direction as **L$_0$** (see Fig. 10-3). In the process of bringing the field up to its final value, we will induce an electric field (Faraday's law) which will on the average be opposite in direction to the velocity of the electron. This will cause the electron to speed up (because of its negative charge) and will increase the magnitude of its magnetic moment (opposite in direction to **B**). If we take a time Δt to bring the magnetic field up, then we expect an average component of electric field along the electron's path given by

$$2\pi r_0 \overline{E} = -\frac{1}{c}\frac{\Delta\Phi}{\Delta t} = -\frac{1}{c}\pi r_0^2 \frac{B}{\Delta t}$$

Solving for $\overline{E}$, we have

$$\overline{E} = -\frac{r_0 B}{2c\,\Delta t} \tag{10-5-2}$$

In this time interval, $\overline{E}$ changes the momentum of the electron by an amount

$$\Delta p = e\overline{E}\,\Delta t = -\frac{er_0 B}{2c} \tag{10-5-3}$$

Before we can evaluate the change in the magnetic moment, we must find if the radius of the orbit will be altered by the introduction of the magnetic field. Remembering that F_0 is the centripetal force excited on the electron before the field is turned on, we have

$$F_0 = \frac{mv_0^2}{r_0} = \frac{p_0^2}{mr_0} \tag{10-5-4}$$

The additional centripetal force *needed* to keep the electron in the *same* orbit after the field has been turned on is just

$$\Delta F = \frac{2p_0\,\Delta p}{mr_0} \tag{10-5-5}$$

But, using the result from Eq. (10-5-3) for Δp, we find

$$\Delta F = -\frac{e}{c}v_0 B \tag{10-5-6}$$

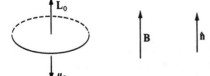

Fig. 10-3 We apply a magnetic field **B** to a classical atom having magnetic moment μ_0 as shown. This causes a change in magnetic moment opposite in direction to **B**.

This is exactly the extra force which is supplied by the magnetic field B acting on the electron as it moves with velocity v_0 around the nucleus. Hence the radius r_0 remains unchanged, and the change in magnetic moment is given by

$$\Delta\mu = \frac{e\,\Delta\mathbf{L}}{2mc} = \frac{er_0\,\Delta p}{2mc}\,\hat{\mathbf{n}}$$

$$= -\frac{e^2r_0{}^2\mathbf{B}}{4mc^2} \tag{10-5-7}$$

In the above example, the applied field caused the (negative) electron to speed up, inducing an increase in the (negative) magnetic moment. If indeed we had taken $\mathbf{L}_0$ opposite in direction to $\mathbf{B}$, the electron would have started with its magnetic moment pointing along $\mathbf{B}$. It would have slowed down upon application of the field, and, again, the change in magnetic moment would be opposite in direction to $\mathbf{B}$. Indeed, retracing our steps, we see that $\Delta\mu$ is again given, exactly as before, by Eq. (10-5-7).

Suppose now that we had set our atom with its magnetic moment μ_0 at right angles to the applied field $\mathbf{B}$. As we learned back in Chap. 4 the angular momentum and magnetic moment would precess about $\mathbf{B}$ as shown in Fig. 10-4. Looking along the direction of $\mathbf{B}$, we would see the plane of the atom rotating clockwise. This clockwise rotation would look to us like an added magnetic moment pointing opposite in direction to $\mathbf{B}$. The angular frequency associated with this precession is [see Eq. (4-10-2)]

$$\omega_L = -\frac{e\mathbf{B}}{2mc} \tag{10-5-8}$$

To find the magnetic moment $\Delta\mu$ associated with ω_L we must first find R^2, the mean squared *distance* of the electron from the axis of precession. Remembering that the magnetic moment of a current loop is equal to the current times the area of the loop, we have

$$|\Delta\mu| = \frac{e}{c}\,\frac{\omega_L}{2\pi}\,\pi\overline{R^2} \tag{10-5-9}$$

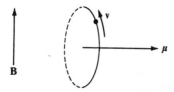

Fig. 10-4 The magnetic moment μ precesses about the magnetic field $\mathbf{B}$ giving rise to an additional component of magnetic moment opposite in direction to $\mathbf{B}$.

A simple geometrical evaluation will demonstrate that

$$\overline{R^2} = \frac{r_0^2}{2} \tag{10-5-10}$$

Substituting into Eq. (10-5-9), we have finally

$$\Delta\mu = -\frac{e^2 r_0^2 \mathbf{B}}{8mc^2} \tag{10-5-11}$$

In fact we can combine the results expressed in Eqs. (10-5-7) and (10-5-11) into one equation which works for any orientation:

$$\Delta\mu = -\frac{e^2 \overline{R^2} \mathbf{B}}{4mc^2} \tag{10-5-12}$$

where again $\overline{R^2}$ is the mean square distance of the electron from a line through the nucleus and parallel to $\mathbf{B}$.

It is illuminating to get some feeling as to the magnitude of $\Delta\mu$ and compare it with the magnetic moment of the electron itself. If we take r_0 as 10^{-8} cm, we obtain

$$|\Delta\mu| \cong 3.5 \times 10^{-28} \, B \text{ esu-cm}$$

The electron's magnetic moment, $g \, eL/2mc$ [see Eq. (4-10-4)], can be calculated if we let L take on the quantized value of $h/4\pi$ where h is Planck's constant. For an electron, $g = 2$, and hence

$$\mu_0 = \frac{eh}{4\pi mc} \cong 9.2 \times 10^{-21} \text{ esu-cm.}$$

We note then that even in fields as high as 30 kilogauss the induced magnetic moment per atom is at least three orders of magnitude less than the magnetic moment of the electron itself. Small wonder that diamagnetic effects are relatively small and easily masked by paramagnetic effects, especially at low temperatures.

10-6 PARAMAGNETISM AND FERROMAGNETISM

As we have mentioned earlier, paramagnetism results from the fact that an atomic or molecular system may have a residual magnetic moment after all energetically feasible "cancellations" among electrons have taken place. For example, if the molecule has an odd number of electrons, then there is no way in which the effect of the intrinsic magnetic moment of the last odd electron can be fully eliminated. Hence, just as when we were dealing with an electric field and polar molecules, there will be a predisposition for the residual magnetic moments to align themselves with the field. Again the extent of alignment will depend on the temperature and will be calculable by means of the Boltzmann distribution function $e^{-W/kt}$. The only problem we have is what to use for W because, as we discovered in Chap. 4, the variation in the energy of the system as we rotate our dipole around will depend on just exactly what we are keeping constant—the current in the dipole, or the flux through the dipole, or perhaps something else. Indeed it would seem at first glance that we would have to understand the detailed dynamics of the atom itself to make some evaluation of W.

Fortunately, this is not the case. As we learned when we studied diamagnetism, the change in the magnetic moment of our atom, even for a fairly large applied field, is small and can be ignored. Hence, as far as an outside observer is concerned, the dipole always has a torque equal to $\mu_0 \times \mathbf{B}_a$ acting on it, where μ_0 is the magnetic moment in the absence of field and $\mathbf{B}_a$ is the applied field. This torque is such as to try to get μ_0 and $\mathbf{B}_a$ in the *same* direction. Thus the externally observable *mechanical* behavior of the magnetic dipole in a magnetic field is exactly the same as if we replaced its magnetic moment by an equal electric dipole moment and the applied magnetic field by an equal applied electric field. We have then

$$W = -\mu_0 \cdot \mathbf{B}_a \tag{10-6-1}$$

This leads us [see Eq. (10-4-9)] to a result for the average value of the component of μ along $\mathbf{B}$ given by

$$\bar{\mu} = \mu_0 \left(\frac{e^a + e^{-a}}{e^a - e^{-a}} - \frac{1}{a} \right) \tag{10-6-2}$$

where

$$a = \frac{\mu_0 B_a}{kT} \tag{10-6-3}$$

For small values of a we can expand the Langevin function as before with the result

$$\bar{\mu} \cong \frac{1}{3} \frac{\mu_0^2 B_a}{kT} \tag{10-6-4}$$

It is again instructive to have a look at the order of magnitude of this polarization and compare it with diamagnetism. We can evaluate $\bar{\mu}/\mu_0$ in terms of B_a and T if we allow μ_0 to be equal to the electron's intrinsic moment

$$\frac{\bar{\mu}}{\mu_0} \cong \frac{1}{3} \frac{\mu_0 B_a}{kT} \cong 0.22 \times 10^{-4} \frac{B_a}{T}$$

For comparison. we find for the case of diamagnetism that

$$\left(\frac{\bar{\mu}}{\mu_0}\right)_{\text{diamag}} \cong -\frac{e^2 \overline{R^2} B_a}{4mc^2 \mu_0} \cong -0.38 \times 10^{-7} B_a$$

We see then that at room temperature paramagnetic effects are not orders of magnitude larger than diamagnetic effects. We might just determine the magnitude of the force on samples of a typical paramagnetic material at room temperature in a reasonable magnetic field gradient. Let us set up a magnetic field along the z direction of our coordinate system of the form

$$\mathbf{B} = \mathbf{B}(z) = (B_0 + \kappa z)\hat{\mathbf{k}} \qquad (10\text{-}6\text{-}5)$$

That is to say, $\mathbf{B}$ increases linearly with increasing z in our region of interest. Let us now place our sample, consisting of N paramagnetic atoms, at the origin of our coordinate system. It will develop a total magnetic moment of $N\bar{\mu}\hat{\mathbf{k}}$:

$$N\bar{\mu}\hat{\mathbf{k}} \cong \frac{1}{3} \frac{N\mu_0^2 B_0 \hat{\mathbf{k}}}{kT} \qquad (10\text{-}6\text{-}6)$$

Making use of Eq. (4-7-20), we find the force on the sample:

$$\mathbf{F} = N\bar{\mu} \frac{\partial}{\partial z} \mathbf{B} = \frac{1}{3} \frac{N\mu_0^2 B_0 \kappa \hat{\mathbf{k}}}{kT} \qquad (10\text{-}6\text{-}7)$$

We take the following typical values (for 1 gram of sample):

$N \cong \frac{1}{4} \times 10^{23}$

$B_0 = 10,000$ gauss

$\kappa = 1,000$ gauss/cm

$T = 300°$K

Inserting these values into Eq. (10-6-7), we find the magnitude of $\mathbf{F}$ to be about 175 dynes. This is not very much at all, not even 20 percent of the weight of the sample. Clearly then, paramagnetism alone is not responsible for the large forces on iron in the vicinity of a magnet.

Before we go on to understand the origin of ferromagnetism, we should make one final observation related to paramagnetism. When we develop the quantum theory of atoms, we will discover that the possible orientations of μ_0 with respect to $\mathbf{B}_a$ are quantized, whereas in our derivation we assumed that any orientation was possible. This does not make a large qualitative difference to our result. For example, in the case of an atom having only a single electron spin contributing to its moment, there are two orientations possible, parallel and antiparallel to the applied field.

In this case the Langevin equation becomes

$$\frac{\bar{\mu}}{\mu_0} = \tanh \frac{\mu_0 B_a}{kT}$$

$$\cong \frac{\mu_0 B_a}{kT} \qquad \text{for low field or high temperature} \qquad (10\text{-}6\text{-}8)$$

Comparison with Eq. (10-6-4) indicates only a factor of 3 difference in the low-field or high-temperature approximation. The agreement of quantum theory with classical theory improves as we deal with larger atomic angular momenta and a correspondingly larger number of possible orientations relative to the applied field.

None of what we have done so far would give us any clue to the remarkable phenomena of ferromagnetism. If we take an iron atom all by itself, then we would find that 11 of its 26 electrons would have their spins lined up in one direction and 15 would have their spins lined up in the opposite direction. This would imply an excess of 4 Bohr magnetons of magnetic moment for the atom as a whole. (The unit of electron magnetic moment, $e\hbar/2mc$, where $\hbar$ is Planck's constant divided by 2π and m is the electron mass, is called the **Bohr magneton.**) The excess moment would actually occur in the next-to-last atomic shell which is not completely filled. Naturally we can calculate how much magnetic field this magnetic moment could lead to in its immediate vicinity and we could then estimate how much correlation should exist between neighboring atoms. It is not hard to estimate the magnitude of this effect; needless to say it is nowhere near what is needed to establish the high degree of correlation which exists between neighboring atoms in a crystal of iron or nickel. There must then be a new type of "force" which is acting here to establish the strong tendency of neighboring atoms to have parallel spin.

To really understand this force even qualitatively requires a study of quantum mechanics beyond the scope of this book. One of the fundamental rules governing the behavior of particles with intrinsic angular momentum equal to a half-integer times Planck's constant divided by 2π [spin = $(n/2)\hbar$] is the Pauli exclusion principle. This principle forbids two such particles from occupying identical quantum-mechanical states at exactly the same time. (That is why the two and only two electrons in the innermost shell of an atom have their spins pointing in opposite directions.) One result of this principle is the fact that two highly overlapping electron clouds tend to have opposite magnetic moments. Now two neighboring iron atoms in an iron crystal are bound together as a result of the sharing of the electrons in their outermost shells. These electrons in turn act on the two sets of aligned electrons in the unfilled inner shells and cause them to line up together It is this so-called **exchange force** between the two atoms that gives rise to ferromagnetism. If we were to examine unmagnetized iron microscopically, we would find that it was made up of almost macroscopic *domains*, each with typical volume of about 10^{-3} mm^3. Within a given domain the moments of the individual atoms all point in the same direction with an average moment per atom of about 2.2 Bohr magnetons. As we apply an external field to the iron the domain boundaries begin to move. Those domains whose moments are along the direction of the applied field grow at the expense of domains whose alignment is not as fortunate. Of course, the more the applied field, the more overall alignment we obtain until we reach the point of saturation. At this point, typically 15,000 gauss in iron, the leading domains have conquered all there is to conquer and no more magnetization can be induced.

Needless to say, if we raise the temperature of our iron we reach the point where the high correlation between neighboring atoms is destroyed. Above this temperature, called the **Curie point,** iron is more or less a paramagnetic solid.

PROBLEM

10-1. Estimate the magnetic field which one iron atom produces at the site of its neighbor in an iron crystal. Compare the magnetic interaction energy of two neighboring iron atoms with kT for $T = 300°$K.

Tables

CONVERSION TO "PRACTICAL" UNITS

Quantity	Symbol	Gaussian unit	Practical unit	Relation
Distance	s	cm	meter	1 meter $=$ 100 cm
Mass	m	gram	kg	1 kg $=$ 1000 grams
Time	t	sec	sec	———
Velocity	v	cm/sec	meter/sec	1 meter/sec $=$ 100 cm/sec
Work, energy	W	erg	joule	1 joule $= 10^7$ ergs
Charge	q	esu	coulomb	1 coulomb $= 2.998 \times 10^9$ esu
Current	I	emu/sec (abampere)	ampere	1 ampere $= 10^{-1}$ emu/sec
Electric field	**E**	gauss	volt/cm	1 volt/cm $= \dfrac{1}{299.8}$ gauss
Electric potential	φ	statvolt	volt	1 volt $= \dfrac{1}{299.8}$ statvolt
Magnetic field	**B**	gauss	weber/m^2	1 weber/$m^2 = 10^4$ gauss
Resistance	**R**	statvolt/ abampere	ohm	1 ohm $= \dfrac{1}{29.98}$ statvolt/abampere

FUNDAMENTAL CONSTANTS

Quantity	Symbol	Value
Speed of light in vacuum	c	2.998×10^{10} cm/sec
Fundamental charge	e	4.803×10^{-10} esu
Planck's constant	h	6.626×10^{-27} erg-sec
Rest mass of electron	m	0.911×10^{-27} gram
Rest mass of proton	M_p	1.672×10^{-24} gram
Avogadro's number	N_0	6.022×10^{23} mole^{-1}
Boltzmann's constant	k	1.38×10^{-16} erg/°K
Electron volt	eV	1.602×10^{-12} erg
Rest energy of electron	mc^2	0.511×10^6 eV
Rest energy of proton	$M_p c^2$	0.938×10^9 eV
Bohr radius	a_0	0.529×10^{-8} cm

SOME USEFUL VECTOR RELATIONS

$$\nabla(uv) = u\nabla v + v\nabla u$$

$$\nabla(\mathbf{A} \cdot \mathbf{B}) = (\mathbf{A} \cdot \nabla)\mathbf{B} + (\mathbf{B} \cdot \nabla)\mathbf{A} + \mathbf{A} \times (\nabla \times \mathbf{B}) + \mathbf{B} \times (\nabla \times \mathbf{A})$$

$$\nabla \cdot (u\mathbf{A}) = u\nabla \cdot \mathbf{A} + \nabla u \cdot \mathbf{A}$$

$$\nabla \cdot (\mathbf{A} \times \mathbf{B}) = \mathbf{B} \cdot (\nabla \times \mathbf{A}) - \mathbf{A} \cdot (\nabla \times \mathbf{B})$$

$$\nabla \times (u\mathbf{A}) = u(\nabla \times \mathbf{A}) + \nabla u \times \mathbf{A}$$

$$\nabla \times (\mathbf{A} \times \mathbf{B}) = (\mathbf{B} \cdot \nabla)\mathbf{A} - (\mathbf{A} \cdot \nabla)\mathbf{B} + \mathbf{A}(\nabla \cdot \mathbf{B}) - \mathbf{B}(\nabla \cdot \mathbf{A})$$

$$\nabla \times (\nabla \times \mathbf{A}) = \nabla(\nabla \cdot \mathbf{A}) - \nabla^2 \mathbf{A}$$

$$\nabla \times (\nabla u) = 0$$

$$\nabla \cdot (\nabla \times \mathbf{A}) = 0$$

$$\int_V \nabla \cdot \mathbf{F} \, dV = \int_S \mathbf{F} \cdot \hat{\mathbf{n}} \, dA$$

$$\int_V \nabla \times \mathbf{F} \, dV = \int_S \hat{\mathbf{n}} \times \mathbf{F} \, dA$$

$$\int_V \nabla\varphi \, dV = \int_S \hat{\mathbf{n}}\varphi \, dA$$

$$\oint_C \mathbf{F} \cdot d\mathbf{l} = \int_S (\nabla \times \mathbf{F}) \cdot \hat{\mathbf{n}} \, dA$$

Index

A CATALOG OF SELECTED
DOVER BOOKS
IN SCIENCE AND MATHEMATICS

A CATALOG OF SELECTED
DOVER BOOKS
IN SCIENCE AND MATHEMATICS

QUALITATIVE THEORY OF DIFFERENTIAL EQUATIONS, V.V. Nemytskii and V.V. Stepanov. Classic graduate-level text by two prominent Soviet mathematicians covers classical differential equations as well as topological dynamics and ergodic theory. Bibliographies. 523pp. 5⅜ × 8½.   65954-2 Pa. $10.95

MATRICES AND LINEAR ALGEBRA, Hans Schneider and George Phillip Barker. Basic textbook covers theory of matrices and its applications to systems of linear equations and related topics such as determinants, eigenvalues and differential equations. Numerous exercises. 432pp. 5⅜ × 8½.   66014-1 Pa. $9.95

QUANTUM THEORY, David Bohm. This advanced undergraduate-level text presents the quantum theory in terms of qualitative and imaginative concepts, followed by specific applications worked out in mathematical detail. Preface. Index. 655pp. 5⅜ × 8½.   65969-0 Pa. $13.95

ATOMIC PHYSICS (8th edition), Max Born. Nobel laureate's lucid treatment of kinetic theory of gases, elementary particles, nuclear atom, wave-corpuscles, atomic structure and spectral lines, much more. Over 40 appendices, bibliography. 495pp. 5⅜ × 8½.   65984-4 Pa. $12.95

ELECTRONIC STRUCTURE AND THE PROPERTIES OF SOLIDS: The Physics of the Chemical Bond, Walter A. Harrison. Innovative text offers basic understanding of the electronic structure of covalent and ionic solids, simple metals, transition metals and their compounds. Problems. 1980 edition. 582pp. 6⅛ × 9¼.   66021-4 Pa. $14.95

BOUNDARY VALUE PROBLEMS OF HEAT CONDUCTION, M. Necati Özisik. Systematic, comprehensive treatment of modern mathematical methods of solving problems in heat conduction and diffusion. Numerous examples and problems. Selected references. Appendices. 505pp. 5⅜ × 8½.   65990-9 Pa. $11.95

A SHORT HISTORY OF CHEMISTRY (3rd edition), J.R. Partington. Classic exposition explores origins of chemistry, alchemy, early medical chemistry, nature of atmosphere, theory of valency, laws and structure of atomic theory, much more. 428pp. 5⅜ × 8½. (Available in U.S. only)   65977-1 Pa. $10.95

A HISTORY OF ASTRONOMY, A. Pannekoek. Well-balanced, carefully reasoned study covers such topics as Ptolemaic theory, work of Copernicus, Kepler, Newton, Eddington's work on stars, much more. Illustrated. References. 521pp. 5⅜ × 8½.   65994-1 Pa. $12.95

PRINCIPLES OF METEOROLOGICAL ANALYSIS, Walter J. Saucier. Highly respected, abundantly illustrated classic reviews atmospheric variables, hydrostatics, static stability, various analyses (scalar, cross-section, isobaric, isentropic, more). For intermediate meteorology students. 454pp. 6⅛ × 9¼. 65979-8 Pa. $12.95

RELATIVITY, THERMODYNAMICS AND COSMOLOGY, Richard C. Tolman. Landmark study extends thermodynamics to special, general relativity; also applications of relativistic mechanics, thermodynamics to cosmological models. 501pp. 5⅜ × 8½. 65383-8 Pa. $12.95

APPLIED ANALYSIS, Cornelius Lanczos. Classic work on analysis and design of finite processes for approximating solution of analytical problems. Algebraic equations, matrices, harmonic analysis, quadrature methods, much more. 559pp. 5⅜ × 8½. 65656-X Pa. $12.95

SPECIAL RELATIVITY FOR PHYSICISTS, G. Stephenson and C.W. Kilmister. Concise elegant account for nonspecialists. Lorentz transformation, optical and dynamical applications, more. Bibliography. 108pp. 5⅜ × 8½. 65519-9 Pa. $4.95

INTRODUCTION TO ANALYSIS, Maxwell Rosenlicht. Unusually clear, accessible coverage of set theory, real number system, metric spaces, continuous functions, Riemann integration, multiple integrals, more. Wide range of problems. Undergraduate level. Bibliography. 254pp. 5⅜ × 8½. 65038-3 Pa. $7.95

INTRODUCTION TO QUANTUM MECHANICS With Applications to Chemistry, Linus Pauling & E. Bright Wilson, Jr. Classic undergraduate text by Nobel Prize winner applies quantum mechanics to chemical and physical problems. Numerous tables and figures enhance the text. Chapter bibliographies. Appendices. Index. 468pp. 5⅜ × 8½. 64871-0 Pa. $11.95

ASYMPTOTIC EXPANSIONS OF INTEGRALS, Norman Bleistein & Richard A. Handelsman. Best introduction to important field with applications in a variety of scientific disciplines. New preface. Problems. Diagrams. Tables. Bibliography. Index. 448pp. 5⅜ × 8½. 65082-0 Pa. $11.95

MATHEMATICS APPLIED TO CONTINUUM MECHANICS, Lee A. Segel. Analyzes models of fluid flow and solid deformation. For upper-level math, science and engineering students. 608pp. 5⅜ × 8½. 65369-2 Pa. $13.95

ELEMENTS OF REAL ANALYSIS, David A. Sprecher. Classic text covers fundamental concepts, real number system, point sets, functions of a real variable, Fourier series, much more. Over 500 exercises. 352pp. 5⅜ × 8½. 65385-4 Pa. $9.95

PHYSICAL PRINCIPLES OF THE QUANTUM THEORY, Werner Heisenberg. Nobel Laureate discusses quantum theory, uncertainty, wave mechanics, work of Dirac, Schroedinger, Compton, Wilson, Einstein, etc. 184pp. 5⅜ × 8½. 60113-7 Pa. $5.95

INTRODUCTORY REAL ANALYSIS, A.N. Kolmogorov, S.V. Fomin. Translated by Richard A. Silverman. Self-contained, evenly paced introduction to real and functional analysis. Some 350 problems. 403pp. 5⅜ × 8½. 61226-0 Pa. $9.95

PROBLEMS AND SOLUTIONS IN QUANTUM CHEMISTRY AND PHYSICS, Charles S. Johnson, Jr. and Lee G. Pedersen. Unusually varied problems, detailed solutions in coverage of quantum mechanics, wave mechanics, angular momentum, molecular spectroscopy, scattering theory, more. 280 problems plus 139 supplementary exercises. 430pp. 6½ × 9¼. 65236-X Pa. $12.95

ASYMPTOTIC METHODS IN ANALYSIS, N.G. de Bruijn. An inexpensive, comprehensive guide to asymptotic methods—the pioneering work that teaches by explaining worked examples in detail. Index. 224pp. 5⅜ × 8½.   64221-6 Pa. $6.95

OPTICAL RESONANCE AND TWO-LEVEL ATOMS, L. Allen and J.H. Eberly. Clear, comprehensive introduction to basic principles behind all quantum optical resonance phenomena. 53 illustrations. Preface. Index. 256pp. 5⅜ × 8½.
65533-4 Pa. $7.95

COMPLEX VARIABLES, Francis J. Flanigan. Unusual approach, delaying complex algebra till harmonic functions have been analyzed from real variable viewpoint. Includes problems with answers. 364pp. 5⅜ × 8½.   61388-7 Pa. $8.95

ATOMIC SPECTRA AND ATOMIC STRUCTURE, Gerhard Herzberg. One of best introductions; especially for specialist in other fields. Treatment is physical rather than mathematical. 80 illustrations. 257pp. 5⅜ × 8½.   60115-3 Pa. $5.95

APPLIED COMPLEX VARIABLES, John W. Dettman. Step-by-step coverage of fundamentals of analytic function theory—plus lucid exposition of five important applications: Potential Theory; Ordinary Differential Equations; Fourier Transforms; Laplace Transforms; Asymptotic Expansions. 66 figures. Exercises at chapter ends. 512pp. 5⅜ × 8½.   64670-X Pa. $11.95

ULTRASONIC ABSORPTION: An Introduction to the Theory of Sound Absorption and Dispersion in Gases, Liquids and Solids, A.B. Bhatia. Standard reference in the field provides a clear, systematically organized introductory review of fundamental concepts for advanced graduate students, research workers. Numerous diagrams. Bibliography. 440pp. 5⅜ × 8½.   64917-2 Pa. $11.95

UNBOUNDED LINEAR OPERATORS: Theory and Applications, Seymour Goldberg. Classic presents systematic treatment of the theory of unbounded linear operators in normed linear spaces with applications to differential equations. Bibliography. 199pp. 5⅜ × 8½.   64830-3 Pa. $7.95

LIGHT SCATTERING BY SMALL PARTICLES, H.C. van de Hulst. Comprehensive treatment including full range of useful approximation methods for researchers in chemistry, meteorology and astronomy. 44 illustrations. 470pp. 5⅜ × 8½.   64228-3 Pa. $10.95

CONFORMAL MAPPING ON RIEMANN SURFACES, Harvey Cohn. Lucid, insightful book presents ideal coverage of subject. 334 exercises make book perfect for self-study. 55 figures. 352pp. 5⅜ × 8¼.   64025-6 Pa. $9.95

OPTICKS, Sir Isaac Newton. Newton's own experiments with spectroscopy, colors, lenses, reflection, refraction, etc., in language the layman can follow. Foreword by Albert Einstein. 532pp. 5⅜ × 8½.   60205-2 Pa. $9.95

GENERALIZED INTEGRAL TRANSFORMATIONS, A.H. Zemanian. Graduate-level study of recent generalizations of the Laplace, Mellin, Hankel, K. Weierstrass, convolution and other simple transformations. Bibliography. 320pp. 5⅜ × 8½.   65375-7 Pa. $7.95

THE ELECTROMAGNETIC FIELD, Albert Shadowitz. Comprehensive undergraduate text covers basics of electric and magnetic fields, builds up to electromagnetic theory. Also related topics, including relativity. Over 900 problems. 768pp. 5⅜ × 8¼. 65660-8 Pa. $17.95

FOURIER SERIES, Georgi P. Tolstov. Translated by Richard A. Silverman. A valuable addition to the literature on the subject, moving clearly from subject to subject and theorem to theorem. 107 problems, answers. 336pp. 5⅜ × 8½. 63317-9 Pa. $8.95

THEORY OF ELECTROMAGNETIC WAVE PROPAGATION, Charles Herach Papas. Graduate-level study discusses the Maxwell field equations, radiation from wire antennas, the Doppler effect and more. xiii + 244pp. 5⅜ × 8½. 65678-0 Pa. $6.95

DISTRIBUTION THEORY AND TRANSFORM ANALYSIS: An Introduction to Generalized Functions, with Applications, A.H. Zemanian. Provides basics of distribution theory, describes generalized Fourier and Laplace transformations. Numerous problems. 384pp. 5⅜ × 8½. 65479-6 Pa. $9.95

THE PHYSICS OF WAVES, William C. Elmore and Mark A. Heald. Unique overview of classical wave theory. Acoustics, optics, electromagnetic radiation, more. Ideal as classroom text or for self-study. Problems. 477pp. 5⅜ × 8½. 64926-1 Pa. $11.95

CALCULUS OF VARIATIONS WITH APPLICATIONS, George M. Ewing. Applications-oriented introduction to variational theory develops insight and promotes understanding of specialized books, research papers. Suitable for advanced undergraduate/graduate students as primary, supplementary text. 352pp. 5⅜ × 8½. 64856-7 Pa. $8.95

A TREATISE ON ELECTRICITY AND MAGNETISM, James Clerk Maxwell. Important foundation work of modern physics. Brings to final form Maxwell's theory of electromagnetism and rigorously derives his general equations of field theory. 1,084pp. 5⅜ × 8½. 60636-8, 60637-6 Pa., Two-vol. set $19.90

AN INTRODUCTION TO THE CALCULUS OF VARIATIONS, Charles Fox. Graduate-level text covers variations of an integral, isoperimetrical problems, least action, special relativity, approximations, more. References. 279pp. 5⅜ × 8½. 65499-0 Pa. $7.95

HYDRODYNAMIC AND HYDROMAGNETIC STABILITY, S. Chandrasekhar. Lucid examination of the Rayleigh-Benard problem; clear coverage of the theory of instabilities causing convection. 704pp. 5⅜ × 8¼. 64071-X Pa. $14.95

CALCULUS OF VARIATIONS, Robert Weinstock. Basic introduction covering isoperimetric problems, theory of elasticity, quantum mechanics, electrostatics, etc. Exercises throughout. 326pp. 5⅜ × 8½. 63069-2 Pa. $7.95

DYNAMICS OF FLUIDS IN POROUS MEDIA, Jacob Bear. For advanced students of ground water hydrology, soil mechanics and physics, drainage and irrigation engineering and more. 335 illustrations. Exercises, with answers. 784pp. 6⅛ × 9¼. 65675-6 Pa. $19.95

CATALOG OF DOVER BOOKS

NUMERICAL METHODS FOR SCIENTISTS AND ENGINEERS, Richard Hamming. Classic text stresses frequency approach in coverage of algorithms, polynomial approximation, Fourier approximation, exponential approximation, other topics. Revised and enlarged 2nd edition. 721pp. 5⅜ × 8½.
65241-6 Pa. $14.95

THEORETICAL SOLID STATE PHYSICS, Vol. I: Perfect Lattices in Equilibrium; Vol. II: Non-Equilibrium and Disorder, William Jones and Norman H. March. Monumental reference work covers fundamental theory of equilibrium properties of perfect crystalline solids, non-equilibrium properties, defects and disordered systems. Appendices. Problems. Preface. Diagrams. Index. Bibliography. Total of 1,301pp. 5⅜ × 8½. Two volumes.
Vol. I 65015-4 Pa. $14.95
Vol. II 65016-2 Pa. $12.95

OPTIMIZATION THEORY WITH APPLICATIONS, Donald A. Pierre. Broadspectrum approach to important topic. Classical theory of minima and maxima, calculus of variations, simplex technique and linear programming, more. Many problems, examples. 640pp. 5⅜ × 8½.
65205-X Pa. $14.95

THE MODERN THEORY OF SOLIDS, Frederick Seitz. First inexpensive edition of classic work on theory of ionic crystals, free-electron theory of metals and semiconductors, molecular binding, much more. 736pp. 5⅜ × 8½.
65482-6 Pa. $15.95

ESSAYS ON THE THEORY OF NUMBERS, Richard Dedekind. Two classic essays by great German mathematician: on the theory of irrational numbers; and on transfinite numbers and properties of natural numbers. 115pp. 5⅜ × 8½.
21010-3 Pa. $4.95

THE FUNCTIONS OF MATHEMATICAL PHYSICS, Harry Hochstadt. Comprehensive treatment of orthogonal polynomials, hypergeometric functions, Hill's equation, much more. Bibliography. Index. 322pp. 5⅜ × 8½. 65214-9 Pa. $9.95

NUMBER THEORY AND ITS HISTORY, Oystein Ore. Unusually clear, accessible introduction covers counting, properties of numbers, prime numbers, much more. Bibliography. 380pp. 5⅜ × 8½. 65620-9 Pa. $9.95

THE VARIATIONAL PRINCIPLES OF MECHANICS, Cornelius Lanczos. Graduate level coverage of calculus of variations, equations of motion, relativistic mechanics, more. First inexpensive paperbound edition of classic treatise. Index. Bibliography. 418pp. 5⅜ × 8½. 65067-7 Pa. $10.95

MATHEMATICAL TABLES AND FORMULAS, Robert D. Carmichael and Edwin R. Smith. Logarithms, sines, tangents, trig functions, powers, roots, reciprocals, exponential and hyperbolic functions, formulas and theorems. 269pp. 5⅜ × 8½.
60111-0 Pa. $6.95

THEORETICAL PHYSICS, Georg Joos, with Ira M. Freeman. Classic overview covers essential math, mechanics, electromagnetic theory, thermodynamics, quantum mechanics, nuclear physics, other topics. First paperback edition. xxiii + 885pp. 5⅜ × 8½. 65227-0 Pa. $18.95

HANDBOOK OF MATHEMATICAL FUNCTIONS WITH FORMULAS, GRAPHS, AND MATHEMATICAL TABLES, edited by Milton Abramowitz and Irene A. Stegun. Vast compendium: 29 sets of tables, some to as high as 20 places. 1,046pp. 8 × 10½. 61272-4 Pa. $22.95

MATHEMATICAL METHODS IN PHYSICS AND ENGINEERING, John W. Dettman. Algebraically based approach to vectors, mapping, diffraction, other topics in applied math. Also generalized functions, analytic function theory, more. Exercises. 448pp. 5⅜ × 8¼. 65649-7 Pa. $9.95

A SURVEY OF NUMERICAL MATHEMATICS, David M. Young and Robert Todd Gregory. Broad self-contained coverage of computer-oriented numerical algorithms for solving various types of mathematical problems in linear algebra, ordinary and partial, differential equations, much more. Exercises. Total of 1,248pp. 5⅜ × 8½. Two volumes. Vol. I 65691-8 Pa. $14.95
Vol. II 65692-6 Pa. $14.95

TENSOR ANALYSIS FOR PHYSICISTS, J.A. Schouten. Concise exposition of the mathematical basis of tensor analysis, integrated with well-chosen physical examples of the theory. Exercises. Index. Bibliography. 289pp. 5⅜ × 8½. 65582-2 Pa. $7.95

INTRODUCTION TO NUMERICAL ANALYSIS (2nd Edition), F.B. Hildebrand. Classic, fundamental treatment covers computation, approximation, interpolation, numerical differentiation and integration, other topics. 150 new problems. 669pp. 5⅜ × 8½. 65363-3 Pa. $14.95

INVESTIGATIONS ON THE THEORY OF THE BROWNIAN MOVEMENT, Albert Einstein. Five papers (1905–8) investigating dynamics of Brownian motion and evolving elementary theory. Notes by R. Fürth. 122pp. 5⅜ × 8½. 60304-0 Pa. $4.95

NUMERICAL METHODS FOR SCIENTISTS AND ENGINEERS, Richard Hamming. Classic text stresses frequency approach in coverage of algorithms, polynomial approximation, Fourier approximation, exponential approximation, other topics. Revised and enlarged 2nd edition. 721pp. 5⅜ × 8½. 65241-6 Pa. $14.95

AN INTRODUCTION TO STATISTICAL THERMODYNAMICS, Terrell L. Hill. Excellent basic text offers wide-ranging coverage of quantum statistical mechanics, systems of interacting molecules, quantum statistics, more. 523pp. 5⅜ × 8½. 65242-4 Pa. $11.95

ELEMENTARY DIFFERENTIAL EQUATIONS, William Ted Martin and Eric Reissner. Exceptionally clear, comprehensive introduction at undergraduate level. Nature and origin of differential equations, differential equations of first, second and higher orders. Picard's Theorem, much more. Problems with solutions. 331pp. 5⅜ × 8½. 65024-3 Pa. $8.95

STATISTICAL PHYSICS, Gregory H. Wannier. Classic text combines thermodynamics, statistical mechanics and kinetic theory in one unified presentation of thermal physics. Problems with solutions. Bibliography. 532pp. 5⅜ × 8½. 65401-X Pa. $11.95

ORDINARY DIFFERENTIAL EQUATIONS, Morris Tenenbaum and Harry Pollard. Exhaustive survey of ordinary differential equations for undergraduates in mathematics, engineering, science. Thorough analysis of theorems. Diagrams. Bibliography. Index. 818pp. 5⅜ × 8½. 64940-7 Pa. $16.95

STATISTICAL MECHANICS: Principles and Applications, Terrell L. Hill. Standard text covers fundamentals of statistical mechanics, applications to fluctuation theory, imperfect gases, distribution functions, more. 448pp. 5⅜ × 8½. 65390-0 Pa. $9.95

ORDINARY DIFFERENTIAL EQUATIONS AND STABILITY THEORY: An Introduction, David A. Sánchez. Brief, modern treatment. Linear equation, stability theory for autonomous and nonautonomous systems, etc. 164pp. 5⅜ × 8¼. 63828-6 Pa. $5.95

THIRTY YEARS THAT SHOOK PHYSICS: The Story of Quantum Theory, George Gamow. Lucid, accessible introduction to influential theory of energy and matter. Careful explanations of Dirac's anti-particles, Bohr's model of the atom, much more. 12 plates. Numerous drawings. 240pp. 5⅜ × 8½. 24895-X Pa. $6.95

THEORY OF MATRICES, Sam Perlis. Outstanding text covering rank, non-singularity and inverses in connection with the development of canonical matrices under the relation of equivalence, and without the intervention of determinants. Includes exercises. 237pp. 5⅜ × 8½. 66810-X Pa. $7.95

GREAT EXPERIMENTS IN PHYSICS: Firsthand Accounts from Galileo to Einstein, edited by Morris H. Shamos. 25 crucial discoveries: Newton's laws of motion, Chadwick's study of the neutron, Hertz on electromagnetic waves, more. Original accounts clearly annotated. 370pp. 5⅜ × 8½. 25346-5 Pa. $9.95

INTRODUCTION TO PARTIAL DIFFERENTIAL EQUATIONS WITH AP-PLICATIONS, E.C. Zachmanoglou and Dale W. Thoe. Essentials of partial differential equations applied to common problems in engineering and the physical sciences. Problems and answers. 416pp. 5⅜ × 8½. 65251-3 Pa. $10.95

BURNHAM'S CELESTIAL HANDBOOK, Robert Burnham, Jr. Thorough guide to the stars beyond our solar system. Exhaustive treatment. Alphabetical by constellation: Andromeda to Cetus in Vol. 1; Chamaeleon to Orion in Vol. 2; and Pavo to Vulpecula in Vol. 3. Hundreds of illustrations. Index in Vol. 3. 2,000pp. 6⅛ × 9¼. 23567-X, 23568-8, 23673-0 Pa., Three-vol. set $41.85

ASYMPTOTIC EXPANSIONS FOR ORDINARY DIFFERENTIAL EQUA-TIONS, Wolfgang Wasow. Outstanding text covers asymptotic power series, Jordan's canonical form, turning point problems, singular perturbations, much more. Problems. 384pp. 5⅜ × 8½. 65456-7 Pa. $9.95

AMATEUR ASTRONOMER'S HANDBOOK, J.B. Sidgwick. Timeless, comprehensive coverage of telescopes, mirrors, lenses, mountings, telescope drives, micrometers, spectroscopes, more. 189 illustrations. 576pp. 5⅜ × 8¼. (USO)24034-7 Pa. $9.95

SPECIAL FUNCTIONS, N.N. Lebedev. Translated by Richard Silverman. Famous Russian work treating more important special functions, with applications to specific problems of physics and engineering. 38 figures. 308pp. 5⅜ × 8½.
60624-4 Pa. $7.95

OBSERVATIONAL ASTRONOMY FOR AMATEURS, J.B. Sidgwick. Mine of useful data for observation of sun, moon, planets, asteroids, aurorae, meteors, comets, variables, binaries, etc. 39 illustrations. 384pp. 5⅜ × 8¼. (Available in U.S. only)
24033-9 Pa. $8.95

INTEGRAL EQUATIONS, F.G. Tricomi. Authoritative, well-written treatment of extremely useful mathematical tool with wide applications. Volterra Equations, Fredholm Equations, much more. Advanced undergraduate to graduate level. Exercises. Bibliography. 238pp. 5⅜ × 8½.
64828-1 Pa. $7.95

CELESTIAL OBJECTS FOR COMMON TELESCOPES, T.W. Webb. Inestimable aid for locating and identifying nearly 4,000 celestial objects. 77 illustrations. 645pp. 5⅜ × 8½.
20917-2, 20918-0 Pa., Two-vol. set $12.00

MODERN NONLINEAR EQUATIONS, Thomas L. Saaty. Emphasizes practical solution of problems; covers seven types of equations. ". . . a welcome contribution to the existing literature. . . ."—*Math Reviews.* 490pp. 5⅜ × 8½. 64232-1 Pa. $9.95

FUNDAMENTALS OF ASTRODYNAMICS, Roger Bate et al. Modern approach developed by U.S. Air Force Academy. Designed as a first course. Problems, exercises. Numerous illustrations. 455pp. 5⅜ × 8½.
60061-0 Pa. $8.95

INTRODUCTION TO LINEAR ALGEBRA AND DIFFERENTIAL EQUATIONS, John W. Dettman. Excellent text covers complex numbers, determinants, orthonormal bases, Laplace transforms, much more. Exercises with solutions. Undergraduate level. 416pp. 5⅜ × 8½.
65191-6 Pa. $9.95

INCOMPRESSIBLE AERODYNAMICS, edited by Bryan Thwaites. Covers theoretical and experimental treatment of the uniform flow of air and viscous fluids past two-dimensional aerofoils and three-dimensional wings; many other topics. 654pp. 5⅜ × 8½.
65465-6 Pa. $16.95

INTRODUCTION TO DIFFERENCE EQUATIONS, Samuel Goldberg. Exceptionally clear exposition of important discipline with applications to sociology, psychology, economics. Many illustrative examples; over 250 problems. 260pp. 5⅜ × 8½.
65084-7 Pa. $7.95

LAMINAR BOUNDARY LAYERS, edited by L. Rosenhead. Engineering classic covers steady boundary layers in two- and three-dimensional flow, unsteady boundary layers, stability, observational techniques, much more. 708pp. 5⅜ × 8½.
65646-2 Pa. $15.95

LECTURES ON CLASSICAL DIFFERENTIAL GEOMETRY, Second Edition, Dirk J. Struik. Excellent brief introduction covers curves, theory of surfaces, fundamental equations, geometry on a surface, conformal mapping, other topics. Problems. 240pp. 5⅜ × 8½.
65609-8 Pa. $7.95

ROTARY-WING AERODYNAMICS, W.Z. Stepniewski. Clear, concise text covers aerodynamic phenomena of the rotor and offers guidelines for helicopter performance evaluation. Originally prepared for NASA. 537 figures. 640pp. 6⅛ × 9¼.
64647-5 Pa. $15.95

DIFFERENTIAL GEOMETRY, Heinrich W. Guggenheimer. Local differential geometry as an application of advanced calculus and linear algebra. Curvature, transformation groups, surfaces, more. Exercises. 62 figures. 378pp. 5⅜ × 8½.
63433-7 Pa. $7.95

INTRODUCTION TO SPACE DYNAMICS, William Tyrrell Thomson. Comprehensive, classic introduction to space-flight engineering for advanced undergraduate and graduate students. Includes vector algebra, kinematics, transformation of coordinates. Bibliography. Index. 352pp. 5⅜ × 8½. 65113-4 Pa. $8.95

A SURVEY OF MINIMAL SURFACES, Robert Osserman. Up-to-date, in-depth discussion of the field for advanced students. Corrected and enlarged edition covers new developments. Includes numerous problems. 192pp. 5⅜ × 8½.
64998-9 Pa. $8.95

ANALYTICAL MECHANICS OF GEARS, Earle Buckingham. Indispensable reference for modern gear manufacture covers conjugate gear-tooth action, gear-tooth profiles of various gears, many other topics. 263 figures. 102 tables. 546pp. 5⅜ × 8½. 65712-4 Pa. $11.95

SET THEORY AND LOGIC, Robert R. Stoll. Lucid introduction to unified theory of mathematical concepts. Set theory and logic seen as tools for conceptual understanding of real number system. 496pp. 5⅜ × 8¼. 63829-4 Pa. $10.95

A HISTORY OF MECHANICS, René Dugas. Monumental study of mechanical principles from antiquity to quantum mechanics. Contributions of ancient Greeks, Galileo, Leonardo, Kepler, Lagrange, many others. 671pp. 5⅜ × 8½.
65632-2 Pa. $14.95

FAMOUS PROBLEMS OF GEOMETRY AND HOW TO SOLVE THEM, Benjamin Bold. Squaring the circle, trisecting the angle, duplicating the cube: learn their history, why they are impossible to solve, then solve them yourself. 128pp. 5⅜ × 8½. 24297-8 Pa. $3.95

MECHANICAL VIBRATIONS, J.P. Den Hartog. Classic textbook offers lucid explanations and illustrative models, applying theories of vibrations to a variety of practical industrial engineering problems. Numerous figures. 233 problems, solutions. Appendix. Index. Preface. 436pp. 5⅜ × 8½. 64785-4 Pa. $9.95

CURVATURE AND HOMOLOGY, Samuel I. Goldberg. Thorough treatment of specialized branch of differential geometry. Covers Riemannian manifolds, topology of differentiable manifolds, compact Lie groups, other topics. Exercises. 315pp. 5⅜ × 8½. 64314-X Pa. $8.95

HISTORY OF STRENGTH OF MATERIALS, Stephen P. Timoshenko. Excellent historical survey of the strength of materials with many references to the theories of elasticity and structure. 245 figures. 452pp. 5⅜ × 8½. 61187-6 Pa. $10.95

GEOMETRY OF COMPLEX NUMBERS, Hans Schwerdtfeger. Illuminating, widely praised book on analytic geometry of circles, the Moebius transformation, and two-dimensional non-Euclidean geometries. 200pp. 5⅜ × 8¼.
63830-8 Pa. $6.95

MECHANICS, J.P. Den Hartog. A classic introductory text or refresher. Hundreds of applications and design problems illuminate fundamentals of trusses, loaded beams and cables, etc. 334 answered problems. 462pp. 5⅜ × 8½. 60754-2 Pa. $8.95

TOPOLOGY, John G. Hocking and Gail S. Young. Superb one-year course in classical topology. Topological spaces and functions, point-set topology, much more. Examples and problems. Bibliography. Index. 384pp. 5⅜ × 8¼.
65676-4 Pa. $8.95

STRENGTH OF MATERIALS, J.P. Den Hartog. Full, clear treatment of basic material (tension, torsion, bending, etc.) plus advanced material on engineering methods, applications. 350 answered problems. 323pp. 5⅜ × 8½. 60755-0 Pa. $8.95

ELEMENTARY CONCEPTS OF TOPOLOGY, Paul Alexandroff. Elegant, intuitive approach to topology from set-theoretic topology to Betti groups; how concepts of topology are useful in math and physics. 25 figures. 57pp. 5⅜ × 8½.
60747-X Pa. $3.50

ADVANCED STRENGTH OF MATERIALS, J.P. Den Hartog. Superbly written advanced text covers torsion, rotating disks, membrane stresses in shells, much more. Many problems and answers. 388pp. 5⅜ × 8½. 65407-9 Pa. $9.95

COMPUTABILITY AND UNSOLVABILITY, Martin Davis. Classic graduate-level introduction to theory of computability, usually referred to as theory of recurrent functions. New preface and appendix. 288pp. 5⅜ × 8½. 61471-9 Pa. $7.95

GENERAL CHEMISTRY, Linus Pauling. Revised 3rd edition of classic first-year text by Nobel laureate. Atomic and molecular structure, quantum mechanics, statistical mechanics, thermodynamics correlated with descriptive chemistry. Problems. 992pp. 5⅜ × 8½. 65622-5 Pa. $19.95

AN INTRODUCTION TO MATRICES, SETS AND GROUPS FOR SCIENCE STUDENTS, G. Stephenson. Concise, readable text introduces sets, groups, and most importantly, matrices to undergraduate students of physics, chemistry, and engineering. Problems. 164pp. 5⅜ × 8½. 65077-4 Pa. $6.95

THE HISTORICAL BACKGROUND OF CHEMISTRY, Henry M. Leicester. Evolution of ideas, not individual biography. Concentrates on formulation of a coherent set of chemical laws. 260pp. 5⅜ × 8½. 61053-5 Pa. $6.95

THE PHILOSOPHY OF MATHEMATICS: An Introductory Essay, Stephan Körner. Surveys the views of Plato, Aristotle, Leibniz & Kant concerning propositions and theories of applied and pure mathematics. Introduction. Two appendices. Index. 198pp. 5⅜ × 8½. 25048-2 Pa. $6.95

THE DEVELOPMENT OF MODERN CHEMISTRY, Aaron J. Ihde. Authoritative history of chemistry from ancient Greek theory to 20th-century innovation. Covers major chemists and their discoveries. 209 illustrations. 14 tables. Bibliographies. Indices. Appendices. 851pp. 5⅜ × 8½. 64235-6 Pa. $17.95

CATALOG OF DOVER BOOKS

DE RE METALLICA, Georgius Agricola. The famous Hoover translation of greatest treatise on technological chemistry, engineering, geology, mining of early modern times (1556). All 289 original woodcuts. 638pp. 6¾ × 11.
60006-8 Pa. $17.95

SOME THEORY OF SAMPLING, William Edwards Deming. Analysis of the problems, theory and design of sampling techniques for social scientists, industrial managers and others who find statistics increasingly important in their work. 61 tables. 90 figures. xvii + 602pp. 5⅜ × 8½.
64684-X Pa. $15.95

THE VARIOUS AND INGENIOUS MACHINES OF AGOSTINO RAMELLI: A Classic Sixteenth-Century Illustrated Treatise on Technology, Agostino Ramelli. One of the most widely known and copied works on machinery in the 16th century. 194 detailed plates of water pumps, grain mills, cranes, more. 608pp. 9 × 12. (EBE)
25497-6 Clothbd. $34.95

LINEAR PROGRAMMING AND ECONOMIC ANALYSIS, Robert Dorfman, Paul A. Samuelson and Robert M. Solow. First comprehensive treatment of linear programming in standard economic analysis. Game theory, modern welfare economics, Leontief input-output, more. 525pp. 5⅜ × 8½.
65491-5 Pa. $13.95

ELEMENTARY DECISION THEORY, Herman Chernoff and Lincoln E. Moses. Clear introduction to statistics and statistical theory covers data processing, probability and random variables, testing hypotheses, much more. Exercises. 364pp. 5⅜ × 8½.
65218-1 Pa. $9.95

THE COMPLEAT STRATEGYST: Being a Primer on the Theory of Games of Strategy, J.D. Williams. Highly entertaining classic describes, with many illustrated examples, how to select best strategies in conflict situations. Prefaces. Appendices. 268pp. 5⅜ × 8½.
25101-2 Pa. $6.95

MATHEMATICAL METHODS OF OPERATIONS RESEARCH, Thomas L. Saaty. Classic graduate-level text covers historical background, classical methods of forming models, optimization, game theory, probability, queueing theory, much more. Exercises. Bibliography. 448pp. 5⅜ × 8¼.
65703-5 Pa. $12.95

CONSTRUCTIONS AND COMBINATORIAL PROBLEMS IN DESIGN OF EXPERIMENTS, Damaraju Raghavarao. In-depth reference work examines orthogonal Latin squares, incomplete block designs, tactical configuration, partial geometry, much more. Abundant explanations, examples. 416pp. 5⅜ × 8¼.
65685-3 Pa. $10.95

THE ABSOLUTE DIFFERENTIAL CALCULUS (CALCULUS OF TENSORS), Tullio Levi-Civita. Great 20th-century mathematician's classic work on material necessary for mathematical grasp of theory of relativity. 452pp. 5⅜ × 8½.
63401-9 Pa. $9.95

VECTOR AND TENSOR ANALYSIS WITH APPLICATIONS, A.I. Borisenko and I.E. Tarapov. Concise introduction. Worked-out problems, solutions, exercises. 257pp. 5⅜ × 8¼.
63833-2 Pa. $6.95

CATALOG OF DOVER BOOKS

THE FOUR-COLOR PROBLEM: Assaults and Conquest, Thomas L. Saaty and Paul G. Kainen. Engrossing, comprehensive account of the century-old combinatorial topological problem, its history and solution. Bibliographies. Index. 110 figures. 228pp. 5⅜ × 8½. 65092-8 Pa. $6.95

CATALYSIS IN CHEMISTRY AND ENZYMOLOGY, William P. Jencks. Exceptionally clear coverage of mechanisms for catalysis, forces in aqueous solution, carbonyl- and acyl-group reactions, practical kinetics, more. 864pp. 5⅜ × 8½. 65460-5 Pa. $19.95

PROBABILITY: An Introduction, Samuel Goldberg. Excellent basic text covers set theory, probability theory for finite sample spaces, binomial theorem, much more. 360 problems. Bibliographies. 322pp. 5⅜ × 8½. 65252-1 Pa. $8.95

LIGHTNING, Martin A. Uman. Revised, updated edition of classic work on the physics of lightning. Phenomena, terminology, measurement, photography, spectroscopy, thunder, more. Reviews recent research. Bibliography. Indices. 320pp. 5⅜ × 8¼. 64575-4 Pa. $8.95

PROBABILITY THEORY: A Concise Course, Y.A. Rozanov. Highly readable, self-contained introduction covers combination of events, dependent events, Bernoulli trials, etc. Translation by Richard Silverman. 148pp. 5⅜ × 8¼. 63544-9 Pa. $5.95

THE CEASELESS WIND: An Introduction to the Theory of Atmospheric Motion, John A. Dutton. Acclaimed text integrates disciplines of mathematics and physics for full understanding of dynamics of atmospheric motion. Over 400 problems. Index. 97 illustrations. 640pp. 6 × 9. 65096-0 Pa. $17.95

STATISTICS MANUAL, Edwin L. Crow, et al. Comprehensive, practical collection of classical and modern methods prepared by U.S. Naval Ordnance Test Station. Stress on use. Basics of statistics assumed. 288pp. 5⅜ × 8½. 60599-X Pa. $6.95

DICTIONARY/OUTLINE OF BASIC STATISTICS, John E. Freund and Frank J. Williams. A clear concise dictionary of over 1,000 statistical terms and an outline of statistical formulas covering probability, nonparametric tests, much more. 208pp. 5⅜ × 8½. 66796-0 Pa. $6.95

STATISTICAL METHOD FROM THE VIEWPOINT OF QUALITY CONTROL, Walter A. Shewhart. Important text explains regulation of variables, uses of statistical control to achieve quality control in industry, agriculture, other areas. 192pp. 5⅜ × 8½. 65232-7 Pa. $7.95

THE INTERPRETATION OF GEOLOGICAL PHASE DIAGRAMS, Ernest G. Ehlers. Clear, concise text emphasizes diagrams of systems under fluid or containing pressure; also coverage of complex binary systems, hydrothermal melting, more. 288pp. 6½ × 9¼. 65389-7 Pa. $10.95

STATISTICAL ADJUSTMENT OF DATA, W. Edwards Deming. Introduction to basic concepts of statistics, curve fitting, least squares solution, conditions without parameter, conditions containing parameters. 26 exercises worked out. 271pp. 5⅜ × 8½. 64685-8 Pa. $8.95

CATALOG OF DOVER BOOKS

TENSOR CALCULUS, J.L. Synge and A. Schild. Widely used introductory text covers spaces and tensors, basic operations in Riemannian space, non-Riemannian spaces, etc. 324pp. 5⅜ × 8¼. 63612-7 Pa. $7.95

A CONCISE HISTORY OF MATHEMATICS, Dirk J. Struik. The best brief history of mathematics. Stresses origins and covers every major figure from ancient Near East to 19th century. 41 illustrations. 195pp. 5⅜ × 8½. 60255-9 Pa. $7.95

A SHORT ACCOUNT OF THE HISTORY OF MATHEMATICS, W.W. Rouse Ball. One of clearest, most authoritative surveys from the Egyptians and Phoenicians through 19th-century figures such as Grassman, Galois, Riemann. Fourth edition. 522pp. 5⅜ × 8½. 20630-0 Pa. $10.95

HISTORY OF MATHEMATICS, David E. Smith. Nontechnical survey from ancient Greece and Orient to late 19th century; evolution of arithmetic, geometry, trigonometry, calculating devices, algebra, the calculus. 362 illustrations. 1,355pp. 5⅜ × 8½. 20429-4, 20430-8 Pa., Two-vol. set $23.90

THE GEOMETRY OF RENÉ DESCARTES, René Descartes. The great work founded analytical geometry. Original French text, Descartes' own diagrams, together with definitive Smith-Latham translation. 244pp. 5⅜ × 8½. 60068-8 Pa. $6.95

THE ORIGINS OF THE INFINITESIMAL CALCULUS, Margaret E. Baron. Only fully detailed and documented account of crucial discipline: origins; development by Galileo, Kepler, Cavalieri; contributions of Newton, Leibniz, more. 304pp. 5⅜ × 8½. (Available in U.S. and Canada only) 65371-4 Pa. $9.95

THE HISTORY OF THE CALCULUS AND ITS CONCEPTUAL DEVELOPMENT, Carl B. Boyer. Origins in antiquity, medieval contributions, work of Newton, Leibniz, rigorous formulation. Treatment is verbal. 346pp. 5⅜ × 8½. 60509-4 Pa. $7.95

THE THIRTEEN BOOKS OF EUCLID'S ELEMENTS, translated with introduction and commentary by Sir Thomas L. Heath. Definitive edition. Textual and linguistic notes, mathematical analysis. 2,500 years of critical commentary. Not abridged. 1,414pp. 5⅜ × 8½. 60088-2, 60089-0, 60090-4 Pa., Three-vol. set $29.85

GAMES AND DECISIONS: Introduction and Critical Survey, R. Duncan Luce and Howard Raiffa. Superb nontechnical introduction to game theory, primarily applied to social sciences. Utility theory, zero-sum games, n-person games, decision-making, much more. Bibliography. 509pp. 5⅜ × 8½. 65943-7 Pa. $11.95

THE HISTORICAL ROOTS OF ELEMENTARY MATHEMATICS, Lucas N.H. Bunt, Phillip S. Jones, and Jack D. Bedient. Fundamental underpinnings of modern arithmetic, algebra, geometry and number systems derived from ancient civilizations. 320pp. 5⅜ × 8½. 25563-8 Pa. $8.95

CALCULUS REFRESHER FOR TECHNICAL PEOPLE, A. Albert Klaf. Covers important aspects of integral and differential calculus via 756 questions. 566 problems, most answered. 431pp. 5⅜ × 8½. 20370-0 Pa. $8.95